Introduction to Fuzzy Logic using MATLAB

S.N. Sivanandam, S. Sumathi and S.N. Deepa

Introduction to Fuzzy Logic using MATLAB

With 304 Figures and 37 Tables

 Springer

Dr. S.N. Sivanandam
Professor and Head
Department of Computer
Science and Engineering
PSG College of Technology
Coimbatore 641 004
Tamil Nadu, India

E-mail: snsivanandam@yahoo.co.in

S.N. Deepa
Faculty
Department of Computer
Science and Engineering
PSG College of Technology
Coimbatore 641 004
Tamil Nadu, India

E-mail: deepanand1999@yahoo.co.in

Dr. S. Sumathi
Assistant Professor
Department of Electrical
and Electronics Engineering
PSG College of Technology
Coimbatore 641 004
Tamil Nadu, India

E-mail: ss_eeein@yahoo.com

Library of Congress Control Number: 2006930099

ISBN-10 3-540-35780-7 Springer Berlin Heidelberg New York
ISBN-13 978-3-540-35780-3 Springer Berlin Heidelberg New York

Springer is a part of Springer Science+Business Media.

springer.com

© Springer-Verlag Berlin Heidelberg 2007

Typesetting by the authors and SPi
Cover design: Erich Kirchner, Heidelberg

Printed on acid-free paper   SPIN 11764601   89/3100/SPi   5 4 3 2 1 0

Preface

The world we live in is becoming ever more reliant on the use of electronics and computers to control the behavior of real-world resources. For example, an increasing amount of commerce is performed without a single banknote or coin ever being exchanged. Similarly, airports can safely land and send off airplanes without ever looking out of a window. Another, more individual, example is the increasing use of electronic personal organizers for organizing meetings and contacts. All these examples share a similar structure; multiple parties (e.g., airplanes or people) come together to co-ordinate their activities in order to achieve a common goal. It is not surprising, then, that a lot of research is being done into how a lot of mechanics of the co-ordination process can be automated using computers.

Fuzzy logic means approximate reasoning, information granulation, computing with words and so on.

Ambiguity is always present in any realistic process. This ambiguity may arise from the interpretation of the data inputs and in the rules used to describe the relationships between the informative attributes. Fuzzy logic provides an inference structure that enables the human reasoning capabilities to be applied to artificial knowledge-based systems. Fuzzy logic provides a means for converting linguistic strategy into control actions and thus offers a high-level computation.

Fuzzy logic provides mathematical strength to the emulation of certain perceptual and linguistic attributes associated with human cognition, whereas the science of neural networks provides a new computing tool with learning and adaptation capabilities. The theory of fuzzy logic provides an inference mechanism under cognitive uncertainty, computational neural networks offer exciting advantages such as learning, adaptation, fault tolerance, parallelism, and generalization.

About the Book

This book is meant for a wide range of readers, especially college and university students wishing to learn basic as well as advanced processes and techniques in fuzzy systems. It can also be meant for programmers who may be involved in programming based on the soft computing applications.

The principles of fuzzy systems are dealt in depth with the information and the useful knowledge available for computing processes. The various algorithms and the solutions to the problems are well balanced pertinent to the fuzzy systems' research projects, labs, and for college- and university-level studies.

Modern aspects of soft computing have been introduced from the first principles and discussed in an easy manner, so that a beginner can grasp the concept of fuzzy systems with minimal effort.

The solutions to the problems are programmed using Matlab 6.0 and the simulated results are given. The fuzzy logic toolbox are also provided in the Appendix for easy reference of the students and professionals.

The book contains solved example problems, review questions, and exercise problems.

This book is designed to give a broad, yet in-depth overview of the field of fuzzy systems. This book can be a handbook and a guide for students of computer science, information technology, EEE, ECE, disciplines of engineering, students in master of computer applications, and for professionals in the information technology sector, etc.

This book will be a very good compendium for almost all readers — from students of undergraduate to postgraduate level and also for researchers, professionals, etc. — who wish to enrich their knowledge on fuzzy systems' principles and applications with a single book in the best manner.

This book focuses mainly on the following academic courses:

- Master of Computer Applications (MCA)
- Master of Computer and Information Technology
- Master of Science (Software)-Integrated
- Engineering students of computer science, electrical and electronics engineering, electronics and communication engineering and information technology both at graduate and postgraduate levels
- Ph.D research scholars who work in this field

Fuzzy systems, at present, is a hot topic among academicians as well as among program developers. As a result, this book can be recommended not only for students, but also for a wide range of professionals and developers who work in this area.

This book can be used as a ready reference guide for fuzzy system research scholars. Most of the algorithms, solved problems, and applications for a wide variety of areas covered in this book can fulfill as an advanced academic book.

In conclusion, we hope that the reader will find this book a truly helpful guide and a valuable source of information about the fuzzy system principles for their numerous practical applications.

Organization of the Book

The book covers 9 chapters altogether. It starts with introduction to the fuzzy system techniques. The application case studies are also discussed.

The chapters are organized as follows:

- Chapter 1 gives an introduction to fuzzy logic and Matlab.
- Chapter 2 discusses the definition, properties, and operations of classical and fuzzy sets. It contains solved sample problems related to the classical and fuzzy sets.
- The Cartesian product of the relation along with the cardinality, operations, properties, and composition of classical and fuzzy relations is discussed in chapter 3.
- Chapter 4 gives details on the membership functions. It also adds features of membership functions, classification of fuzzy sets, process of fuzzification, and various methods by means of which membership values are assigned.
- The process and the methods of defuzzification are described in chapter 5. The lambda cut method for fuzzy set and relation along with the other methods like centroid method, weighted average method, etc. are discussed with solved problems inside.
- Chapter 6 describes the fuzzy rule-based system. It includes the aggregation, decomposition, and the formation of rules. Also the methods of fuzzy inference system, mamdani, and sugeno methods are described here.
- Chapter 7 provides the information regarding various decision-making processes like fuzzy ordering, individual decision making, multiperson decision making, multiobjective decision making, and fuzzy Bayesian decision-making method.
- The application of fuzzy logic in various fields along with case studies and adaptive fuzzy in image segmentation is given in chapter 8.
- Chapter 9 gives information regarding a few projects implemented using the fuzzy logic technique.
- The appendix includes fuzzy Matlab tool box.
- The bibliography is given at the end after the appendix chapter.

Salient Features of Fuzzy Logic

The salient features of this book include

- Detailed description on fuzzy logic techniques
- Variety of solved examples

- Review questions and exercise problems
- Simulated results obtained for the fuzzy logic techniques using Matlab version 6.0
- Application case studies and projects on fuzzy logic in various fields.

S.N. Sivanandam completed his B.E (Electrical and Electronics Engineering) in 1964 from Government College of Technology, Coimbatore, and M.Sc (Engineering) in Power System in 1966 from PSG College of Technology, Coimbatore. He acquired PhD in Control Systems in 1982 from Madras University. His research areas include modeling and simulation, neural networks, fuzzy systems and genetic algorithm, pattern recognition, multidimensional system analysis, linear and nonlinear control system, signal and image processing, control system, power system, numerical methods, parallel computing, data mining, and database security. He received "Best Teacher Award" in 2001, "Dhakshina Murthy Award for Teaching Excellence" from PSG College of Technology, and "The Citation for Best Teaching and Technical Contribution" in 2002 from Government College of Technology, Coimbatore. He is currently working as a Professor and Head of Computer Science and Engineering Department, PSG College of Technology, Coimbatore. He has published nine books and is a member of various professional bodies like IE (India). ISTE, CSI, ACS, etc. He has published about 600 papers in national and international journals.

S. Sumathi completed B.E. (Electronics and Communication Engineering), M.E. (Applied Electronics) at Government College of Technology, Coimbatore, and Ph.D. in data mining. Her research interests include neural networks, fuzzy systems and genetic algorithms, pattern recognition and classification, data warehousing and data mining, operating systems, parallel computing, etc. She received the prestigious gold medal from the Institution of Engineers Journal Computer Engineering Division for the research paper titled "Development of New Soft Computing Models for Data Mining" and also best project award for UG Technical Report titled "Self-Organized Neural Network Schemes as a Data Mining Tool." Currently, she is working as Lecturer in the Department of Electrical and Electronics Engineering, PSG College of Technology, Coimbatore. Sumathi has published several research articles in national and international journals and conferences.

Deepa has completed her B.E. from Government College of Technology, Coimbatore, and M.E. from PSG College of Technology, Coimbatore. She was a gold medallist in her B.E. exams. She has published two books and articles in national and international journals and conferences. She was a recipient of national award in the year 2004 from ISTE and Larsen & Toubro Limited. Her research areas include neural network, fuzzy logic, genetic algorithm, digital control, adaptive and nonlinear control.

Coimbatore, India   S.N. Sivanandam
2006–2007   S. Sumathi
                                                                                             S.N. Deepa

Acknowledgments

The authors are always thankful to the Almighty for perseverance and achievements. They wish to thank Shri. G. Rangasamy, Managing Trustee, PSG Institutions; Shri. C.R. Swaminathan, Chief Executive; and Dr. R. Rudramoorthy, Principal, PSG College of Technology, Coimbatore, for their whole-hearted cooperation and great encouragement given in this successful endeavor. Sumathi owes much to her daughter Priyanka and to the support rendered by her husband, brother and family. Deepa wishes to thank her husband Anand and her daughter Nivethitha, and her parents for their support.

Contents

1

Introduction

1.1 Fuzzy Logic

In the literature sources, we can find different kinds of justification for fuzzy systems theory. Human knowledge nowadays becomes increasingly important – we gain it from experiencing the world within which we live and use our ability to reason to create order in the mass of information (i.e., to formulate human knowledge in a systematic manner). Since we are all limited in our ability to perceive the world and to profound reasoning, we find ourselves everywhere confronted by *uncertainty* which is a result of lack of information (lexical impression, incompleteness), in particular, inaccuracy of measurements. The other limitation factor in our desire for precision is a natural language used for describing/sharing knowledge, communication, etc. We understand core meanings of word and are able to communicate accurately to an acceptable degree, but generally we cannot precisely agree among ourselves on the single word or terms of common sense meaning. In short, natural languages are *vague*.

Our perception of the real world is pervaded by concepts which do not have sharply defined boundaries – for example, *many, tall, much larger than, young,* etc. are true only to some degree and they are false to some degree as well. These concepts (facts) can be called *fuzzy* or *gray (vague)* concepts – a human brain works with them, while computers may not do it (they reason with strings of 0s and 1s). Natural languages, which are much higher in level than programming languages, are fuzzy whereas programming languages are not. The door to the development of fuzzy computers was opened in 1985 by the design of the first logic chip by Masaki Togai and Hiroyuki Watanabe at Bell Telephone Laboratories. In the years to come fuzzy computers will employ both *fuzzy hardware* and *fuzzy software*, and they will be much closer in structure to the human brain than the present-day computers are.

The entire real world is complex; it is found that the complexity arises from uncertainty in the form of ambiguity. According to Dr. Lotfi Zadeh, Principle of Compatability, the complexity, and the imprecision are correlated and adds,

The closer one looks at a real world problem, the fuzzier becomes its solution (Zadeh 1973)

The Fuzzy Logic tool was introduced in 1965, also by Lotfi Zadeh, and is a mathematical tool for dealing with uncertainty. It offers to a soft computing partnership the important concept of computing with words'. It provides a technique to deal with imprecision and information granularity. The fuzzy theory provides a mechanism for representing linguistic constructs such as "many," "low," "medium," "often," "few." In general, the fuzzy logic provides an inference structure that enables appropriate human reasoning capabilities. On the contrary, the traditional binary set theory describes crisp events, events that either do or do not occur. It uses probability theory to explain if an event will occur, measuring the chance with which a given event is expected to occur. The theory of fuzzy logic is based upon the notion of relative graded membership and so are the functions of mentation and cognitive processes. The utility of fuzzy sets lies in their ability to model uncertain or ambiguous data, Fig. 1.1, so often encountered in real life.

It is important to observe that there is an *intimate connection* between *Fuzziness* and *Complexity*. As the complexity of a task (problem), or of a system for performing that task, exceeds a certain threshold, the system must necessarily become fuzzy in nature. Zadeh, originally an engineer and systems scientist, was concerned with the rapid decline in information afforded by traditional mathematical models as the complexity of the target system increased. As he stressed, with the increasing of complexity our ability to make precise and yet significant statements about its behavior diminishes. Real-world problems (situations) are too complex, and the *complexity involves the degree of uncertainty* – as uncertainty increases, so does the complexity of the problem. Traditional system modeling and analysis techniques are too precise for such problems (systems), and in order to make complexity less daunting we introduce appropriate simplifications, assumptions, etc. (i.e., *degree of uncertainty* or *Fuzziness*) to achieve a satisfactory compromise between the information we have and the amount of uncertainty we are willing to accept. In this aspect, fuzzy systems theory is similar to other engineering theories, because almost all of them characterize the real world in an approximate manner.

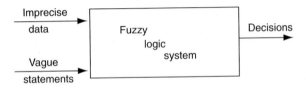

Fig. 1.1. A fuzzy logic system which accepts imprecise data and vague statements such as low, medium, high and provides decisions

Fuzzy sets provide means to model the uncertainty associated with vagueness, imprecision, and lack of information regarding a problem or a plant, etc. Consider the meaning of a "short person." For an individual X, the short person may be one whose height is below 4′25″. For other individual Y, the short person may be one whose height is below or equal to 3′90″. This "short" is called as a linguistic descriptor. The term "short" informs the same meaning to the individuals X and Y, but it is found that they both do not provide a unique definition. The term "short" would be conveyed effectively, only when a computer compares the given height value with the preassigned value of "short." This variable "short" is called as *linguistic variable*, which represents the imprecision existing in the system.

The uncertainty is found to arise from ignorance, from chance and randomness, due to lack of knowledge, from vagueness (unclear), like the fuzziness existing in our natural language. Lotfi Zadeh proposed the *set membership* idea to make suitable decisions when uncertainty occurs. Consider the "short" example discussed previously. If we take "short" as a height equal to or less than 4 feet, then 3′90″ would easily become the member of the set "short" and 4′25″ will not be a member of the set "short." The membership value is "1" if it belongs to the set or "0" if it is not a member of the set. Thus membership in a set is found to be binary i.e., the element is a member of a set or not.

It can be indicated as,

$$\chi_A(x) = \left\{ \begin{array}{ll} 1 & , x \in A \\ 0 & , x \notin A \end{array} \right\},$$

where $\chi_A(x)$ is the membership of element x in set A and A is the entire set on the universe.

This membership was extended to possess various "degree of membership" on the real continuous interval [0,1]. Zadeh formed *fuzzy sets* as the sets on the universe X which can accommodate "degrees of membership." The concept of a fuzzy set contrasts with a classical concept of a bivalent set (crisp set), whose boundary is required to be precise, i.e., a crisp set is a collection of things for which it is known whether any given thing is inside it or not. Zadeh generalized the idea of a crisp set by extending a valuation set {1,0} (definitely in/definitely out) to the interval of real values (degrees of membership) between 1 and 0 denoted as [0,1]. We can say that the degree of membership of any particular element of a fuzzy set express the degree of compatibility of the element with a concept represented by fuzzy set. It means that a fuzzy set A contains an object x to degree $a(x)$, i.e., $a(x) = Degree(x \in A)$, and the map $a : X \rightarrow \{Membership\ Degrees\}$ is called a *set function* or *membership function*. The fuzzy set A can be expressed as $A = \{(x, a(x))\}, x \in X$, and it imposes an elastic constrain of the possible values of elements $x \in X$ called the *possibility distribution*. Fuzzy sets tend to capture vagueness exclusively via membership functions that are mappings from a given universe of discourse

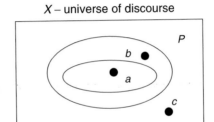

Fig. 1.2. Boundary region of a fuzzy set

X to a unit interval containing membership values. It is important to note that membership can take values between 0 and 1.

Fuzziness describes the ambiguity of an event and randomness describes the uncertainty in the occurrence of an event. It can be generally seen in classical sets that there is no uncertainty, hence they have crisp boundaries, but in the case of a fuzzy set, since uncertainty occurs, the boundaries may be ambiguously specified.

From the Fig. 1.2, it can be noted that a is clearly a member of fuzzy set P, c is clearly not a member of fuzzy set P, but the membership of b is found to be vague. Hence a can take membership value 1, c can take membership value 0 and b can take membership value between 0 and 1 [0 to 1], say 0.4, 0.7, etc. This is set to be a partial member ship of fuzzy set P.

The membership function for a set maps each element of the set to a membership value between 0 and 1 and uniquely describes that set. The values 0 and 1 describe "not belonging to" and "belonging to" a conventional set respectively; values in between represent "fuzziness." Determining the membership function is subjective to varying degrees depending on the situation. It depends on an individual's perception of the data in question and does not depend on randomness. This is important, and distinguishes fuzzy set theory from probability theory (Fig. 1.3).

In practice fuzzy logic means computation of words. Since computation with words is possible, computerized systems can be built by embedding human expertise articulated in daily language. Also called a fuzzy inference engine or fuzzy rule-base, such a system can perform approximate reasoning somewhat similar to but much more primitive than that of the human brain. Computing with words seems to be a slightly futuristic phrase today since only certain aspects of natural language can be represented by the calculus of fuzzy sets, but still fuzzy logic remains one of the most practical ways to mimic human expertise in a realistic manner. The fuzzy approach uses a premise that humans do not represent classes of objects (e.g. *class of bald men*, or the *class of numbers which are much greater than 50*) as fully disjoint but rather as sets in which there may be grades of membership intermediate between full

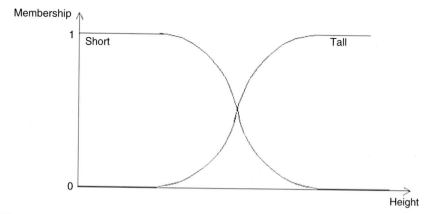

Fig. 1.3. *The fuzzy sets "tall" and "short."* The classification is subjective – it depends on what height is measured relative to. At the extremes, the distinction is clear, but there is a large amount of overlap in the middle

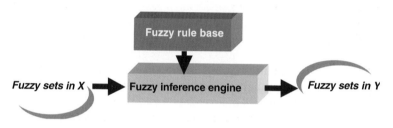

Fig. 1.4. Configuration of a pure fuzzy system

membership and non-membership. Thus, a fuzzy set works as a concept that makes it possible to *treat fuzziness in a quantitative manner.*

Fuzzy sets form the building blocks for fuzzy *IF–THEN* rules which have the general form "IF X is A THEN Y is B," where A and B are fuzzy sets. The term "fuzzy systems" refers mostly to systems that are governed by fuzzy IF–THEN rules. The IF part of an implication is called the *antecedent* whereas the second, THEN part is a *consequent*. A fuzzy system is a set of fuzzy rules that converts inputs to outputs. The basic configuration of a pure fuzzy system is shown in Fig. 1.4. The fuzzy inference engine (algorithm) combines fuzzy *IF–THEN* rules into a mapping from fuzzy sets in the input space X to fuzzy sets in the output space Y based on fuzzy logic principles. From a knowledge representation viewpoint, a fuzzy *IF–THEN* rule is a scheme for capturing knowledge that involves imprecision. The main feature of reasoning using these rules is its *partial matching* capability, which enables an inference to be made from a fuzzy rule even when the rule's condition is only partially satisfied.

Fuzzy systems, on one hand, are rule-based systems that are constructed from a collection of linguistic rules; on the other hand, fuzzy systems are nonlinear mappings of inputs (stimuli) to outputs (responses), i.e., certain types of fuzzy systems can be written as compact nonlinear formulas. The inputs and outputs can be numbers or vectors of numbers. These rule-based systems in theory model represents any system with arbitrary accuracy, i.e., they work as *universal approximators*.

The Achilles' heel of a fuzzy system is its rules; smart rules give smart systems and other rules give smart systems and other rules give less smart or even dumb systems. The *number of rules* increases exponentially with the dimension of the input space (number of system variables). This rule explosion is called the *principle of dimensionality* and is a general problem for mathematical models. For the last five years several approaches based on decomposition (cluster) merging and fusing have been proposed to overcome this problem.

Hence, Fuzzy models are not replacements for probability models. The fuzzy models sometimes found to work better and sometimes they do not. But mostly fuzzy is evidently proved that it provides better solutions for complex problems.

1.2 Mat LAB – An Overview

Dr Cleve Moler, Chief scientist at MathWorks, Inc., originally wrote Matlab, to provide easy access to matrix software developed in the LINPACK and EISPACK projects. The very first version was written in the late 1970s for use in courses in matrix theory, linear algebra, and numerical analysis. Matlab is therefore built upon a foundation of sophisticated matrix software, in which the basic data element is a matrix that does not require predimensioning.

Matlab is a product of The Math works, Inc. and is an advanced interactive software package specially designed for scientific and engineering computation. The Matlab environment integrates graphic illustrations with precise numerical calculations, and is a powerful, easy-to-use, and comprehensive tool for performing all kinds of computations and scientific data visualization. Matlab has proven to be a very flexible and usable tool for solving problems in many areas. Matlab is a high-performance language for technical computing. It integrates computation, visualization, and programming in an easy-to-use environment where problems and solutions are expressed in familiar mathematical notation. Typical use includes:

– Math and computation
– Algorithm development
– Modeling, simulation, and prototyping
– Data analysis, exploration, and visualization
– Scientific and engineering graphics
– Application development, including graphical user interface building

Matlab is an interactive system whose basic elements are an array that does not require dimensioning. This allows solving many computing problems, especially those with matrix and vector formulations, in a fraction of the time it would take to write a program in a scalar noninteractive language such as C or FORTRAN. Mathematics is the common language of science and engineering. Matrices, differential equations, arrays of data, plots, and graphs are the basic building blocks of both applied mathematics and Matlab. It is the underlying mathematical base that makes Matlab accessible and powerful. Matlab allows expressing the entire algorithm in a few dozen lines, to compute the solution with great accuracy in about a second.

Matlab is both an environment and programming language, and the major advantage of the Matlab language is that it allows building our own reusable tools. Our own functions and programs (known as M-files) can be created in Matlab code. The toolbox is a specialized collection of M-files for working on particular classes of problems. The Matlab documentation set has been written, expanded, and put online for ease of use. The set includes online help, as well as hypertext-based and printed manuals. The commands in Matlab are expressed in a notation close to that used in mathematics and engineering. There is a very large set of commands and functions, known as Matlab M-files. As a result solving problems in Matlab is faster than the other traditional programming. It is easy to modify the functions since most of the M-files can be open. For high performance, the Matlab software is written in optimized C and coded in assembly language.

Matlab's two- and three-dimensional graphics are object oriented. Matlab is thus both an environment and a matrix/vector-oriented programming language, which enables the use to build own required tools.
The main features of Matlab are:

- Advance algorithms for high-performance numerical computations, especially in the field of matrix algebra.
- A large collection of predefined mathematical functions and the ability to define one's own functions.
- Two- and three-dimensional graphics for plotting and displaying data.
- A complete help system online.
- Powerful matrix/vector-oriented high-level programming language for individual applications.
- Ability to cooperate with programs written in other languages and for importing and exporting formatted data.
- Toolboxes available for solving advanced problems in several application areas.

Figure 1.5 shows the main features and capabilities of Matlab.

SIMULINK is a Matlab toolbox designed for the dynamic simulation of linear and nonlinear systems as well as continuous and discrete-time systems. It can also display information graphically. Matlab is an interactive package for numerical analysis, matrix computation, control system design, and linear system analysis and design available on most CAEN platforms (Macintosh,

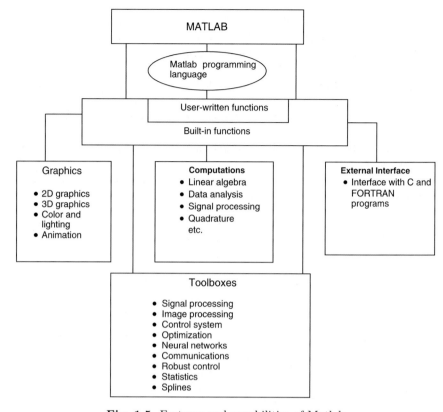

Fig. 1.5. Features and capabilities of Matlab

PCs, Sun, and Hewlett-Packard). In addition to the standard functions provided by Matlab, there exist large set of toolboxes, or collections of functions and procedures, available as part of the Matlab package. The toolboxes are:

- *Control system.* Provides several features for advanced control system design and analysis
- *Communications.* Provides functions to model the components of a communication system's physical layer
- *Signal processing.* Contains functions to design analog and digital filters and apply these filters to data and analyze the results
- *System identification.* Provides features to build mathematical models of dynamical systems based on observed system data
- *Robust control.* Allows users to create robust multivariable feedback control system designs based on the concept of the singular value Bode plot
- *Simulink.* Allows you to model dynamic systems graphically
- *Neural network.* Allows you to simulate neural networks
- *Fuzzy logic.* Allows for manipulation of fuzzy systems and membership functions

- *Image processing.* Provides access to a wide variety of functions for reading, writing, and filtering images of various kinds in different ways
- *Analysis.* Includes a wide variety of system analysis tools for varying matrices
- *Optimization.* Contains basic tools for use in constrained and unconstrained minimization problems
- *Spline.* Can be used to find approximate functional representations of data sets
- *Symbolic.* Allows for symbolic (rather than purely numeric) manipulation of functions
- *User interface utilities.* Includes tools for creating dialog boxes, menu utilities, and other user interaction for script files

Matlab has been used as an efficient tool, all over this text to develop the applications based on neural net, fuzzy systems and genetic algorithm.

Review Questions

1) Define uncertainty and vagueness
2) Compare – precision an impression
3) Explain the concept of fuzziness a said by Lotfi A. Zadeh
4) What is a membership function?
5) Describe in detail about fuzzy system with basic configuration
6) Write short note on "degree of uncertainty"
7) Write an over view of Mat Lab

2

Classical Sets and Fuzzy Sets

2.1 Introduction

The theory on classical sets and the basic ideas of the fuzzy sets are discussed in detail in this chapter. The various operations, laws and properties of fuzzy sets are introduced along with that of the classical sets. The classical set we are going to deal is defined by means of the definite or crisp boundaries. This means that there is no uncertainty involved in the location of the boundaries for these sets. But whereas the fuzzy set, on the other hand is defined by its vague and ambiguous properties, hence the boundaries are specified ambiguously. The crisp sets are sets without ambiguity in their membership. The fuzzy set theory is a very efficient theory in dealing with the concepts of ambiguity. The fuzzy sets are dealt after reviewing the concepts of the classical or crisp sets.

2.2 Classical Set

Consider a classical set where X denotes the universe of discourse or universal sets. The individual elements in the universe X will be denoted as x. The features of the elements in X can be discrete, countable integers, or continuous valued quantities on the real line. Examples of elements of various universes might be as follows.

- The clock speeds of computers CPUs.
- The operating temperature of an air conditioner.
- The operating currents of an electronic motor or a generator set.
- The integers 1–100.

Choosing a universe that is discrete and finite or one that it continuous and infinite is a modeling choice, the choice does not alter the characterization of sets defined on the universe. If the universe possesses continuous elements, then the corresponding set defined on the universe will also be continuous.

The total number of elements in a universe X is called its *cardinal number* and is denoted by η_x. Discrete universe is composed of countable finite collection of elements and has a finite cardinal number and the continuous universe consists of uncountable or infinite collection of elements and thus has a infinite cardinal number.

As we all know, the collection of elements in the universe are called as sets, and the collections of elements within sets are called as subsets. The collection of all the elements in the universe is called the whole set. The null set $\emptyset$, which has no elements is analogous to an impossible event, and the whole set is analogous to certain event. Power set constitutes all possible sets of X and is denoted by $P(X)$.

Example 2.1. Let universe comprised of four elements $X = \{1, 2, 3, 4\}$ find cardinal number, power set, and cardinality of the power set.

Solution. The cardinal number is the number of elements in the defined set. The defined set X consists of four elements $1, 2, 3$, and 4. Therefore, the Cardinal number $= \eta_x = 4$.

The power set consists of all possible sets of X. It is given by,

Power set $P(x) = \{\emptyset, \{1\}, \{2\}, \{3\}, \{4\}, \{1, 2\}, \{1, 3\}, \{1, 4\}, \{2, 3\}, \{2, 4\}, \{3, 4\}, \{1, 2, 3\}, \{2, 3, 4\}, \{1, 3, 4\}, \{1, 2, 4\}, \{1, 2, 3, 4\}\}$

Cardinality of the power set is given by,

$$\eta_{P(x)} = 2^{\eta_x} = 2^4 = 16.$$

2.2.1 Operations on Classical Sets

There are various operations that can be performed in the classical or crisp sets. The results of the operation performed on the classical sets will be definite. The operations that can be performed on the classical sets are dealt in detail below:

Consider two sets A and B defined on the universe X. The definitions of the operation for classical sets are based on the two sets A and B defined on the universe X.

Union

The Union of two classical sets A and B is denoted by $A \cup B$. It represents all the elements in the universe that reside in either the set A, the set B or both sets A and B. This operation is called the logical OR.

In set theoretic form it is represented as

$$A \cup B = \{x/x \in A \text{ or } x \in B\}.$$

In Venn diagram form it can be represented as shown in Fig. 2.1.

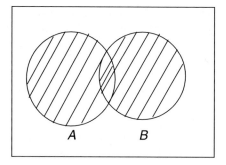

Fig. 2.1. $A \cup B$

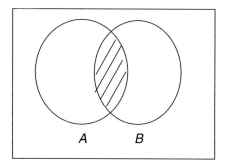

Fig. 2.2. $A \cap B$

Intersection

The intersection of two sets A and B is denoted $A \cap B$. It represents all those elements in the universe X that simultaneously reside in (or belongs to) both sets A and B.

In set theoretic form it is represented as

$$A \cap B = \{x/x \in A \text{ and } x \in B\}.$$

In Venn diagram form it can be represented as shown in Fig. 2.2.

Complement

The complement of set A denoted $\overline{A}$, is defined as the collection of all elements in the universe that do not reside in the set A.

In set theoretic form it is represented as

$$\overline{A} = \{x/x \notin A, x \in X\}.$$

In Venn diagram form it is represented as shown in Fig. 2.3.

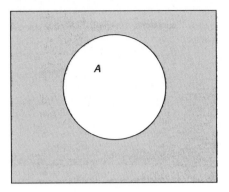

Fig. 2.3. Complement of set A

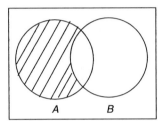

Fig. 2.4. Difference $A|B$

Difference

The difference of a set A with respect to B, denoted $A|B$ is defined as collection of all elements in the universe that reside in A and that do not reside in B simultaneously.

In set theoretic form it is represented as

$$A|B = \{x/x \in A \text{ and } x \notin B\}.$$

In Venn diagram form it is represented as shown in Fig. 2.4.

2.2.2 Properties of Classical Sets

In any mathematical operations the properties plays a major role. Based upon the properties, the solution can be obtained for the problems. The following are the important properties of classical sets:

Commutativity

$$A \cup B = B \cup A,$$
$$A \cap B = B \cap A.$$

Associativity
$$A \cup (B \cup C) = (A \cup B) \cup C,$$
$$A \cap (B \cap C) = (A \cap B) \cap C.$$

Distributivity
$$A \cup (B \cap C) = (A \cup B) \cap (A \cup C),$$
$$A \cap (B \cup C) = A \cap B) \cup (A \cap C).$$

Idempotency
$$A \cup A = A,$$
$$A \cap A = A.$$

Identity
$$A \cup \phi = A$$
$$A \cap X = A$$
$$A \cap \phi = \phi$$
$$A \cup X = X.$$

Transitivity
$$\text{If } A \subseteq B \subseteq C, \text{ then } A \subseteq C.$$

In this case the symbol $\subseteq$ means contained in or equivalent to and $\subset$ means contained in.

Involution
$$\overline{\overline{A}} = A.$$

The other two important special properties include the Excluded middle laws and the Demorgan's law.

Excluded middle law includes the law of excluded middle and the law of contradiction. The excluded middle laws is very important because these are the only set operations that are not valid for both classical and fuzzy sets.

Law of excluded middle. It represents union of a set A and its complement.

$$A \cup \overline{A} = X.$$

Law of contradication. It represents the intersection of a set A and its complement

$$A \cap \overline{A} = \phi.$$

De Morgan's Law

These are very important because of their efficiency in proving the tautologies and contradictions in logic. The demorgan's law are given by

$$\overline{A \cap B} = \overline{A} \cup \overline{B},$$
$$\overline{A \cup B} = \overline{A} \cap \overline{B}.$$

In Venn diagram form it is represented as shown in Fig. 2.5.

The complement of a union or an intersection of two sets is equal to the intersection or union of the respective complements of the two sets. This is the statement made for the demorgan's law.

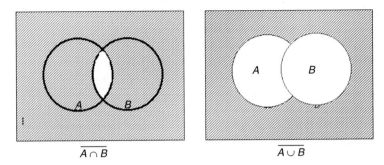

Fig. 2.5. Demorgan's law

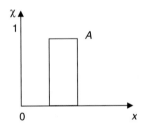

Fig. 2.6. Membership mapping for Crisp Set A

2.2.3 Mapping of Classical Sets to a Function

Mapping of set theoretic forms to function theoretic forms is an important concept. In general it can be used to map elements or subsets on one universe of discourse to elements or sets in another universe. Suppose X and Y are two different universe of discourse. If an element x is contained in X and corresponds to an element y contained in Y, it is generally represented as $f : X \rightarrow Y$, which is said as the *mapping from X to Y*. The characteristic function χ_A is defined by

$$\chi_A(x) = \begin{cases} 1 & x \in A, \\ 0 & x \notin A, \end{cases}$$

where χ_A represents the membership in set a for the elements x in the universe. The membership mapping for the crisp set A is shown in Fig. 2.6.

Let us define two sets A and B on the Universe X.

Union

The union of these two sets in terms of function theoretic form is given as follows:

$$A \cup B \rightarrow \chi_{A \cup B}(x) = \chi_A(x) V \chi_B(x)$$
$$= \max(\chi_A(x), \chi_B(x)).$$

Here V indicates the maximum operator.

Intersection

The intersection of two sets in function theoretic form is given as

$$A \cap B \rightarrow \chi_{A \cap B}(x) = \chi_A(x) \Lambda \chi_B(x)$$
$$= \min(\chi_A(x), \chi_B(x)).$$

Here Λ indicates the minimum operator.

Complement

The complement of single set on universe X, say A is given by

$$\overline{A} \rightarrow \chi_{\overline{A}}(x) = 1 - \chi_A(x).$$

Containment

The two sets A and B in universe, if one set (A) is contained in another set B, then

$$A \subseteq B \rightarrow \chi_A(x) \leq \chi_B(x).$$

Thus, the mapping of classical sets to functions is mentioned here.

2.2.4 Solved Examples

Example 2.2. Given the classical sets,

$$A = \{9, 5, 6, 8, 10\} \qquad B = \{1, 2, 3, 7, 9\} \qquad C = \{1, 0\}$$

defined on universe $X = \{$ Set of all 'n' natural no$\}$
    Prove the classical set properties associativity and distributivity.

Solution. The associative property is given by

1. $A \cup (B \cup C) = (A \cup B) \cup C$.
   LHS
   $A \cup (B \cup C)$

        (a) $B \cup C = \{2, 3, 7, 9, 1, 0\}$.
        (b) $A \cup (B \cup C) = \{5, 6, 8, 10, 2, 3, 7, 9, 1, 0\}$.   (2.1)

  RHS
  $(A \cup B) \cup C$

        (a) $(A \cup B) = \{9, 5, 6, 8, 10, 1, 2, 3, 7\}$.
        (b) $(A \cup B) \cup C = \{9, 5, 6, 10, 8, 1, 2, 3, 7, 0\}$.   (2.2)

From (2.1) and (2.2)

$$\text{LHS} = \text{RHS}$$
$$A \cup (B \cup C) = (A \cup B) \cup C.$$

2. $(A \cap (B \cap C) = (A \cap B) \cap C$
   LHS

        (a) $(B \cap C) = \{1\}.$

        (b) $A \cap (B \cap C) = \{\phi\}.$ (2.3)

RHS
$(A \cap B) \cap C$

        (a) $(A \cap B) = \{9\}.$

        (b) $(A \cap B) \cap C = \{\phi\}.$ (2.4)

From (2.3) and (2.4)

$$\text{LHS} = \text{RHS}$$
$$A \cap (B \cap C) = (A \cap B) \cap C.$$

Thus associative property is proved.
The distributive property is given by,

1. $A \cup (B \cap C) = (A \cup B) \cap (A \cup C)$
   LHS

        (a) $B \cap C = \{1\}.$

        (b) $A \cup (B \cap C) = \{9, 5, 6, 8, 10, 1\}.$ (2.5)

RHS
$$(A \cup B) \cap (A \cup C)$$

        (a) $(A \cup B) = \{9, 5, 6, 8, 10, 1, 2, 3, 7\}.$

        (b) $(A \cup C) = \{9, 5, 6, 8, 10, 1, 0\}.$

        (c) $(A \cup B) \cap (A \cup C) = \{9, 5, 6, 8, 10, 1\}.$ (2.6)

From (2.5) and (2.6)

$$\text{LHS} = \text{RHS}$$
$$A \cup (B \cap C) = (A \cup B) \cap (A \cup C).$$

2. $A \cap (B \cup C) = (A \cap B) \cup (A \cap C)$
   LHS

        (a) $(B \cup C) = \{1, 2, 3, 7, 9, 0\}.$

        (b) $A \cap (B \cup C) = \{9\}.$ (2.7)

RHS

(a) $A \cap B = \{9\}$.

(b) $A \cap C = \{\phi\}$.

(c) $(A \cap B) \cup (A \cap C) = \{9\}$.   (2.8)

From (2.7) and (2.8),

$$LHS = RHS$$
$$A \cap (B \cup C) = (A \cap B) \cup (A \cap B).$$

Hence distributive property is proved.

Example 2.3. Consider, $X = \{a, b, c, d, e, f, g, h\}$.
and the set A is defined as $\{a, d, f\}$. So for this classical set prove the identity property.

Solution. Given, $X = \{a, b, c, d, e, f, g, h\}$,
   $A = \{a, d, f\}$.
   The identity property is given as

1. A $\cup \phi = A$.
2. A $\cap \phi = A$.
   ϕ is going to be a null set, hence it is clearly understood, that, $A \cup \phi$ &
   $A \cap \phi$ will give as the same set A.
3. $A \cap X = A$.

$$A \cap X = \{a, d, f\},$$
$$A = \{a, d, f\}.$$
$$\text{Hence, } A \cap X = A.$$

4. $A \cup X = X$,

$$A \cup X = \{a, b, c, d, e, f, g, h\},$$
$$X = \{a, b, c, d, e, f, g, h\}.$$

Hence, $A \cup X = X$.
This identity property is proved.

2.3 Fuzzy Sets

In the classical set, its characteristic function assigns a value of either 1 or 0 to each individual in the universal set, there by discriminating between members and nonmembers of the crisp set under consideration. The values assigned to

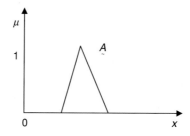

Fig. 2.7. Membership function of fuzzy set A
~

the elements of the universal set fall within a specified range and indicate the membership grade of these elements in the set. Larger values denote higher degrees of set membership such a function is called a membership function and the set is defined by it is a fuzzy set.

A fuzzy set is thus a set containing elements that have varying degrees of membership in the set. This idea is in contrast with classical or crisp, set because members of a crisp set would not be members unless their membership was full or complete, in that set (i.e., their membership is assigned a value of 1). Elements in a fuzzy set, because their membership need not be complete, can also be members of other fuzzy set on the same universe. Fuzzy set are denoted by a set symbol with a tilde understrike. Fuzzy set is mapped to a real numbered value in the interval 0 to 1. If an element of universe, say x, is a member of fuzzy set A, then the mapping is given by $\mu_A(x) \in [0, 1]$. This is the membership mapping and is shown in Fig. 2.7.

2.3.1 Fuzzy Set Operations

Considering three fuzzy sets A, B and C on the universe X. For a given element x of the universe, the following function theoretic operations for the set theoretic operations unions, intersection and complement are defined for A, B and C on X:

Union:

$$\mu_{A \cup B}(x) = \mu_A(x) \vee \mu_B(x).$$

Intersection:

$$\mu_{A \cap B}(x) = \mu_A(x) \wedge \mu_B(x).$$

Complement

$$\mu_{\underset{\sim}{A}}(x) = 1 - \mu_A(x).$$

$$A \subseteq X \rightarrow \mu_{\underset{\sim}{A}}(x) \leq \mu_X(x)$$

for all $x \in X \mu_{\phi}(x) = 0.$

for all $x \in X \mu_x(x) = 1.$

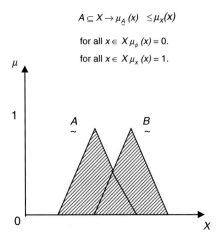

Fig. 2.8. Union of fuzzy sets

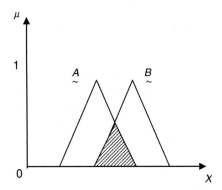

Fig. 2.9. Intersection of fuzzy sets

The venn diagram representation of these operations are shown in Figs. 2.8–2.10.

Any fuzzy set $\underset{\sim}{A}$ defined on a universe x is a subset of that universe. The membership value of any element x in the null set ϕ is 0, and the membership value of any element x in the whole set x is 1. This statement is given by

De Morgan's laws stated for classical sets also hold for fuzzy sets, as denoted by these expressions.

$$\overline{\underset{\sim}{A} \cap \underset{\sim}{B}} = \overline{\underset{\sim}{A}} \cup \overline{\underset{\sim}{B}},$$

$$\overline{\underset{\sim}{A} \cup \underset{\sim}{B}} = \overline{\underset{\sim}{A}} \cap \overline{\underset{\sim}{B}}.$$

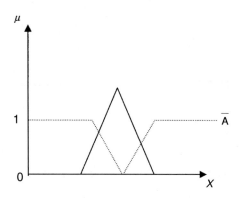

Fig. 2.10. Complement of fuzzy set

All operations on classical sets also hold for the fuzzy set except for the excluded middle laws. These two laws does not hold good for fuzzy sets. Since fuzzy sets can overlap, a set and its complement also can overlap.

The excluded middle law for fuzzy sets is given by

$$A \cup \bar{A} \neq X,$$

$$A \cap \bar{A} \neq \phi.$$

Comparing Venn diagram for classical sets and fuzzy sets for excluded middle law are shown in Figs. 2.11 and 2.12.

2.3.2 Properties of Fuzzy Sets

The properties of the classical set also suits for the properties of the fuzzy sets. The important properties of fuzzy set includes:

Commutativity

$$A \cup B = B \cup A,$$
$$A \cap B = B \cap A.$$

Associativity

$$A \cup (B \cup C) = (A \cup B) \cup C,$$
$$A \cap (B \cap C) = (A \cap B) \cap C.$$

Distributivity

$$A \cup (B \cap C) = (A \cup B) \cap (A \cup C),$$
$$A \cap (B \cup C) = (A \cap B) \cup (A \cap C).$$

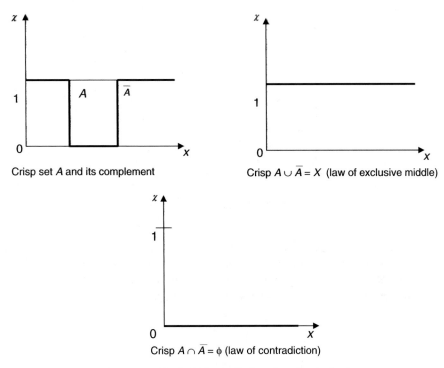

Fig. 2.11. Excluded middle law for classical sets

Idempotency

$$A \cup A = A,$$
$$A \cap A = A.$$

Identity

$$A \cup \phi = A \quad \text{and} \quad A \cap X = A,$$
$$A \cap \phi = \phi \quad \text{and} \quad A \cup X = X.$$

Transtivity

$$\text{If } A \subset B \subset C \text{ then } A \subset C$$

Involution

$$\overline{\overline{A}} = A.$$

These are the important properties of the fuzzy set.

2.3.3 Solved Examples

Example 2.4. Consider two fuzzy sets A and B find Complement, Union, Intersection, Difference, and De Morgan's law.

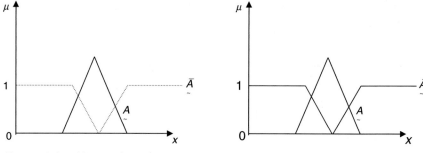

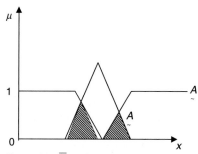

Fuzzy $A \cup \bar{A} \neq \phi$ (law of contradiction)

Fig. 2.12. Excluded middle law for fuzzy set

$$\underset{\sim}{A} = \left\{ \frac{1}{2} + \frac{0.5}{3} + \frac{0.6}{4} + \frac{0.2}{5} + \frac{0.6}{6} \right\},$$
$$\underset{\sim}{B} = \left\{ \frac{0.5}{2} + \frac{0.8}{3} + \frac{0.4}{4} + \frac{0.7}{5} + \frac{0.3}{6} \right\}.$$

Solution.
Complement

$$\underset{\sim}{\bar{A}} = \left\{ \frac{0}{2} + \frac{0.5}{3} + \frac{0.4}{4} + \frac{0.8}{5} + \frac{0.4}{6} \right\},$$
$$\underset{\sim}{\bar{B}} = \left\{ \frac{0.5}{2} + \frac{0.2}{3} + \frac{0.6}{4} + \frac{0.3}{5} + \frac{0.7}{6} \right\}.$$

Union

$$\underset{\sim}{A} \cup \underset{\sim}{B} = \left\{ \frac{1}{2} + \frac{0.8}{3} + \frac{0.6}{4} + \frac{0.7}{5} + \frac{0.6}{6} \right\}.$$

Comparing the membership values and writing maximum of the two values
determine Union of the fuzzy set.

Intersection

$$\underset{\sim}{A} \cap \underset{\sim}{B} = \left\{ \frac{0.5}{2} + \frac{0.5}{3} + \frac{0.4}{4} + \frac{0.2}{5} + \frac{0.3}{6} \right\}.$$

Comparing the membership values and writing minimum of the two values determine intersection of the fuzzy set.

Difference

$$A/B = A \cap \bar{B} = \left\{ \frac{0.5}{2} + \frac{0.2}{3} + \frac{0.6}{4} + \frac{0.2}{5} + \frac{0.6}{6} \right\},$$

$$B/A = B \cap \bar{A} = \left\{ \frac{0}{2} + \frac{0.5}{3} + \frac{0.4}{4} + \frac{0.7}{5} + \frac{0.3}{6} \right\}.$$

De Morgan's Laws

$$\overline{A \cup B} = \bar{A} \cap \bar{B} = \left\{ \frac{0}{2} + \frac{0.2}{3} + \frac{0.4}{4} + \frac{0.3}{5} + \frac{0.4}{6} \right\},$$

$$\overline{A \cap B} = \bar{A} \cup \bar{B} = \left\{ \frac{0.5}{2} + \frac{0.5}{3} + \frac{0.6}{4} + \frac{0.8}{5} + \frac{0.7}{6} \right\}.$$

Example 2.5. We want to compare two sensors based upon their detection levels and gain settings. The following table of gain settings and sensor detection levels with a standard item being monitored provides typical membership values to represents the detection levels for each of the sensors.

Gain setting	Sensor 1 detection levels	Sensor 2 detection levels
0	0	0
20	0.5	0.35
40	0.65	0.5
60	0.85	0.75
80	1	0.90
100	1	1

The universe of discourse is $X = \{0, 20, 40, 60, 80, 100\}$. Find the membership function for the two sensors: Find the following membership functions using standard set operations:

(a) $\mu_{S_1 \cup S_2}(x)$ (b) $\mu_{S_1 \cap S_2}(x)$

(c) $\mu_{\bar{S}_1}(x)$ (d) $\mu_{\bar{S}_2}(x)$

(e) $\mu_{\overline{S_1 \cup S_1}}(x)$ (f) $\mu_{\overline{S_1 \cup S_1}}(x)$

(g) $\mu_{S_1 \cup \bar{S}_1}(x)$ (h) $\mu_{S_1 \cup \bar{S}_1}(x)$

Solution. The membership functions for the two sensors in standard discrete form are

$$S_1 = \left\{ \frac{0.5}{20} + \frac{0.65}{40} + \frac{0.85}{60} + \frac{1}{80} + \frac{1}{100} \right\}$$

$$S_2 = \left\{ \frac{0.35}{20} + \frac{0.5}{40} + \frac{0.75}{60} + \frac{0.90}{80} + \frac{1}{100} \right\}$$

(a) $\mu_{S_1 \cup S_2}(x) = \mu_{s_s}(x) \vee \mu_{s_2}(x)$

$$= \left\{ \frac{0.5}{20} + \frac{0.65}{40} + \frac{0.85}{60} + \frac{1}{80} + \frac{1}{100} \right\}$$

(b) $\mu_{S_1 \cap S_2}(x) = \mu_{s_s}(x) \wedge \mu_{s_2}(x)$

$$= \left\{ \frac{0.35}{20} + \frac{0.5}{40} + \frac{0.75}{60} + \frac{0.9}{80} + \frac{1}{100} \right\}$$

(c) $\mu_{\bar{S}_1}(x) = \left\{ \frac{0.5}{20} + \frac{0.35}{40} + \frac{0.15}{60} + \frac{0}{80} + \frac{0}{100} \right\}$

(d) $\mu_{\bar{S}_2}(x) = \left\{ \frac{0.65}{20} + \frac{0.5}{40} + \frac{0.25}{60} + \frac{0.1}{80} + \frac{0}{100} \right\}$

(e) $\mu_{\overline{S_1 \cup S_1}}(x) = \mu_{\bar{S}_1 \cap \bar{S}_1} = \mu_{\bar{S}_1}(x) \wedge \mu_{\bar{S}_1}(x) = \left\{ \frac{0.5}{20} + \frac{0.35}{40} + \frac{0.15}{60} + \frac{0}{80} + \frac{0}{100} \right\}$

(f) $\mu_{\overline{S_1 \cap S_1}}(x) = \mu_{\bar{S}_1 \cup \bar{S}_1} = \mu_{\bar{S}_1}(x) \vee \mu_{\bar{S}_1}(x) = \left\{ \frac{0.5}{20} + \frac{0.35}{40} + \frac{0.15}{60} + \frac{0}{80} + \frac{0}{100} \right\}$

(g) $\mu_{S_1 \cup \bar{S}_1}(x) = \mu_{S_1}(x) \vee \mu_{\bar{S}_1}(x) = \left\{ \frac{0.5}{20} + \frac{0.65}{40} + \frac{0.85}{60} + \frac{1}{80} + \frac{1}{100} \right\}$

(h) $\mu_{S_1 \cap \bar{S}_1}(x) = \mu_{S_1}(x) \wedge \mu_{\bar{S}_1}(x) = \left\{ \frac{0.5}{20} + \frac{0.35}{40} + \frac{0.15}{60} + \frac{0}{80} + \frac{0}{100} \right\}$

Example 2.6. Let x be the universe of commercial aircraft of interest

$$X = \{\, a10, b52, b117, C5, C130, f4, f14, f15, f16, f111, kc130 \,\}.$$

Let A be the fuzzy set passenger class aircraft

$$A = \left\{ \frac{0.3}{f16} + \frac{0.5}{f4} + \frac{0.4}{a10} + \frac{0.6}{f14} + \frac{0.7}{f111} + \frac{1.0}{b117} + \frac{1.0}{b52} \right\}$$

Let B be the fuzzy set of cargo

$$B = \left\{ \frac{0.4}{b117} + \frac{0.4}{f111} + \frac{0.6}{f4} + \frac{0.8}{f15} + \frac{0.9}{f14} + \frac{1.0}{f16} \right\}$$

Find the values of the operation performed on these fuzzy sets.

Solution. The operation are union, intersection, and complement.

(a) $A \cup B = \left\{ \frac{0.4}{a10} + \frac{1.0}{b52} + \frac{1.0}{b117} + \frac{0}{C5} + \frac{0}{C130} + \frac{0.6}{f4} + \frac{0.9}{f14} + \frac{0.8}{f15} + \frac{1.0}{f16} \right.$
$\left. + \frac{0.7}{f111} + \frac{0}{KC130} \right\}$

(b) $A \cap B = \left\{ \dfrac{0.4}{a10} + \dfrac{1.0}{b52} + \dfrac{0.4}{b117} + \dfrac{0}{C5} + \dfrac{0}{C130} + \dfrac{0.5}{f4} + \dfrac{0.6}{f14} + \dfrac{0.8}{f15} + \dfrac{0.3}{f16} \right.$
$\left. + \dfrac{0.4}{f111} + \dfrac{0}{KC130} \right\}$

(c) $\bar{A} = \left\{ \dfrac{0.7}{f16} + \dfrac{0.5}{f4} + \dfrac{0.6}{a10} + \dfrac{0.4}{f14} + \dfrac{0.7}{f15} + \dfrac{0.3}{f111} + \dfrac{0}{b117} + \dfrac{0}{b52} + \dfrac{1}{C5} \right.$
$\left. + \dfrac{1}{C130} + \dfrac{1}{KC130} \right\}$

(d) $\bar{B} = \left\{ \dfrac{0.6}{b117} + \dfrac{0.6}{f111} + \dfrac{0.4}{f4} + \dfrac{0.2}{f15} + \dfrac{0.1}{f14} + \dfrac{0}{f16} + \dfrac{1}{C5} + \dfrac{1}{C130} + \dfrac{1}{KC130} \right\}$

Example 2.7. For the given fuzzy set

$$A = \left\{ \frac{1}{1.0} + \frac{0.65}{1.5} + \frac{0.4}{2.0} + \frac{0.35}{2.5} + \frac{0}{3.0} \right\},$$

$$B = \left\{ \frac{0}{1.0} + \frac{0.25}{1.5} + \frac{0.6}{2.0} + \frac{0.25}{2.5} + \frac{1}{3.0} \right\},$$

$$C = \left\{ \frac{0.5}{1.0} + \frac{0.25}{1.5} + \frac{0}{2.0} + \frac{0.25}{2.5} + \frac{0.5}{3.0} \right\}.$$

Prove the associativity and the distributivity property for the above given sets.

Solution.
To prove associative property

1. $A \cup \left(B \cup C \right) = \left(A \cup B \right) \cup C$

    LHS

    $$A \cup \left(B \cup C \right)$$

    $$\left(B \cup C \right) = \left\{ \frac{0.5}{1.0} + \frac{0.25}{1.5} + \frac{0.6}{2.0} + \frac{0.25}{2.5} + \frac{1}{3.0} \right\}$$

    $$A \cup \left(B \cup C \right) = \left\{ \frac{1}{1.0} + \frac{0.65}{1.5} + \frac{0.6}{2.0} + \frac{0.35}{2.5} + \frac{1}{3.0} \right\} \qquad (2.9)$$

    RHS

    $$\left(A \cup B \right) \cup C$$

    $$\left(A \cup B \right) = \left\{ \frac{1}{1.0} + \frac{0.65}{1.5} + \frac{0.6}{2.0} + \frac{0.35}{2.5} + \frac{1}{3.0} \right\}$$

    $$\left(A \cup B \right) \cup C = \left\{ \frac{1}{1.0} + \frac{0.65}{1.5} + \frac{0.6}{2.0} + \frac{0.35}{2.5} + \frac{1}{3.0} \right\} \qquad (2.10)$$

From (2.9) and (2.10),

$$\text{LHS} = \text{RHS}$$

$$\underset{\sim}{A} \cup \left(\underset{\sim}{B} \cup \underset{\sim}{C} \right) = \left(\underset{\sim}{A} \cup \underset{\sim}{B} \right) \cup \underset{\sim}{C}$$

Thus associative property is proved.
To prove distribute property

2. $\left(\underset{\sim}{A} \cap \left(\underset{\sim}{B} \cup \underset{\sim}{C} \right) \right) = \left(\underset{\sim}{A} \cap \underset{\sim}{B} \right) \cup \left(\underset{\sim}{A} \cap \underset{\sim}{C} \right)$

LHS

$$\underset{\sim}{A} \cap \left(\underset{\sim}{B} \cup \underset{\sim}{C} \right)$$

$$\left(\underset{\sim}{B} \cup \underset{\sim}{C} \right) = \left\{ \frac{0.5}{1.5} + \frac{0.25}{1.5} + \frac{0.6}{2.0} + \frac{0.25}{2.5} + \frac{1}{3.0} \right\}$$

$$\underset{\sim}{A} \cap \left(\underset{\sim}{B} \cup \underset{\sim}{C} \right) = \left\{ \frac{0.5}{1.0} + \frac{0.25}{1.5} + \frac{0.4}{2.0} + \frac{0.25}{2.5} + \frac{0}{3.0} \right\} \qquad (2.11)$$

RHS

$$\left(\underset{\sim}{A} \cap \underset{\sim}{B} \right) = \left\{ \frac{0}{1.0} + \frac{0.25}{1.5} + \frac{0.4}{2.0} + \frac{0.25}{2.5} + \frac{0}{3.0} \right\}$$

$$\left(\underset{\sim}{A} \cap \underset{\sim}{C} \right) = \left\{ \frac{0.5}{1.0} + \frac{0.25}{1.5} + \frac{0}{2.0} + \frac{0.25}{2.5} + \frac{0}{3.0} \right\}$$

$$\left(\underset{\sim}{A} \cap \underset{\sim}{B} \right) \cup \left(\underset{\sim}{A} \cap \underset{\sim}{C} \right) = \left\{ \frac{0.5}{1.0} + \frac{0.25}{1.5} + \frac{0.4}{2.0} + \frac{0.25}{2.5} + \frac{0}{3.0} \right\} \quad (2.12)$$

From (2.11) and (2.12), LHS = RHS proving distributive property.

Example 2.8. Consider the following fuzzy sets

$$A = \left\{ \frac{1}{2} + \frac{0.5}{3} + \frac{0.3}{4} + \frac{0.2}{5} \right\},$$

$$B = \left\{ \frac{0.5}{2} + \frac{0.7}{3} + \frac{0.2}{4} + \frac{0.4}{5} \right\}.$$

Calculate, $A \cup B$, $A \cap B$, $\bar{A}$, $\bar{B}$ by a Matlab program.

Solution. The Matlab program for the union, intersection, and complement is

Program

```
% enter the two matrix
u=input('enter the first matrix');
v=input('enter the second matrix');
```

```
option=input('enter the option');
%option 1 Union
%option 2 intersection
%option 3 complement
if (option==1)
        w=max(u,v)
end
if (option==2)
        p=min(u,v)
end
if (option==3)
        option1=input('enter whether to find complement for first matrix
            or second matrix');
        if (option1==1)
                [m,n]=size(u);
                q=ones(m)-u;
        else
                q=ones(m)-v;
        end
end
```

Output

(1) To find union of A and B
    enter the first matrix[1 0.5 0.2 0.3]
    enter the second matrix[0.5 0.7 0.2 0.4]
    enter the option1
    w =
        1.0000 0.7000 0.2000 0.4000

(2) To find Intersection of A and B is
    enter the first matrix[1 0.5 0.2 0.3]
    enter the second matrix[0.5 0.7 0.2 0.4]
    enter the option2
    p =
        0.5000 0.5000 0.2000 0.3000

(3) To find complement of A
    enter the first matrix[1 0.5 0.2 0.3]
    enter the second matrix[0.5 .7 .2 .4]
    enter the option3
    enter the whether to find complement for first matrix or second matrix
    1
    q =
        0 0.5000 0.8000 0.7000

(4) To find complement of B
    enter the first matrix[1 0.5 0.2 0.3]
    enter the second matrix[0.5 .7 .2 .4]
    enter the option3
    enter the whether to find complement for first matrix or second matrix
    2
    q =
      0.5000 0.3000 0.8000 0.6000

Example 2.9. Consider the following fuzzy sets

$$A = \left\{ \frac{0.1}{2} + \frac{0.6}{3} + \frac{0.4}{4} + \frac{0.3}{5} + \frac{0.8}{6} \right\},$$

$$B = \left\{ \frac{0.5}{2} + \frac{0.8}{3} + \frac{0.4}{4} + \frac{0.6}{5} + \frac{0.4}{6} \right\}.$$

Calculate $A \cap \bar{B}$ (difference), $B \cap \bar{A}$ by writing an M-file

Solution. The Matlab program for the difference of A and B is

Program

```
% enter the two matrix
u=input('enter the first matrix');
v=input('enter the second matrix');
option=input('enter the option');
%option 1 u|v
%option 2 v|u
%to find difference of u and v
if option==1
    %to find v complement
        [m,n]=size(v);
        vcomp=ones(m)-v;
        r=min(u,vcomp);
end
%to find difference v and u
if option==2
    %to find u complement
        [m,n]=size(u);
        ucomp=ones(m)-u;
        r=min(ucomp,v);
end
fprintf('output result')
printf(r)
```

Output of the Matlab program

(1) to find A difference B is
    enter the first matrix[0.1 0.6 0.4 0.3 0.8]
    enter the second matrix[0.5 0.8 0.4 0.6 0.4]
    enter the option1
    output result
    r =
      0.1000 0.2000 0.4000 0.3000 0.6000
(2) to find B difference A is
    enter the first matrix[0.1 0.6 0.4 0.3 0.8]
    enter the second matrix[0.5 0.8 0.4 0.6 0.4]
    enter the option2
    output result
    r =
      0.5000 0.4000 0.4000 0.6000 0.2000

Example 2.10. Consider the following fuzzy sets

$$A = \left\{ \frac{0.8}{10} + \frac{0.3}{15} + \frac{0.6}{20} + \frac{0.2}{25} \right\},$$

$$B = \left\{ \frac{0.4}{10} + \frac{0.2}{15} + \frac{0.9}{20} + \frac{0.1}{25} \right\}.$$

Calculate the Demorgan's law $\overline{A \cup B} = \bar{A} \cap \bar{B}$, and $\overline{A \cap B} = \bar{A} \cup \bar{B}$ using a matlab program.

Solution. The Matlab program for the demorgan's law for A and B is

Program

```
% Demorgan's law
% enter the two matrix
u=input('enter the first matrix');
v=input('enter the second matrix');
% first find u's complement
[m,n]=size(u);
ucomp=ones(m)-u;
% second to find v's complement
[a,b]=size(v);
vcomp=ones(a)-v;
p=min(ucomp,vcomp)
q=max(ucomp,vcomp)
fprintf(p)
fprintf(q)
```

Output

enter the first matrix[0.8 0.3 0.6 0.2]
enter the second matrix[0.4 0.2 0.9 0.1]
p =
   0.2000 0.7000 0.1000 0.8000
q =
   0.6000 0.8000 0.4000 0.9000

Summary

In this chapter, we have defined on the classical sets and the fuzzy sets. The various operations and properties of these sets were also dealt in detail. It was found that the variation of the classical and the fuzzy set was in the excluded middle law. The demorgan's law discussed helps in determining some tautologies while some operations are performed. Except for the excluded middle law all other operations and properties are common for the crisp set and the fuzzy set.

Review Questions

1. Define classical set.
2. How is the power set formed from the existing universe?
3. What is the difference between the whole set and the power set?
4. Give a few examples for classical set.
5. What are the operations that can be performed on the classical set?
6. Write the expressions involved for the operations of classical set in function – theoretic form and set- theoretic form.
7. State the properties of classical sets.
8. What is the cardinal number of a set?
9. How is the cardinality defined for a power set?
10. What are the additional properties added with the existing properties of classical set?
11. What is the variation between the law of excluded middle and law of contradiction?
12. State Demorgan's law. Explain the law with the help of Venn diagram representation.
13. Discuss in detail how classical sets are mapped to functions.
14. Define fuzzy set.
15. What is the membership function for fuzzy set $\underset{\sim}{A}$?
16. State the reason for the membership function to be in the interval 0 to 1.
17. What are the operations that can be performed by a fuzzy set?

18. Explain about the properties present in the fuzzy set.

19. Write the expressions for the fuzzy set operation in set-theoretic form and function theoretic form.

20. How is the excluded middle law different for the fuzzy set and the classical set?

21. Discuss about the Demorgan's law for the fuzzy sets. Say whether it is similar to that of classical sets.

Exercise Problems

1. Consider two fuzzy sets one representing a scooter and other van.

$$\underset{\sim}{\text{Scooter}} = \left\{ \frac{0.6}{\text{van}} + \frac{0.3}{\text{motor cycle}} + \frac{0.8}{\text{boat}} + \frac{0.9}{\text{scooter}} + \frac{0.1}{\text{house}} \right\},$$

$$\underset{\sim}{\text{Van}} = \left\{ \frac{1}{\text{Van}} + \frac{0.2}{\text{motor cycle}} + \frac{0.5}{\text{boat}} + \frac{0.3}{\text{scooter}} + \frac{0.2}{\text{house}} \right\}.$$

Find the following:

(a) $\underset{\sim}{\text{Scooter}} \cup \underset{\sim}{\text{Van}}$ (b) $\underset{\sim}{\text{Scooter}} / \underset{\sim}{\text{Van}}$ (c) $\underset{\sim}{\text{Scooter}} \cap \overline{\underset{\sim}{\text{Scooter}}}$

(d) $\overline{\underset{\sim}{\text{Scooter}} \cup \underset{\sim}{\text{Scooter}}}$ (e) $\overline{\underset{\sim}{\text{Scooter}} \cap \underset{\sim}{\text{Scooter}}}$

(f) $\underset{\sim}{\text{Scooter}} \cup \overline{\underset{\sim}{\text{Van}}}$ (g) $\underset{\sim}{\text{Van}} \cup \overline{\underset{\sim}{\text{Van}}}$ (h) $\underset{\sim}{\text{Van}} \cap \overline{\underset{\sim}{\text{Van}}}$

2. Consider flight simulator data, the determination of certain changes in creating conditions of the aircraft is made on the basis of hard breakpoint in the mach region. Let us define a fuzzy set to represent the condition of near a match number of 0.644. A second fuzzy sets in the region of mach number 0.74

$$\underset{\sim}{A} = \text{near mach } 0.64.$$

$$= \left\{ \frac{0.1}{0.630} + \frac{0.6}{0.635} + \frac{1}{0.64} + \frac{0.8}{0.645} + \frac{0.2}{0.650} \right\}.$$

$$\underset{\sim}{B} = \text{near mach } 0.64.$$

$$= \left\{ \frac{0}{0.630} + \frac{0.5}{0.635} + \frac{0.8}{0.64} + \frac{1}{0.645} + \frac{0.4}{0.650} \right\}.$$

Find the following:

(a) $\underset{\sim}{A} \cup \underset{\sim}{B}$ (b) $\underset{\sim}{A} \cap \underset{\sim}{B}$ (c) $\overline{\underset{\sim}{A}}$ (d) $\overline{\underset{\sim}{B}}$ (e) $\underset{\sim}{A} / \underset{\sim}{B}$ (f) $\overline{\underset{\sim}{A} \cup \underset{\sim}{B}}$ (g) $\overline{\underset{\sim}{A} \cap \underset{\sim}{B}}$

3. The continuous form of MOSFET and a transistor are shown in figure below. The discretized membership functions are given by the following equations:

$$\mu_{\underset{\sim}{m}} = \left\{ \frac{0}{0} + \frac{0.4}{2} + \frac{0.6}{4} + \frac{0.7}{6} + \frac{0.8}{8} + \frac{0.9}{10} \right\},$$

$$\mu_{\underset{\sim}{m}} = \left\{ \frac{0}{0} + \frac{0.1}{2} + \frac{0.2}{4} + \frac{0.3}{6} + \frac{0.4}{8} + \frac{0.5}{10} \right\}.$$

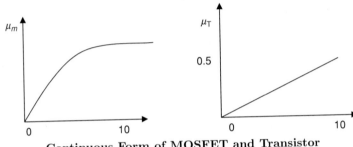

Continuous Form of MOSFET and Transistor

For these two fuzzy calculate the following:

(a) $\mu_{\underset{\sim}{m}} \cup \mu_{\underset{\sim}{T}}$

(b) $\mu_{\underset{\sim}{m}} \cap \mu_{\underset{\sim}{T}}$

(c) $\bar{\mu}_{\underset{\sim}{T}} = 1 - \mu_{\underset{\sim}{T}}$

(d) $\bar{\mu}_{\underset{\sim}{M}} = 1 - \mu_{\underset{\sim}{M}}$

(e) De Morgan's law

4. Samples of new microprocessors IC chip are to be sent to several customers for beta testing. The chips are sorted to meet certain maximum electrical characteristics say frequency, and temperature rating, so that the "best" chips are distributed to preferred customer 1. Suppose that each sample chip is screened and all chips are found to have a maximum operating frequency in the range 7–15 MHz at 20°C. Also the maximum operating temperature range $(20°C \pm \Delta T)$ at 8 MHz is determined. Suppose there are eight sample chips with the following electrical characteristics:

Chip number	1	2	3	4	5	6	7	8
F_{max} (MHz)	6	7	8	9	10	11	12	13
$\Delta T_{max}(°C)$	0	0	20	40	30	50	40	60

The following fuzzy sets are defined.

$$A = \text{Set of "Fast" chips} = \text{chips with } f_{max} \geq 12\,\text{MHz}$$

$$= \left\{ \frac{0}{1} + \frac{0}{2} + \frac{0.1}{3} + \frac{0.1}{4} + \frac{0.2}{5} + \frac{0.8}{6} + \frac{1}{7} + \frac{1}{8} \right\},$$

$$B = \text{Set of "Fast" chips} = \text{chips with } f_{max} \geq 8\,\text{MHz}$$

$$= \left\{ \frac{0.1}{1} + \frac{0.5}{2} + \frac{1}{3} + \frac{1}{4} + \frac{1}{5} + \frac{1}{6} + \frac{1}{7} + \frac{1}{8} \right\},$$

$$C = \text{Set of "Fast" chips} = \text{chips with } T_{max} \geq 10°C$$

$$= \left\{ \frac{0}{1} + \frac{0}{2} + \frac{1}{3} + \frac{1}{4} + \frac{1}{5} + \frac{1}{6} + \frac{1}{7} + \frac{1}{8} \right\},$$

$$D = \text{Set of "Fast" chips} = \text{chips with } T_{max} \geq 50°C$$

$$= \left\{ \frac{0}{1} + \frac{0.6}{2} + \frac{0.1}{3} + \frac{0.2}{4} + \frac{0.5}{5} + \frac{0.8}{6} + \frac{1}{7} + \frac{1}{8} \right\}.$$

Using fuzzy set illustrate various set operations possible.

5. Consider two fuzzy sets A and B as shown in figure below. Write the fuzzy set using membership definition and find the following properties:

(a) $\underset{\sim}{A} \cup \underset{\sim}{B}$ (b) $\underset{\sim}{A} \cap \underset{\sim}{B}$ (c) $\overline{\underset{\sim}{A}}$ (d) $\overline{\underset{\sim}{B}}$ (e) $\underset{\sim}{A} \Big/ \underset{\sim}{B}$ (f) $\overline{\underset{\sim}{A} \cup \underset{\sim}{B}}$

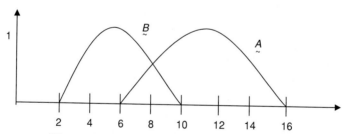

Figure to define the membership function

6. Consider two fuzzy sets $\underset{\sim}{A}$ and $\underset{\sim}{B}$ as shown

$$\underset{\sim}{A} = \left\{ \frac{0}{1} + \frac{0.5}{2} + \frac{0.3}{3} + \frac{0.7}{4} + \frac{0.9}{5} \right\},$$

$$\underset{\sim}{B} = \left\{ \frac{0.2}{1} + \frac{0.4}{2} + \frac{0.6}{3} + \frac{0.9}{4} + \frac{0.4}{8} \right\}.$$

Find (a) $\underset{\sim}{A} \cup \underset{\sim}{B}$ (b) $\underset{\sim}{A} \cap \underset{\sim}{B}$ (c) $\overline{\underset{\sim}{A}}$ (d) $\overline{\underset{\sim}{B}}$ (e) $\underset{\sim}{A} \Big/ \underset{\sim}{B}$ (f) $\overline{\underset{\sim}{A} \cup \underset{\sim}{B}}$

7. Prove why law of excluded middle law and contradiction does not hold good for fuzzy.
8. Consider the universe with two elements $X = \{a, b\}$ and consider Y with $Y = \{0, 1\}$. Find the power set.
9. Consider a universe of four elements $x = \{1, 2, 3, 4, 5, 6\}$. Find the cardinal number power set and cardinality.
10. Consider the following fuzzy sets:

$$A = \left\{ \frac{1}{2} + \frac{0.1}{3} + \frac{0.8}{4} + \frac{0.6}{5} \right\},$$

$$B = \left\{ \frac{0.3}{2} + \frac{0.9}{3} + \frac{0}{4} + \frac{0.4}{5} \right\}.$$

Calculate, $A \cup B$, $A \cap B$, $\bar{A}$, $\bar{B}$ by a Matlab program.
11. For the above problem perform the De Morgan's law by writing an M-file.

3

Classical and Fuzzy Relations

3.1 Introduction

A relation is of fundamental importance in all-engineering, science, and mathematically based fields. It is associated with graph theory, a subject of wide impact in design and data manipulation. Relations are intimately involved in logic, approximate reasoning, classification, rule-based systems, pattern recognition, and control. Relations represent the mapping of the sets. In the case of crisp relation there are only two degrees of relationship between the elements of sets in a crisp relation, i.e., "completely related" and "not related". But fuzzy relations have infinite number of relationship between the extremes of completely related and not related between the elements of two or more sets considered. A crisp relation represents the presence or absence of association, interaction, or interconnectedness between the elements of two or more sets. Degrees of association can be represented by membership grades in a fuzzy relation by membership grades in a fuzzy relation in the same way as degrees of set membership are represented in the fuzzy set. Crisp set can be viewed as a restricted case of the more general fuzzy set concept. In this chapter the classical and fuzzy relation are dealt in detail.

3.2 Cartesian Product of Relation

An ordered sequence of n elements is called as ordered n-tuple. The ordered sequence is in the form of $a_1, a_2, \ldots, a_n$. An unordered sequence is that it is a collection of n elements without restrictions in the order. The $n-$tuple is called as an ordered pair when $n = 2$. For the crisp sets $A_1, A_2, \ldots, A_n$, the set of n-tuples $a_1, a_2, \ldots, a_n$, where $a_1 \in A_1, a_2 \in A_2, \ldots, a_n \in A_n$, is called the Cartesian product of $A_1, A_2, \ldots, A_n$. The Cartesian product is denoted by $A_1 \times A_2 \times \cdots \times A_n$. In Cartesian product the first element in each pair is a member of x and the second element is a member of y formally,

$$x \times y = \{(x, y)/x \in X \text{ and } y \in Y\},$$
$$\text{if } x \neq y \text{ then } x \times y \neq y \times x.$$

If all the A_n are identical and equal to A, then the cartesian product of $A_1, A_2, \ldots, A_n$ becomes A^n.

Example 3.1. The elements in two sets A and B are given as $A = \{0, 1\}$ and $B = \{e, f, g\}$ find the Cartesian product $A \times B, B \times A, A \times A, B \times B$.

Solution. The Cartesian product for the given sets is as follows:

$$A \times B = \{(0, e), (0, f), (0, g), (1, e), (1, f), (1, g)\},$$
$$B \times A = \{(e, 0), (e, 1), (e, 1), (f, 1), (g, 1)\},$$
$$A \times A = A^2 = \{(0, 0), (0, 1), (1, 0), (1, 1)\},$$
$$B \times B = B^2 = \{(e, e), (e, f), (e, g), (f, e), (f, f), (f, g), (g, e), (g, f), (g, g)\}.$$

3.3 Classical Relations

A relation among classical sets $x_1, x_2, \ldots, x_n$ and $y_1, y_2, \ldots, y_n$ is a subset of the Cartesian product. It is denoted either by R or by the abbreviated form

$$X \times Y = \{(x, y)/x \in X, y \in Y.$$

In the case of an ordered pair, the relation is a subset of the Cartesian product $A_1 \times A_2$. This subset of the full Cartesian product is called as the binary relation from A_1 into A_2. If it consists of three, four, or five sets are the subsets of the full Cartesian product, then the relationship is termed as ternary, quaternary, and quinary. But mostly we are into deal with that of the binary relation only.

The strength of the relationship between ordered pairs of elements in each universe is measured by the characteristic function denoted by χ, where a value of unity is associated with complete relationship and a value of zero is associated with no relationship, i.e.,

$$\chi_{x \times y}(x, y) = \begin{cases} 1 & (x, y) \in X \times Y, \\ 0 & (x, y) \notin X \times Y. \end{cases}$$

When the universe or the set are finite, a matrix called as relation matrix can conveniently represent the relation. A two-dimensional matrix represents the binary relation. If $X = \{2, 4, 6\}$ and $Y = \{p, q, r\}$, if they both are related to each other entirely, then the relation between them can be given by:

$$R = \begin{array}{c c c c} & p & q & r \\ 2 & 1 & 1 & 1 \\ 4 & 1 & 1 & 1 \\ 6 & 1 & 1 & 1 \end{array}$$

Example 3.2. Let R be a relation among the three sets
$X= \{$Hindi, English$\}$, $Y = \{$Dollar, Euro, Pound, Rupees$\}$, and $Z = \{$India, Nepal, United States, Canada$\}$

$$R\,(x,y,z) = \{\text{Hindi, Rupees, India}\}$$
$$\{\text{Hindi, Rupees, Nepal}\}$$
$$\{\text{English, Dollar, Canada}\}$$
$$\{\text{English, Dollar, United States}\}.$$

Solution. The relation can be represented as follows:

	India	Nepal	US	Canada
Dollar	0	0	0	0
Euro	0	0	0	0
Pound	0	0	0	0
Rupees	1	1	0	0

Hindi

	India	Nepal	US	Canada
Dollar	0	0	1	1
Euro	0	0	0	0
Pound	0	0	0	0
Rupees	0	0	0	0

English

3.3.1 Cardinality of Crisp Relation

Suppose n elements of the universe X are related to m elements of the universe Y. If the cardinality of X is n_x and the cardinality of Y is n_y, then the cardinality of the relation R, between these two universe $n_{x\times y} = n_x \times n_y$. The cardinality of the power set describing this relation, $P(X \times Y)$ in then $n_p(x \times y) = 2^{n_x n_y}$.

3.3.2 Operations on Crisp Relation

The following are the function – theoretic operations for the two crisp sets (R, S):

Union

$$R \cup S = \chi_{R \cup S}(x,y) : \chi_{R \cup S}(x,y) = \max[\chi_R(x,y), \chi_S(x,y)].$$

Intersection

$$R \cap S = \chi_{R \cap S}(x,y) : \chi_{R \cap S}(x,y) = \min[\chi_R(x,y), \chi_S(x,y)].$$

Complement

$$\bar{R} = \chi_{\bar{R}}(x, y) : \chi_{\bar{R}}(x, y) = 1 - \chi_{R(x,y)}.$$

Containment

$$R \subset S = \chi_R(x, y) : \chi_R(x, y) \leq \chi_S(x, y).$$

These are the various operations that can be performed in a classical relation.

3.3.3 Properties of Crisp Relations

The properties of commutativity, associativity, distributivity, involution, and idempotency as discussed in Sect. 2.2.2 for the classical sets also hold good for crisp relation. This includes DeMorgan's laws and the excluded middle laws too. The null relation O and the complement relation E are given by:

$$O = \begin{bmatrix} 0 & 0 & 0 \\ 0 & 0 & 0 \\ 0 & 0 & 0 \end{bmatrix}, \quad E = \begin{bmatrix} 1 & 1 & 1 \\ 1 & 1 & 1 \\ 1 & 1 & 1 \end{bmatrix}.$$

3.3.4 Composition

Let R be relation that relates elements from universe X to universe Y. Let S be the relation that relates elements from universe Y to universe Z. Let T relates the same element in universe that R contains to the same elements in the universe Z that S contains. The two methods of the composition operations are:

- Max–min composition,
- Max–product composition.

The max–min composition is defined by the set-theoretic and membership function-theoretic expressions:

$$T = R \circ S,$$

$$\chi_T(x, z) = \bigvee_{y \in Y} (\chi_R(x, y) \wedge \chi_S(y, z)).$$

The max–product composition is defined by the set-theoretic and membership function-theoretic expressions:

$$T = R \circ S,$$

$$\chi_T(x, z) = \bigvee_{y \in Y} (\chi_R(x, y) \bullet \chi_S(y, z)).$$

Example 3.3. Using max–min composition find relation between R and S:

$$R = \begin{array}{c} x_1 \\ x_2 \\ x_3 \end{array} \begin{array}{ccc} y_1 & y_2 & y_3 \\ \left[\begin{array}{ccc} 1 & 1 & 0 \\ 0 & 0 & 1 \\ 0 & 1 & 0 \end{array}\right] \end{array}, \quad S = \begin{array}{c} x_1 \\ x_2 \\ x_3 \end{array} \begin{array}{cc} z_1 & z_2 \\ \left[\begin{array}{cc} 0 & 1 \\ 1 & 0 \\ 1 & 1 \end{array}\right] \end{array}.$$

Solution. The max–min composition is given by:

$$\mu_T(x_1, z_1) = \max\,(\min\,(1,\,0),\,\min\,(1,\,1),\,\min\,(0,\,1))$$
$$= \max\,[0,\,1,\,0] = 1,$$
$$\mu_T(x_1, z_2) = \max\,(\min\,(1,\,1),\,\min\,(1,\,0),\,\min\,(1,\,1))$$
$$= \max\,[1,\,0,\,1] = 1,$$
$$\mu_T(x_2, z_1) = \max\,(\min\,(0,\,0),\,\min\,(0,\,1),\,\min\,(1,\,1))$$
$$= \max\,[0,\,0,\,1] = 1,$$
$$\mu_T(x_2, z_2) = \max\,(\min\,(0,\,1),\,\min\,(0,\,0),\,\min\,(1,\,1))$$
$$= \max\,[0,\,0,\,1] = 1,$$
$$\mu_T(x_3, z_1) = \max\,(\min\,(0,\,0),\,\min\,(1,\,1),\,\min\,(0,\,1))$$
$$= \max\,[0,\,1,\,0] = 1,$$
$$\mu_T(x_3, z_2) = \max\,(\min\,(0,\,1),\,\min\,(1,\,0),\,\min\,(0,\,1))$$
$$= \max\,[0,\,0,\,0] = 0,$$

$$R \circ S = \left[\begin{array}{cc} 1 & 1 \\ 1 & 1 \\ 1 & 0 \end{array}\right].$$

3.4 Fuzzy Relations

Fuzzy relations are fuzzy subsets of $X \times Y$, i.e., mapping from $X \to Y$. It maps elements of one universe, X to those of another universe, say Y, through the Cartesian product of the two universes. A fuzzy relation $\underset{\sim}{R}$ is mapping from the Cartesian space $X \times Y$ to the interval $[0, 1]$ where the strength of the mapping is expressed by the membership function of the relation for ordered pairs from the two expressed as or $\mu_R(x, y)$. This can be expressed as

$$\underset{\sim}{R} = \{((x, y), \mu_{\underset{\sim}{R}}(x, y)) | (x, y) \in X \times Y\}$$

is called a fuzzy relation on $X \times Y$.

3.4.1 Cardinality of Fuzzy Relations

As we know that the cardinality of fuzzy set is infinity, the cardinality of fuzzy relations between two or more universes is also infinity.

3.4.2 Operations on Fuzzy Relations

Let R and T be fuzzy relation on Cartesian space $X \times Y$. Then the following operations apply for the membership values for various set operations:

Union

The union of two fuzzy relation R and T is defined by

$$\mu_{R \cup T}(x, y) = \max\left(\mu_R(x, y), \mu_T(x, y)\right).$$

Intersection

The intersection of two fuzzy relation R and T is defined by

$$\mu_{R \cap T}(x, y) = \min\left(\mu_R(x, y), \mu_T(x, y)\right).$$

Complement

The complement of fuzzy relation R is given by

$$\mu_{\bar{R}}(x, y) = 1 - \mu_R(x, y).$$

Containment

The containment of two fuzzy relation R and T is given by

$$R \subset T \Rightarrow \mu_R(x, y) \leq \mu_T(x, y).$$

These are some of the operations performed on the fuzzy relation.

3.4.3 Properties of Fuzzy Relations

The properties of fuzzy relations include commutativity, associativity, distributivity, idempotency, and involution.

Commutativity

$$\mu_{R \cup S}(x, y) = \mu_{S \cup R}(x, y).$$

Associativity

$$\mu_{\left(R \cup S\right) \cup T}(x, y) = \mu_{R \cup \left(S \cup T\right)}(x, y).$$

Distributivity

$$\mu_{\left(R \cup S\right) \cap T}(x, y) = \mu_{R \cup \left(S \cap T\right)}(x, y).$$

Idempotency
$$\mu_{R \cup R}(x,y) = \mu_R.$$

Involution
$$\mu_{\bar{\bar{R}}}(x,y) = \mu_R(x,y).$$

As in case of fuzzy set, in fuzzy relation also excluded middle law and contradiction law does not holds good. The null relation O and the complete relation E are analogous to the null set and the whole set is in set theoretic form. Since the fuzzy relation R is also a fuzzy set, there is a overlap between a relation and its complement, hence

$$R \cup \bar{R} \neq E,$$
$$R \cap \bar{R} \neq O.$$

3.4.4 Fuzzy Cartesian Product and Composition

Let A be a fuzzy set on universe X and B be a fuzzy set on universe Y, then the Cartesian product between fuzzy sets A and B will result in a fuzzy relation R which is contained with the full Cartesian product space or

$$A \times B = R \subset X \times Y,$$

where the fuzzy relation R has membership function.

$$\mu_R(x,y) = \mu_{A \times B}(x,y) = \min\left(\mu_A(x), \mu_B(y)\right).$$

Each fuzzy set could be thought of as a vector of membership values; each value is associated with a particular element in each set. For example, for fuzzy set A that has four elements, hence column vector size 4×1 and for fuzzy set (vector) B that has five elements, hence a row vector of 1×5. The resulting fuzzy relation R will be represented by a matrix of size 4×5 (i.e.,) R will have four rows and five columns.

Example 3.4. Consider two fuzzy sets A and B. A represents universe of three discrete temperatures $x = \{x_1, x_2, x_3\}$ and B represents universe of two discrete flow $y = \{y_1, y_2\}$. Find the fuzzy Cartesian product between them:

$$A = \frac{0.4}{x_1} + \frac{0.7}{x_2} + \frac{0.1}{x_3} \quad \text{and} \quad B = \frac{0.5}{\gamma_1} + \frac{0.8}{\gamma_2}.$$

Solution. A represents column vector of size 3×1 and B represents column vector of size 1×2. The fuzzy Cartesian product results in a fuzzy relation R of size 3×2:

$$A \times B = R = \begin{array}{c} \\ x_1 \\ x_2 \\ x_3 \end{array} \begin{array}{cc} \gamma_1 & \gamma_2 \\ \left[\begin{array}{cc} 0.4 & 0.4 \\ 0.5 & 0.7 \\ 0.1 & 0.1 \end{array} \right] \end{array}.$$

Composition of Fuzzy Relation

Let R be a fuzzy relation on the Cartesian space $X \times Y$, S be a fuzzy relation on $Y \times Z$, and T be a fuzzy relation on $X \times Z$, then the fuzzy set max–min composition is defined as:

$$T = R \circ S \text{ (set-theoretic notation)}$$

$$= \left\{ (x,z), \max_y \left\{ \min \left(\mu_R (x,y), \mu_S (y,z) \right) \right\} \Big/ x \in X, y \in Y z \in Z \right\}.$$

In function-theoretic form

$$\mu_T (x,z) = \bigvee_{y \in Y} \left(\mu_R (x,y) \wedge \mu_S (y,z) \right).$$

In fuzzy max–product composition is defined in terms of set-theoretic

$$T = R \circ S$$

$$= \left\{ (x,z), \max_y \left(\mu_R (x,y) * \mu_S (y,z) \right) \Big/ x \in X, y \in Y, z \in Z \right\}$$

$$\mu_T (x,z) = \bigvee_{y \in Y} \left(\mu_R (x,y) \bullet \mu_S (y,z) \right).$$

Max–Average Composition

The max–average composition $S \underset{\text{org}}{O} R$ are then defined as follows:

$$S \underset{\text{org}}{O} R (x,z) = \left\{ (x,z), \frac{1}{2} \max \left(\mu_R (x,y) + \mu_S (y,z) \right) \Big/ x \in X, y \in Y, z \in Z \right\}.$$

Example 3.5. Consider fuzzy relations:

$$R = \begin{array}{c} \\ x_1 \\ x_2 \end{array} \begin{array}{cc} y_1 & y_2 \\ \left[\begin{array}{cc} 0.7 & 0.6 \\ 0.8 & 0.3 \end{array} \right] \end{array}, \quad S = \begin{array}{c} \\ y_1 \\ y_2 \end{array} \begin{array}{ccc} z_1 & z_2 & z_2 \\ \left[\begin{array}{ccc} 0.8 & 0.5 & 0.4 \\ 0.1 & 0.6 & 0.7 \end{array} \right] \end{array}.$$

Find the relation $T = R \circ S$ using max–min and max–product composition.

Solution. **Max–Min Composition**

$$T = R \circ S$$
$$\mu_T(x_1, z_1) = \max\left[\min(0.7, 0.8), \min(0.6, 0.1)\right]$$
$$= \max[0.7, 0.1]$$
$$= 0.7,$$
$$\mu_T(x_1, z_2) = \max\left[\min(0.7, 0.5), \min(0.6, 0.6)\right]$$
$$= \max[0.5, 0.6]$$
$$= 0.6,$$
$$\mu_T(x_1, z_3) = \max\left[\min(0.7, 0.4), \min(0.6, 0.7)\right]$$
$$= \max[0.4, 0.7]$$
$$= 0.7,$$
$$\mu_T(x_2, z_1) = \max\left[\min(0.8, 0.8), \min(0.3, 0.1)\right]$$
$$= \max[0.8, 0.1]$$
$$= 0.8,$$
$$\mu_T(x_2, z_2) = \max\left[\min(0.8, 0.5), \min(0.3, 0.6)\right]$$
$$= \max[0.5, 0.3]$$
$$= 0.5,$$
$$\mu_T(x_2, z_3) = \max\left[\min(0.8, 0.4), \min(0.3, 0.7)\right]$$
$$= 0.4,$$

$$\underset{\sim}{S} = \begin{array}{c} \\ x_1 \\ x_2 \end{array} \overset{\begin{array}{ccc} z_1 & z_2 & z_2 \end{array}}{\begin{bmatrix} 0.7 & 0.6 & 0.7 \\ 0.8 & 0.5 & 0.4 \end{bmatrix}}.$$

Max–Product Composition

$$\mu_T(x_1, z_1) = \max\left[\min(0.7 \times 0.8), \min(0.6 \times 0.1)\right]$$
$$= \max[0.56, 0.06]$$
$$= 0.56,$$
$$\mu_T(x_1, z_2) = \max\left[\min(0.7 \times 0.5), \min(0.6 \times 0.6)\right]$$
$$= \max[0.35, 0.36]$$
$$= 0.36,$$
$$\mu_T(x_1, z_3) = \max\left[\min(0.7 \times 0.4), \min(0.5 \times 0.7)\right]$$
$$= \max[0.28, 0.35]$$
$$= 0.35,$$

$$\mu_T (x_2, z_1) = \max \left[\min (0.8 \times 0.8), \min (0.3 \times 0.1)\right]$$
$$= \max \left[0.64, 0.03\right]$$
$$= 0.64,$$
$$\mu_T (x_2, z_2) = \max \left[\min (0.8 \times 0.5), \min (0.3 \times 0.6)\right]$$
$$= \max \left[0.40, 0.18\right]$$
$$= 0.40,$$
$$\mu_T (x_2, z_3) = \max \left[\min (0.8 \times 0.4), \min (0.3 \times 0.7)\right]$$
$$= \max \left[0.32, 0.21\right]$$
$$= 0.32,$$
$$\underset{\sim}{T} = \begin{bmatrix} 0.56 & 0.36 & 0.35 \\ 0.64 & 0.40 & 0.32 \end{bmatrix}.$$

Example 3.6. In the field of computer networking there is an imprecise relationship between the level of use of a network communication bandwidth and the latency experienced in peer-to-peer communication. Let $\underset{\sim}{X}$ be a fuzzy set of use levels (in terms of the percentage of full bandwidth used) and $\underset{\sim}{Y}$ be a fuzzy set of latencies (in milliseconds) with the following membership function:

$$\underset{\sim}{X} = \left\{ \frac{02}{10} + \frac{0.5}{20} + \frac{0.8}{40} + \frac{1.0}{60} + \frac{0.6}{80} + \frac{0.1}{100} \right\},$$

$$\underset{\sim}{Y} = \left\{ \frac{0.3}{0.5} + \frac{0.6}{1} + \frac{0.9}{1.5} + \frac{1.0}{4} + \frac{0.6}{8} + \frac{0.3}{20} \right\}.$$

(a) Find the Cartesian product represented by the relation $\underset{\sim}{R} = \underset{\sim}{X} \times \underset{\sim}{Y}$.

Now, suppose we have second fuzzy set of bandwidth usage given by

$$\underset{\sim}{X} = \left\{ \frac{0.3}{10} + \frac{0.6}{20} + \frac{0.7}{40} + \frac{0.9}{60} + \frac{1}{80} + \frac{0.5}{100} \right\}.$$

(b) Find $\underset{\sim}{S} = \underset{\sim 1 \times 6}{Z} \circ \underset{\sim 6 \times 6}{R}$ using (1) Max–min composition and (2) Using max–product composition.

Solution. (a) Cartesian product

$$\underset{\sim}{R} = \underset{\sim}{X} \times \underset{\sim}{Y}$$

fuzzy relation $\underset{\sim}{R}$ is given by

$$\underset{\sim}{X} = \mu_R (x, y) = \mu_{A \times B} (x, y)$$

$$= \min \left(\mu_A (x), \mu_B (y) \right),$$

$$R = \begin{bmatrix} 0.2 & 0.2 & 0.2 & 0.2 & 0.2 & 0.2 \\ 0.3 & 0.5 & 0.5 & 0.5 & 0.5 & 0.3 \\ 0.3 & 0.6 & 0.8 & 0.8 & 0.6 & 0.3 \\ 0.3 & 0.6 & 0.9 & 1.0 & 0.6 & 0.3 \\ 0.3 & 0.6 & 0.6 & 0.6 & 0.6 & 0.3 \\ 0.1 & 0.1 & 0.1 & 0.1 & 0.1 & 0.1 \end{bmatrix}.$$

(b) (1) Max–min composition

$$z_1 \times 6 = \left\{ \frac{0.3}{10} + \frac{0.6}{20} + \frac{0.7}{40} + \frac{0.9}{60} + \frac{1}{80} + \frac{0.5}{100} \right\},$$

$$S = \begin{bmatrix} 0.3 \ 0.6 \ 0.9 \ 0.9 \ 0.6 \ 0.3 \end{bmatrix}.$$

(2) Max–product composition

$$S(x_1, z_1) = \max(0.06, 0.18, 0.21, 0.27, 0.3, 0.05)$$
$$= 0.27,$$
$$S(x_1, z_2) = \max(0.06, 0.30, 0.42, 0.54, 0.6, 0.05)$$
$$= 0.6,$$
$$S(x_1, z_3) = \max(0.06, 0.30, 0.56, 0.81, 0.6, 0.05)$$
$$= 0.81,$$
$$S(x_1, z_4) = \max(0.06, 0.30, 0.56, 0.9, 0.6, 0.05)$$
$$= 0.9,$$
$$S(x_1, z_5) = \max(0.06, 0.30, 0.42, 0.54, 0.6, 0.05)$$
$$= 0.6,$$
$$S(x_1, z_6) = \max(0.06, 0.18, 0.21, 0.27, 0.3, 0.05)$$
$$= 0.3,$$
$$S = \begin{bmatrix} 0.27 & 0.6 & 0.81 & 0.9 & 0.6 & 0.3 \end{bmatrix}.$$

Example 3.7. Find max–average composition for $R_1(x, y)$ and $R_2(y, z)$ defined by the following relational matrices:

$$R_1 = \begin{array}{c|ccccc} & y_1 & y_2 & y_3 & y_4 & y_5 \\ \hline x_1 & 0.1 & 0.2 & 0 & 1 & 0.7 \\ x_2 & 0.3 & 0.5 & 0 & 0.2 & 1 \\ x_3 & 0.8 & 0 & 1 & 0.4 & 0.3 \end{array}$$

$$S_2 = \begin{array}{c|cccc} & z_1 & z_2 & z_3 & z_4 \\ \hline y_1 & 0.9 & 0 & 0.3 & 0.4 \\ y_2 & 0.2 & 1 & 0.8 & 0 \\ y_3 & 0.8 & 0 & 0.7 & 1 \\ y_4 & 0.4 & 0.2 & 0.3 & 0 \\ y_5 & 0 & 1 & 0 & 0.8 \end{array}$$

Solution. The max–average composition is given by

$$R \circ S = \frac{1}{2} \max \left\{ \mu_R\left(x, y_i\right) + \mu_S\left(y_{ij} z_1\right) \right\}$$

$$= \frac{1}{2}(1.4) = 0.7$$

$$\begin{array}{c|c} i & \mu(x_1, y_i) + \mu(y_i, z_1) \\ \hline 1 & 1 \\ 2 & 0.4 \\ 3 & 0.8 \\ 4 & 1.4 \\ 5 & 0.7 \end{array}$$

$$\frac{1}{2} \max \left\{ \mu_R\left(x_1, y_i\right) \; \mu_S\left(y_i, z_2\right) \right\}$$

$$\begin{array}{c|c} I & \mu(x_1, y_1) + \mu(y_i, z_2) \\ \hline 1 & 0.1 \\ 2 & 1.2 \\ 3 & 0 \\ 4 & 1.2 \\ 5 & 1.7 \end{array}$$

$$= \frac{1}{2}(1.7) = 0.85$$

$$T\left(x_1, z_3\right) = \frac{1}{2} \max \left\{ \mu_R\left(x_1, y_i\right) \mu_S\left(y_i, z_3\right) \right\}$$

$$\begin{array}{c|c} i & \mu(x_1, y_1) + \mu(y_i, z_3) \\ \hline 1 & 0.4 \\ 2 & 1.0 \\ 3 & 0.7 \\ 4 & 1.3 \\ 5 & 0.7 \end{array}$$

$$= \frac{1}{2}(1.3) = 0.65$$

$$T\left(x_1, z_4\right) = \frac{1}{2} \max \left\{ \mu_R\left(x_1, y_i\right) \mu_S\left(y_i, z_4\right) \right\}$$

i	$\mu(x_1, y_1) + \mu(y_i, z_4)$
1	0.5
2	0.2
3	1
4	1
5	1.5

$$= \frac{1}{2}(1.5) = 0.75$$

$$\underset{\sim}{T}(x_2,\ z_1) = \frac{1}{2}\max\left\{\mu_{\underset{\sim}{R}}(x_2,\ y_i)\,\mu_{\underset{\sim}{S}}(y_i,\ z_1)\right\}$$

i	$\mu(x_2, y_1) + \mu(y_i, z_4)$
1	1.2
2	0.4
3	0.8
4	0.6
5	1

$$= \frac{1}{2}(1.2) = 0.6$$

$$\underset{\sim}{T}(x_2,\ z_2) = \frac{1}{2}\max\left\{\mu_{\underset{\sim}{R}}(x_2, y_i)\,\mu_{\underset{\sim}{S}}(y_i, z_2)\right\}$$

i	$\mu(x_2, y_1) + \mu(y_i, z_2)$
1	0.3
2	1.5
3	0
4	0.4
5	2

$$= \frac{1}{2}(2) = 1$$

$$\underset{\sim}{T}(x_2,\ z_3) = \frac{1}{2}\max\left\{\mu_{\underset{\sim}{R}}(x_2, y_i)\,\mu_{\underset{\sim}{S}}(y_i, z_3)\right\}$$

i	$\mu(x_2, y_1) + \mu(y_i, z_3)$
1	0.6
2	1.3
3	0.7
4	0.5
5	1

$$= \frac{1}{2}(1.3) = 0.65$$

$$\underset{\sim}{T}(x_2,\ z_4) = \frac{1}{2}\max\left\{\mu_{\underset{\sim}{R}}(x_2, y_i)\,\mu_{\underset{\sim}{S}}(y_i, z_4)\right\}$$

i	$\mu(x_2, y_1) + \mu(y_i, z_4)$
1	0.7
2	0.5
3	1
4	0.2
5	1.8

$$= \frac{1}{2}(1.8) = 0.9$$

$$\mathop{T}_{\sim}(x_3, z_1) = \frac{1}{2}\max\left\{\mu_{\mathop{R}_{\sim}}(x_3, y_i)\,\mu_{\mathop{S}_{\sim}}(y_i, z_1)\right\}$$

i	$\mu(x_3, y_1) + \mu(y_i, z_1)$
1	1.7
2	0.2
3	1.8
4	0.8
5	0.3

$$= \frac{1}{2}(1.8) = 0.85$$

$$\mathop{T}_{\sim}(x_3, z_2) = \frac{1}{2}\max\left\{\mu_{\mathop{R}_{\sim}}(x_3, y_i)\,\mu_{\mathop{S}_{\sim}}(y_i, z_2)\right\}$$

i	$\mu(x_3, y_1) + \mu(y_i, z_2)$
1	0.8
2	1
3	1
4	0.6
5	1.3

$$= \frac{1}{2}(1.3) = 0.65$$

$$\mathop{T}_{\sim}(x_3, z_3) = \frac{1}{2}\max\left\{\mu_{\mathop{R}_{\sim}}(x_3, y_i)\,\mu_{\mathop{S}_{\sim}}(y_i, z_3)\right\}$$

i	$\mu(x_3, y_1) + \mu(y_i, z_3)$
1	1.1
2	0.8
3	1.7
4	0.7
5	0.3

$$= \frac{1}{2}(1.7) = 0.85$$

$$\mathop{T}_{\sim}(x_3, z_4) = \frac{1}{2}\max\left\{\mu_{\mathop{R}_{\sim}}(x_3, y_i)\,\mu_{\mathop{S}_{\sim}}(y_i, z_4)\right\}$$

i	$\mu(x_3, y_1) + \mu(y_i, z_4)$
1	1.2
2	0
3	2
4	0.4
5	1.1

$$= \frac{1}{2}(2) = 1,$$

$$\underset{\sim}{T}(x,\ y) = \begin{array}{c} x_1 \\ x_2 \\ x_3 \end{array} \overset{\begin{array}{cccc} z_1 & z_2 & z_3 & z_4 \end{array}}{\begin{bmatrix} 0.7 & 0.85 & 0.65 & 0.75 \\ 0.6 & 1 & 0.65 & 0.9 \\ 0.9 & 0.65 & 0.85 & 1 \end{bmatrix}}.$$

3.5 Tolerance and Equivalence Relations

Relations exhibit various other properties apart from that discussed in Sects. 3.3 and 3.4. It is already said that the relation can be used in graph theory. The various other properties that are dealt here include reflexivity, symmetry, and transitivity. These are discussed in detail for the crisp and fuzzy relations and are called as equivalence relation. Apart from these, tolerance relations of both fuzzy and crisp relations are also described.

3.5.1 Crisp Relation

Crisp Equivalence Relation

A relation R is called as an equivalence relation if it satisfies the following properties. They are (1) reflexivity, (2) symmetry, and (3) transitivity.

Reflexivity

When a relation satisfies the reflexive property then every vertex in the graph originates a single loop. This is shown in Fig. 3.1.

For matrix relation reflexivity is given by,

$$(x_i, x_j) \in R \quad \text{or} \quad \chi_R(x_i, x_j) = 1.$$

Symmetry

When a relation satisfies symmetric property then in the graph for every edge pointing from vertex i to vertex j there is a edge pointing in the opposite direction, i.e., from vertex j to vertex i. This is shown in Fig. 3.2.

For matrix relation symmetry is given by

$$(x_i, x_j) \in R \rightarrow (x_i, x_j) \in R \quad \text{or} \quad \chi_R(x_i, x_j) = \chi_R(x_j, x_i).$$

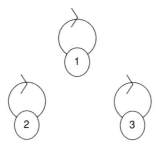

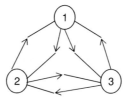

Fig. 3.1. Reflexivity

Fig. 3.2. Symmetry

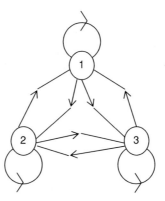

Fig. 3.3. Transitivity

Transitivity

When a relation satisfies transitivity property then every pair of edges in the graph, one pointing from vertex i to vertex j and the other from vertex j to vertex $k(i,j,k = 1,2,3)$ there is an edge pointing from vertex i directly to vertex k. This is shown in Fig. 3.3.

For matrix relation transitivity is given by

$$(x_i, x_j) \in R \quad \text{and} \quad (x_j, x_k) \in R \rightarrow (x_i, x_k) \in R$$

or

$$\chi_R(x_i, x_j) \quad \text{and} \quad \chi_R(x_j, x_k) = 1 \rightarrow \chi_R(x_i, x_k) = 1.$$

Crisp Tolerance Relation

A binary relation R on a universe X that is reflexive and symmetric is called as compatibility relation or a tolerance relation. When a relation is a reflexive and symmetric it is called as proximity relation. A tolerance relation R can be reformed into an equivalence relation by at most $(n-1)$ composition with itself, where n is the cardinal number of the set defining R, in the case X, i.e.,

$$R_1^{n-1} = R_1 \circ R_1 \circ \cdots \circ R_1 = R.$$

3.5.2 Fuzzy Relation

Fuzzy Equivalence Relation

A fuzzy relation $\underset{\sim}{R}$ on a single universe X is also a relation from X to X. It is a fuzzy equivalence relation if all three of the following properties for matrix relations define it; e.g.,

Reflexivity $\mu_{\underset{\sim}{R}}(x_i, x_i) = 1,$

Symmetry $\mu_{\underset{\sim}{R}}(x_i, x_j) = \mu_{\underset{\sim}{R}}(x_j, x_i),$

Transitivity $\mu_{\underset{\sim}{R}}(x_i, x_j) = \lambda_1$ and $\mu_{\underset{\sim}{R}}(x_j, x_k) = \lambda_2$

$\rightarrow \mu_{\underset{\sim}{R}}(x_i, x_k) = \lambda$

where $\lambda \geq \min[\lambda_1\ \lambda_2]$.

Fuzzy Tolerance Relation

If the fuzzy relation R_1 satisfies both reflexivity and symmetry then it is called as fuzzy tolerance relation R_1. A fuzzy tolerance relation can be reformed into fuzzy equivalence relation at most $(n-1)$ compositions. This is given by:

$$\underset{\sim}{R_1^{n-1}} = \underset{\sim}{R_1} \circ \underset{\sim}{R_1} \circ \cdots \circ \underset{\sim}{R_1} = \underset{\sim}{R}.$$

3.5.3 Solved Examples

Example 3.8. Consider fuzzy relation

$$R = \begin{bmatrix} 1 & 0.6 & 0 & 0.2 & 0.3 \\ 0.6 & 1 & 0.4 & 0 & 0.8 \\ 0 & 0.4 & 1 & 0 & 0 \\ 0.2 & 0 & 0 & 1 & 0.5 \\ 0.3 & 0.8 & 0 & 0.5 & 1 \end{bmatrix}$$

is reflexive and symmetric. However it is not transitive, e.g.,

$$\mu_R\widetilde{(x_1, x_2)} = 0.6, \quad \mu_R (x_2, x_5) = 0.8,$$

$$\mu_R\widetilde{(x_1, x_5)} = 0.3 \leq \min(0.8, 0.6).$$

One composition results in the following relation:

$$R^2 = R \circ R = \begin{bmatrix} 1 & 0.6 & 0.4 & 0.3 & 0.6 \\ 0.6 & 1 & 0.4 & 0.5 & 0.8 \\ 0.4 & 0.4 & 1 & 0 & 0.4 \\ 0.3 & 0.5 & 0 & 1 & 0.5 \\ 0.6 & 0.8 & 0.4 & 0.5 & 1 \end{bmatrix},$$

where transitivity still does not result for example

$$\mu_R (x_1, x_2) = 0.6, \quad \mu_R (x_2, x_4) = 0.8,$$

$$\mu_R\widetilde{(x_1, x_4)} = 0.3 \leq \min(0.6, \ 0.5).$$

Finally after one or two more compositions, transitivity results

$$R^3 = \begin{bmatrix} 1 & 0.6 & 0.4 & 0.5 & 0.6 \\ 0.6 & 1 & 0.4 & 0.5 & 0.8 \\ 0.4 & 0.4 & 1 & 0.4 & 0.4 \\ 0.5 & 0.5 & 0.4 & 1 & 0.5 \\ 0.6 & 0.8 & 0.4 & 0.5 & 1 \end{bmatrix},$$

$$R^3(x_1, x_2) = 0.6 \geq 0.5,$$
$$R^3(x_2, x_4) = 0.3 \geq 0.5,$$
$$R^3(x_1, x_4) = 0.5 \geq 0.5.$$

The transitivity is satisfied hence equivalence relation is also satisfied.

Example 3.9. Find whether the given matrix is reflexive or not

$$R = \begin{bmatrix} 1 & 1 & 0 & 0 & 0 \\ 1 & 1 & 0 & 0 & 1 \\ 0 & 0 & 1 & 0 & 0 \\ 0 & 0 & 0 & 1 & 0 \\ 0 & 1 & 0 & 0 & 1 \end{bmatrix}$$ by writing an M-file.

Solution. The Matlab program to find the reflexivity is

Program

```
% to find reflexvity
r=input('enter the matrix');
sum1=0;
[m,n]=size(r);
if(m==n)
        for i=1:m
        %to find the reflexivity
                if(r(1,1)==r(i,i))
                else
                fprintf('the given matrix is irreflexive');
                sum1=1;
                break;
                end
        end
        if(sum1 ~= 1)
                fprintf('the given matrix is reflexive');
        end
end
```

Output

enter the matrix[1 1 0 0 0;1 1 0 0 1;0 0 1 0 0;0 0 0 1 0;0 1 0 0 1]
the given matrix is reflexive.

Example 3.10. Find whether the given matrix is symmetry or not

$$R = \begin{bmatrix} 1 & 0.5 & 0.3 & 0.6 & 0 \\ 0.5 & 1 & 0.7 & 0.5 & 0.9 \\ 0.3 & 0.7 & 1 & 0.6 & 0 \\ 0.6 & 0.5 & 0.6 & 1 & 0.5 \\ 0 & 0.9 & 0 & 0.5 & 1 \end{bmatrix}$$ by a Matlab program.

Solution. The Matlab program is

Program

```
% to find the symmetric
r=input('enter the matrix');
sum=0;
for i=1:m
        for j=1:n
                if(r(i,j)==r(j,i))
                else
                        fprintf('matrix is not symmetry');
                        sum=1;
```

```
                    break;
                end
            end
        if(sum==1)
            break;
        end
    end
    if(sum~=1)
        fprintf('The given matrix is symmetry');
    end
```

Output

enter the matrix[1 0.5 0.3 0.6 0;0.5 1 0.7 0.5 0.9;0.3 0.7 1 0.6 0;0.6 0.5 0.6 1 0.5;0 0.9 0 0.5 1]

r =

1.0000	0.5000	0.3000	0.6000	0
0.5000	1.0000	0.7000	0.5000	0.9000
0.3000	0.7000	1.0000	0.6000	0
0.6000	0.5000	0.6000	1.0000	0.5000
0	0.9000	0	0.5000	1.0000

The given matrix is symmetry.

Example 3.11. Find whether the given matrix is a tolerance matrix or not

$$R = \begin{bmatrix} 1 & 1 & 0 & 0 & 0 \\ 1 & 1 & 0 & 0 & 1 \\ 0 & 0 & 1 & 0 & 0 \\ 0 & 0 & 0 & 1 & 0 \\ 0 & 1 & 0 & 0 & 1 \end{bmatrix}$$ by writing a Matlab file.

Solution. The Matlab program is given

Program

```
% to find tolerance
r=input('enter the matrix')
sum=0;
sum1=0;
[m,n]=size(r);
if(m==n)
    for i=1:m
        if(r(1,1)==r(i,i))
        else
            fprintf('the given matrix is irreflexive and');
            sum1=1;
            break;
        end
    end
end
```

```
    if(sum1 ~= 1)
        fprintf('the given matrix is reflexive and');
    end
    for i=1:m
        for j=1:n
            if(r(i,j)==r(j,i))
            else
                fprintf('not symmetry hence');
                sum=1;
                break;
            end
        end
        if(sum==1)
            break;
        end
    end
    if(sum~=1)
        fprintf('symmetry hence');
    end
end
if(sum1~=1)
    if(sum~=1)
        fprintf('the given matrix tolerance matrix');
    else
        fprintf('the given matrix is not tolerance matrix');
    end
else
    fprintf('the given matrix is not tolerance matrix');
end
```

Output

enter the matrix[1 1 0 0 0;1 1 0 0 1;0 0 1 0 0;0 0 0 1 0;0 1 0 0 1]

r =

1	1	0	0	0
1	1	0	0	1
0	0	1	0	0
0	0	0	1	0
0	1	0	0	1

The given matrix is reflexive and symmetry hence the given matrix is tolerance matrix.

Example 3.12. To find whether the given matrix is transitivity or not

$$R = \begin{bmatrix} 1 & 1 & 0 & 0 & 1 \\ 1 & 1 & 0 & 0 & 1 \\ 0 & 0 & 1 & 0 & 0 \\ 0 & 0 & 0 & 1 & 0 \\ 1 & 1 & 0 & 0 & 1 \end{bmatrix}$$ using a Matlab program.

Solution. The Matlab program is

Program

```
% to find transitvity matrix
r=input('enter the matrix')
sum2=0;
[m,n]=size(r);
for i=1:m
   for j=1:n
      for k=n:1
         lambda1=r(i,j);
         lambda2=r(j,k);
         lambda3=r(i,k);
         p=min(lambda1,lambda2)
         if(lambda3 <= p)
            fprintf('The given matrix is not transitivity');
            sum2=1;
            break;
         end
      end
      if(sum2==1)
         break;
      end
   end
   if(sum2==1)
      break;
   end
end
if(sum~=2)
   fprintf('The given matrix is transitivity');
end
```

Output
enter the matrix[1 1 0 0 1;1 1 0 0 1;0 0 1 0 0;0 0 0 1 0;1 1 0 0 1]

r =
1 1 0 0 1
1 1 0 0 1
0 0 1 0 0
0 0 0 1 0
1 1 0 0 1

The given matrix is transitivity.

Example 3.13. To find whether the given matrix is equivalence or not

$$R = \begin{bmatrix} 1 & 0.8 & 0.4 & 0.5 & 0.8 \\ 0.8 & 1 & 0.4 & 0.5 & 0.9 \\ 0.4 & 0.4 & 1 & 0.4 & 0.4 \\ 0.5 & 0.5 & 0.4 & 1 & 0.5 \\ 0.8 & 0.9 & 0.4 & 0.5 & 1 \end{bmatrix} \text{ by means of a Matlab program.}$$

Solution. The Matlab program is

Program

```
% to find equivalence matrix
r=input('enter the matrix')
sum=0;
sum1=0;
sum2=0;
[m,n]=size(r);
if(m==n)
    for i=1:m
        if(r(1,1)==r(i,i))
        else
            fprintf('the given matirx is irreflexive');
            sum1=1;
            break;
        end
    end
    if(sum1 ~= 1)
        fprintf('the given matrix is reflexive');
    end
    for i=1:m
        for j=1:n
            if(r(i,j)==r(j,i))
            else
                fprintf('and not symmetry');
                sum=1;
                break;
            end
        end
    end
```

```
      if(sum==1)
         break;
      end
   end
   if(sum~=1)
      fprintf('and symmetry');
   end
end
for i=1:m
   for j=1:n
      for k=n:1
         lambda1=r(i,j);
         lambda2=r(j,k);
         lambda3=r(i,k);
         p=min(lambda1,lambda2)
         if(lamda3 <= p)
            fprintf('and not transitivity hence');
            sum2=1;
            break;
         end
      end
      if(sum2==1)
         break;
      end
   end
   if(sum2==1)
      break;
   end
end
if(sum~=2)
      fprintf('and transitivity hence');
   end
if(sum1~=1)
   if(sum~=1)
      if(sum~=2)
      fprintf('the given matrix is equivalence matrix');
   else
      fprintf('the given matrix is not equivalence matrix');
   end
else
   fprintf('not equivalence matrix');
end
else
   fprintf('not equivalence matrix');
end
```

Output
enter the matrix[1 0.8 0.4 0.5 0.8;0.8 1 0.4 0.5 0.9;0.4 0.4 1 0.4 0.4;0.5 0.5 0.4 1 0.5;0.8 0.9 0.4 0.5 1]

r =
        1.0000 0.8000 0.4000 0.5000 0.8000
        0.8000 1.0000 0.4000 0.5000 0.9000
        0.4000 0.4000 1.0000 0.4000 0.4000
        0.5000 0.5000 0.4000 1.0000 0.5000
        0.8000 0.9000 0.4000 0.5000 1.0000

The given matrix is reflexive, symmetry, and transitivity hence the given matrix is equivalence matrix.

Example 3.14. To find whether the given matrix is equivalence or not

$$
R = \begin{bmatrix} 1 & 0.8 & 0 & 0.1 & 0.2 \\ 0.8 & 1 & 0.4 & 0 & 0.9 \\ 0. & 0.4 & 1 & 0 & 0 \\ 0.1 & 0 & 0 & 1 & 0.5 \\ 0.2 & 0.9 & 0 & 0.5 & 1 \end{bmatrix} \text{ using a Matlab program.}
$$

Solution. The Matlab program is

Program

```
% to find equivalence matrix
r=input('enter the matrix')
sum=0;
sum1=0;
sum2=0;
sum3=0;
[m,n]=size(r);
l=m;
if(m==n)
   for i=1:m
      if(r(1,1)==r(i,i))
      else
         fprintf('the given matrix is irreflexive');
         sum1=1;
         break;
      end
   end
   if(sum1 ~= 1)
      fprintf('the given matrix is reflexive');
   end
   m
   n
   [m,n]=size(r)
```

```
    for i=1:m
        for j=1:n
            if(r(i,j)==r(j,i))
            else
                fprintf(',not symmetry');
                sum=1;
                break;
            end
        end
        if(sum==1)
            break;
        end
    end
    if(sum~=1)
        fprintf(',symmetry');
    end
for i=1:m
    for j=1:n
        for k=l:-1:1
            lambda1=r(i,j);
            lambda2=r(j,k);
            lambda3=r(i,k);
            p=min(lambda1,lambda2);
            if(lambda3 >= p)
            else
                sum2=1;
                break;
            end
        end
    end
end
if(sum2 ~= 1)
    fprintf('and transitivity hence');
else
    fprintf('and not transitivity hence');
end
if(sum1~=1)
    if(sum~=1)
        if(sum2~=1)
        fprintf('the given matrix is equivalence matrix');
    else
        fprintf('the given matrix is not equivalence matrix');
    end
else
    fprintf('not equivalence matrix');
```

```
end
else
    fprintf('not equivalence matrix');
end
end
```

Output

enter the matrix[1 0.8 0 0.1 0.2;0.8 1 0.4 0 0.9;0 0.4 1 0 0;0.1 0 0 1 0.5;0.2 .9 0 0.5 1]

r =

1.0000	0.8000	0	0.1000	0.2000
0.8000	1.0000	0.4000	0	0.9000
0	0.4000	1.0000	0	0
0.1000	0	0	1.0000	0.5000
0.2000	0.9000	0	0.5000	1.0000

The given matrix is reflexive, symmetry and not transitivity hence the given matrix is not equivalence matrix.

Example 3.15. Find the fuzzy relation between two vectors R and S

$$R =$$
$$\begin{array}{cc} 0.70 & 0.50 \\ 0.80 & 0.40 \end{array}$$

$$S =$$
$$\begin{array}{ccc} 0.90 & 0.60 & 0.20 \\ 0.10 & 0.70 & 0.50 \end{array}$$

Using max–product method by a Matlab program.

Solution. The Matlab program for the max–product method is shown below

Program

```
%enter the two input vectors
R=input('enter the first vector')
S=input('enter the second vector')
%find the size of the two vector
[m,n]=size(R)
[a,b]=size(S)
if(n==a)
            for i=1:m
                    for j=1:b
                    c=R(i,:);
                    d=S(:,j);
                    [f,g]=size(c);
                    [h,q]=size(d);
```

```
                    %finding product
                         for l=1:g
                              e(1,l)=c(1,l)*d(1,1);
                    end
                    %finding maximum
                    t(i,j)=max(e);
                    end
          end
else
          display('cannot be find min-max');
end
```

Output

enter the first vector[0.7 0.5;0.8 0.4]

R =

   0.7000 0.5000
   0.8000 0.4000

enter the second vector[0.9 0.6 0.2;0.1 0.7 0.5]

S =

   0.9000 0.6000 0.2000
   0.1000 0.7000 0.5000

the final max–product answer is

t =

   0.6300 0.4200 0.2500
   0.7200 0.4800 0.2000

Example 3.16. Find the fuzzy relation using fuzzy max–min Method for the given using Matlab program

$$
R = \begin{bmatrix} 1 & 0 & 1 & 0 \\ 0 & 0 & 0 & 1 \\ 0 & 0 & 0 & 0 \end{bmatrix} \quad \text{and} \quad S = \begin{bmatrix} 0 & 1 \\ 0 & 0 \\ 0 & 1 \\ 0 & 0 \end{bmatrix}.
$$

Solution. The Matlab program for finding fuzzy relation using fuzzy max–min method is

Program

```
%enter the two vectors whose relation is to be found
R=input('enter the first vector')
S=input('enter the second vector')
% find the size of two vectors
[m,n]=size(R)
[a,b]=size(S)
if(n==a)
          for i=1:m
```

```
           for j=1:b
           c=R(i,:)
           d=S(:,j)
           f=d'
           %find the minimum of two vectors
           e=min(c,f)
           %find the maximum of two vectors
           h(i,j)=max(e);
           end
      end
      %print the result
      display('the fuzzy relation between two vectors is');
      display(h)
else
      display('The fuzzy relation cannot be found')
end
```

Output

enter the first vector[1 0 1 0;0 0 0 1;0 0 0 0]

```
   R =
      1 0 1 0
      0 0 0 1
      0 0 0 0
```

enter the second vector[0 1;0 0;0 1;0 0]

```
   S =
      0 1
      0 0
      0 1
      0 0
```

The fuzzy relation between two vectors is

```
   h =
      0 1
      0 0
      0 0
```

Summary

In this chapter, we have studied about the operations and the properties of the crisp and the fuzzy relation. The basic concept of relation has to be efficient in order to perform the problems of classification, rule base, control system, etc. The composition of the relation was also introduced and discussed. The special properties related to that of the crisp and fuzzy equivalence and tolerance relation were also described. These are used in similarity and classification applications.

Review Questions

1. State some of the applications where relation concept plays a major role.
2. What is the degree of relationship for the classical relation and the fuzzy relation? Explain the differences between the two relations in terms of degree of relationship.
3. What is meant by ordered n-tuple?
4. State the Cartesian product of set theory.
5. How are the crisp relations formed based on the Cartesian product of set?
6. On what basis is the relation matrix formed in crisp relations.
7. Define the cardinality of the classical relation.
8. Write function-theoretic operations of crisp relation.
9. Explain with function-theoretic form the operations that can be performed on the classical relation.
10. State some of the properties of the crisp relation.
11. How is the composition operation performed on the classical relations? Briefly describe the types of composition methods.
12. Define fuzzy relation.
13. What are the operations that can be performed on the fuzzy relations?
14. State the properties of fuzzy relations.
15. What are the properties that hold well in crisp relation but not in fuzzy relation? Explain.
16. Define fuzzy Cartesian product.
17. Explain about the composition methods adopted in fuzzy relation. Write the various expressions involved here.
18. State whether $R \circ S = S \circ R$. Explain.
19. Define reflexivity, symmetry, and transitivity properties of relations.
20. When does a relation become an equivalence relation?
21. State the expression for the classical equivalence relation.
22. Define tolerance classical relation.
23. Describe about the fuzzy equivalence and tolerance relation.

Exercise Problems

1. Consider speed control of DC motor. Two variables are speed (in RPM) and load (torque) resulting in the following two fuzzy membership functions.

$$S_{\sim} = \left\{ \frac{0.2}{x_1} + \frac{0.6}{x_2} + \frac{0.8}{x_3} + \frac{0.6}{x_4} + \frac{0.4}{x_5} \right\},$$

$$T_{\sim} = \left\{ \frac{0.3}{y_1} + \frac{0.5}{y_2} + \frac{0.6}{y_3} + \frac{1.0}{y_4} + \frac{0.8}{y_5} + \frac{0.3}{y_6} + \frac{0.2}{y_7} \right\},$$

$T_{\sim}$ is on universe $Y_{\sim}$.

(a) Find fuzzy relation that relates these two variables $R = S \times T$. Now another variable fuzzy armature current I that relates elements in universe Y to element in two as given here

$$I = \begin{array}{c} y_1 \\ y_2 \\ y_3 \\ y_4 \\ y_5 \\ y_6 \\ y_7 \end{array} \begin{bmatrix} z_1 \\ 0.4 \\ 0.5 \\ 0.6 \\ 0.3 \\ 0.7 \\ 0.6 \\ 1.0 \end{bmatrix}.$$

(b) Find $Q = I \circ R$ using max–min composition and max–product composition.

2. The three variables of interest in the MOSFET are the amount of current that can be switched, the voltage that can be switched and the cost. The following membership function for the transistor was developed

$$\text{Current} = I = \left\{ \frac{0.4}{0.8} + \frac{0.7}{0.9} + \frac{1}{1} + \frac{0.8}{1.1} + \frac{0.6}{1.2} \right\},$$

$$\text{Voltage} = V = \left\{ \frac{0.2}{30} + \frac{0.8}{45} + \frac{1}{60} + \frac{0.9}{75} + \frac{0.7}{90} \right\},$$

$$\text{Cost} = \left\{ \frac{0.4}{0.5} + \frac{1}{0.6} + \frac{0.5}{0.7} \right\}.$$

The power is given by $P = VI$.

(a) Find the fuzzy Cartesian product $P = V \times I$.

(b) Find the fuzzy Cartesian product $T = I \times C$.

(c) Using max–min composition find $E = P \circ T$.

(d) Using max–product composition find $E = P \circ T$.

3. Relating earthquake intensity to ground acceleration is an imprecise science. Suppose we have a universe of earthquake intensities $I = \{5, 6, 7, 8, 9\}$ and a universe of accelerations, $A = \{0.2, 0.4, 0.6, 0.8, 1.0, 1.2\}$ in 8s. The following fuzzy relation R exists on confession space $I \times A$.

$$R = \begin{bmatrix} 0.75 & 1 & 0.65 & 0.4 & 0.2 & 0.1 \\ 0.5 & 0.9 & 1 & 0.65 & 0.3 & 0 \\ 0.1 & 0.4 & 0.7 & 1 & 0.6 & 0 \\ 0.1 & 0.2 & 0.4 & 0.9 & 1 & 0.6 \\ 0 & 0.1 & 0.3 & 0.45 & 0.8 & 1 \end{bmatrix}.$$

Fuzzy set "intensing about 7" is defined as:

$$\underset{\sim7}{I} = \left\{ \frac{0.1}{5} + \frac{0.6}{6} + \frac{1}{7} + \frac{0.8}{8} + \frac{0.4}{9} \right\}.$$

Determine the fuzzy membership of $\underset{\sim7}{I}$ on the universe of accelerations, A.

4. The speed of the motor m degrees per second and the voltage in volts. One having fuzzy set.

$$\underset{\sim}{S_2} = \left\{ \frac{1/3}{0} + \frac{2/3}{1} + \frac{1}{2} + \frac{2/3}{3} \right\} = \text{speed about 2},$$

$$\underset{\sim}{V_0} = \left\{ \frac{1}{0} + \frac{3/4}{1} + \frac{1/2}{2} + \frac{1/4}{3} + \frac{0}{5} + \frac{0}{6} \right\} = \text{voltage about } 0''.$$

(a) Find the Cartesian product relation $\underset{\sim}{R}$ between $\underset{\sim}{S_2}$ and $\underset{\sim}{V_0}$

Creating another fuzzy set on universe $\underset{\sim}{V}$ for "voltage about 3" might give

$$\underset{\sim}{V_3} = \left\{ \frac{0}{0} + \frac{1/4}{1} + \frac{1/2}{2} + \frac{1}{3} + \frac{1/2}{5} + \frac{0}{6} \right\}.$$

(b) Use max–min composition to find $\underset{\sim}{V_3} \text{ o } \underset{\sim}{R}$

5. Consider two fuzzy sets $\underset{\sim}{A}$ and $\underset{\sim}{B}$

$$\underset{\sim}{A} = \left\{ \frac{0}{5} + \frac{0.1}{30} + \frac{0.3}{50} + \frac{0.8}{100} + \frac{1.0}{300} \right\},$$

$$\underset{\sim}{B} = \left\{ \frac{0.7}{2} + \frac{0.8}{4} + \frac{0.2}{8} + \frac{0.1}{10} + \frac{0.7}{1.2} \right\}.$$

(a) Find fuzzy relation using the Cartesian product between $\underset{\sim}{A}$ and $\underset{\sim}{B}$

Another fuzzy set $\underset{\sim}{C}$ is defines as

$$\underset{\sim}{C} = \left\{ \frac{1.0}{5} + \frac{0.8}{30} + \frac{0.1}{50} + \frac{0.2}{100} + \frac{0}{300} \right\}.$$

Find a relation between a $\underset{\sim}{C}$ and previously determines relation of part (a)

(b) Using max–min composition
(c) Using max–product composition

6. Using max–min find the relational matrix between $\underset{\sim}{R_1}$ and $\underset{\sim}{R_2}$

$$R_1 = \begin{array}{c} x_1 \\ x_2 \\ x_3 \end{array} \begin{bmatrix} y_1 & y_2 & y_3 & y_4 & y_5 \\ 0.1 & 0.2 & 0.1 & 1 & 0.8 \\ 0.4 & 0.5 & 0 & 0.2 & 1 \\ 0.9 & 0.2 & 0.4 & 0.3 & 0.2 \end{bmatrix},$$

$$R_2 = \begin{array}{c} y_1 \\ y_2 \\ y_3 \\ y_4 \\ y_5 \end{array} \begin{bmatrix} z_1 & z_2 & z_3 & z_4 \\ 0.8 & 0.1 & 0.5 & 0.4 \\ 0.3 & 0.9 & 0.8 & 0.1 \\ 1 & 0.2 & 0.6 & 0.1 \\ 0.4 & 0.2 & 0.3 & 0 \\ 0.1 & 1 & 0.8 & 0.7 \end{bmatrix}.$$

7. Find the relation between two fuzzy sets R_1 and R_2 using

(a) Max–min composition
(b) Max–product composition
(c) Max–average composition

$$R_1 = \begin{array}{c} x_1 \\ x_2 \end{array} \begin{bmatrix} y_1 & y_2 & y_3 & y_4 \\ 0.3 & 0.1 & 0.6 & 0.3 \\ 0.1 & 1 & 0.2 & 0.1 \end{bmatrix},$$

$$R_2 = \begin{array}{c} y_1 \\ y_2 \\ y_3 \\ y_4 \end{array} \begin{bmatrix} z_1 & z_2 & z_3 \\ 0.9 & 0.1 & 1 \\ 0.1 & 0.5 & 0.4 \\ 0.6 & 0.8 & 0.5 \\ 0.1 & 0 & 0 \end{bmatrix}.$$

8. Find the relation between tow fuzzy sets using max–average method

$$A = \begin{array}{c} x_1 \\ x_2 \end{array} \begin{bmatrix} y_1 & y_2 & y_3 \\ 0.1 & 0.5 & 0.6 \\ 0.4 & 0.8 & 0.3 \end{bmatrix}$$

$$B = \begin{array}{c} y_1 \\ y_2 \\ y_3 \end{array} \begin{bmatrix} z_1 & z_2 \\ 0.4 & 0.8 \\ 0.6 & 0.5 \\ 1 & 0.8 \end{bmatrix}$$

9. Discuss the reflexivity property of the following fuzzy relation

$$\begin{bmatrix} 1 & 0.7 & 0.3 \\ 0.4 & 0.5 & 0.8 \\ 0.7 & 0.5 & 1 \end{bmatrix}.$$

10. For each of the following relation on single set state whether the relation is reflexive, symmetric, and transitive
    (a) "is a sibling of,"

(b) "is a parent of,"
(c) "is a smarter than,"
(d) "is the same height as,"
(e) "is at least as tall as."

11. State which of the following are equivalence relations and draw graph for that equivalence relations with appropriate labels.

Set	Relation on the set
(a) People	is the brother of
(b) People	has the same parent
(c) Points and map	as is connected by a road to
(d) Lines in plane geometry	is perpendicular to
(e) Positive integers	10^m times m_{ij} some integer

12. Find whether the given matrix is reflexive or not

$$R = \begin{bmatrix} 1 & 1 & 0 & 0 & 1 \\ 1 & 1 & 1 & 0 & 1 \\ 0 & 0 & 1 & 0 & 0 \\ 0 & 1 & 0 & 1 & 0 \\ 0 & 1 & 1 & 0 & 1 \end{bmatrix}$$ by writing an M-file.

13. For the above given matrix in problem 12, check the symmetry and transitivity property.

14. Find whether the given relation is a tolerance or not

$$R = \begin{bmatrix} 1 & 1 & 0 & 0 & 1 \\ 1 & 1 & 1 & 0 & 1 \\ 0 & 0 & 1 & 0 & 0 \\ 0 & 1 & 0 & 1 & 0 \\ 0 & 1 & 1 & 0 & 1 \end{bmatrix}$$ by writing an M-file.

15. Find whether the given matrix is equivalence or not

$$R = \begin{bmatrix} 1 & 0.8 & 0.4 & 0.5 & 0.6 \\ 0.5 & 1 & 0.4 & 0.3 & 0.9 \\ 0.4 & 0.2 & 1 & 0.4 & 0.4 \\ 0.5 & 0.5 & 0.3 & 1 & 0.3 \\ 0.4 & 0.9 & 0.4 & 0.7 & 1 \end{bmatrix}$$ by means of a Matlab program.

16. Implement neuro-fuzzy NAND using fuzzy propagation algorithm in Matlab.

17. Find the fuzzy relation between two vectors R and S

$$R =$$

$$
\begin{array}{cc}
0.60 & 0.25 \\
0.81 & 0.45
\end{array}
$$

$$S =$$

$$
\begin{array}{ccc}
0.80 & 0.40 & 0.20 \\
0.10 & 0.60 & 0.10
\end{array}
$$

using max–product and max–min method by a Matlab program.

4

Membership Functions

4.1 Introduction

Fuzziness in a fuzzy set is characterized by its membership functions. It classifies the element in the set, whether it is discrete or continuous. The membership functions can also be formed by graphical representations. The graphical representations may include different shapes. There are certain restrictions regarding the shapes used. The rules formed to represent the fuzziness in an application are also fuzzy. The "shape" of the membership function is an important criterion that has to be considered. There are different methods to form membership functions. This chapter discusses on the features and the various methods of arriving membership functions.

4.2 Features of Membership Function

The feature of the membership function is defined by three properties. They are:

(1) Core
(2) Support
(3) Boundary

The Fig. 4.1 shown below defines the properties listed above.

The membership can take value between 0 and 1.

(1) Core

If the region of universe is characterized by full membership (1) in the set A then this gives the core of the membership function of fuzzy at A.

The elements, which have the membership function as 1, are the elements of the core, i.e., here $\mu_A(x) = 1$.

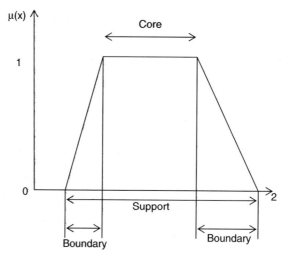

Fig. 4.1. Features of membership function

(2) Support

If the region of universe is characterized by nonzero membership in the set A, this defines the support of a membership function for fuzzy set A.

The support has the elements whose membership is greater than 0. $\mu_A(x) > 0$.

(3) Boundary

If the region of universe has a nonzero membership but not full membership, this defines the boundary of a membership; this defines the boundary of a membership function for fuzzy set A:

The boundary has the elements whose membership is between 0 and 1, $0 < \mu_A(x) < 1$.

These are the standard regions defined in the membership functions.

Defining two important terms.

Crossover point

The crossover point of a membership function is the elements in universe whose membership value is equal to 0.5, $\mu_A(x) = 0.5$.

Height

The height of the fuzzy set A is the maximum value of the membership function,

$$\max\left(\mu_A(x)\right).$$

The membership functions can be symmetrical or asymmetrical. Membership value is between 0 and 1.

4.3 Classification of Fuzzy Sets

The fuzzy sets can be classified based on the membership functions. They are:

Normal fuzzy set. If the membership function has at least one element in the universe whose value is equal to 1, then that set is called as normal fuzzy set.

Subnormal fuzzy set. If the membership function has the membership values less than 1, then that set is called as subnormal fuzzy set.

These two sets are shown in Fig. 4.2.

Convex fuzzy set. If the membership function has membership values those are monotonically increasing, or, monotonically decreasing, or they are monotonically increasing and decreasing with the increasing values for elements in the universe, those fuzzy set A is called convex fuzzy set.

Nonconvex fuzzy set. If the membership function has membership values which are not strictly monotonically increasing or monotonically decreasing or both monotonically increasing and decreasing with increasing values for elements in the universe, then this is called as nonconvex fuzzy set. Figure 4.3 shows convex and nonconvex fuzzy set.

When intersection is performed on two convex fuzzy sets, the intersected portion is also a convex fuzzy set.
This is shown in Fig. 4.4.
The shaded portions show that the intersected portion is also a convex fuzzy set. The membership functions can have different shapes like triangle, trapezoidal, Gaussian, etc.

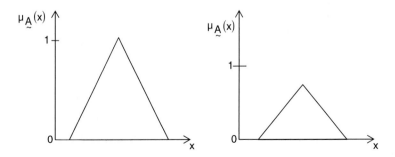

Fig. 4.2. (1) Normal fuzzy set and (2) subnormal fuzzy set

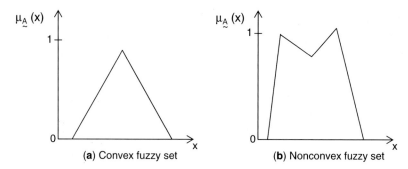

(a) Convex fuzzy set   **(b)** Nonconvex fuzzy set

Fig. 4.3. (a) Convex set and (b) Nonconvex set

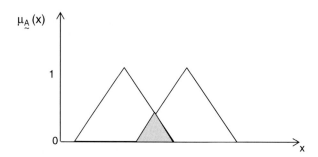

Fig. 4.4. Intersection of two convex sets

4.4 Fuzzification

Fuzzification is an important concept in the fuzzy logic theory. Fuzzification is the process where the crisp quantities are converted to fuzzy (crisp to fuzzy). By identifying some of the uncertainties present in the crisp values, we form the fuzzy values. The conversion of fuzzy values is represented by the membership functions.

In any practical applications, in industries, etc., measurement of voltage, current, temperature, etc., there might be a negligible error. This causes imprecision in the data. This imprecision can be represented by the membership functions. Hence fuzzification is performed.

Thus fuzzification process may involve assigning membership values for the given crisp quantities.

4.5 Membership Value Assignments

There are various methods to assign the membership values or the membership functions to fuzzy variables. The assignment can be just done by intuition or by using some algorithms or logical procedures. The methods for assigning the membership values are listed as follows:

- Intuition,
- Inference,
- Rank ordering,
- Angular fuzzy sets,
- Neural networks,
- Genetic algorithms, and
- Inductive seasoning

All these methods are discussed in detail in the following sections.

4.5.1 Intuition

Intuition is based on the human's own intelligence and understanding to develop the membership functions. The thorough knowledge of the problem has to be known, the knowledge regarding the linguistic variable should also be known. Figure 4.5 shows membership function for imprecision in crisp temperature reading.

For example, consider the speed of a dc-motor. The shape of the universe of speed given in rpm is shown in Fig. 4.6.

The curves represent membership function corresponding to various fuzzy variables. The range of speed is splitted into low, medium, and high. The

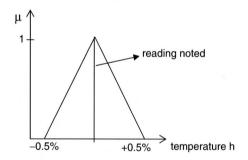

Fig. 4.5. Membership functions representing imprecision in crisp temperature reading

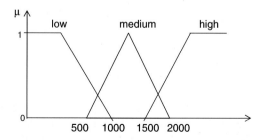

Fig. 4.6. Membership for fuzzy variable "speed" in rpm

curves differentiate the ranges, said by humans. The placement of curves is approximate over the universe of discourse; the number of curves and the overlapping of curves is an important criteria to be considered while defining membership functions.

4.5.2 Inference

This method involves the knowledge to perform deductive reasoning. The membership function is formed from the facts known and knowledge.

Let us use inference method for the identification of the triangle. Let U be universe of triangles and A, B, and C be the inner angles of the triangles. Also $A \geq B \geq C \geq 0$. Therefore the universe is given by:

$$U = \{(A, B, C), A \geq B \geq C \geq 0, A + B + C = 180°\}.$$

There are various types of triangles, for identifying, we define three types of triangles:

$\underset{\sim}{I}$ Appropriate isosceles triangle

$\underset{\sim}{R}$ Appropriate right triangle

$\underset{\sim}{O}$ Other triangles

The membership vales can be inferred to all of these triangle types through the method of inference, as we know the knowledge about the geometry of the triangles.

The membership for the approximate isosceles triangle, for the given conditions $A \geq B \geq C \geq 0$ and $A + B + C = 180°$, is given as,

$$\mu_{\underset{\sim}{I}}(A, B, C) = 1 - \frac{1}{60°} \min(A - B, B - C).$$

The membership for the appropriate right triangle, for the same conditions, is

$$\mu_{\underset{\sim}{R}}(A, B, C) = 1 - \frac{1}{90°}(A - 90°).$$

The membership for the other triangles can be given as the complement of the logical union of the two already defined membership functions

$$\mu_{\underset{\sim}{O}}(A, B, C) = \overline{\underset{\sim}{I} \cup \underset{\sim}{R}} \quad \text{(or)}$$

by using demorgans law, it is,

$$\mu_{\underset{\sim}{O}}(A, B, C) = \underset{\sim}{I} \cap \underset{\sim}{R} = \min \left\{ \begin{array}{c} 1 - \mu_{\underset{\sim}{I}}(A, B, C) \\ 1 - \mu_{\underset{\sim}{R}}(A, B, C) \end{array} \right\}.$$

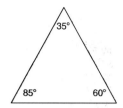

Fig. 4.7. A given triangle

Example 4.1. Define the triangle for the figure shown in Fig. 4.7 with the three given angles.

Solution.
The condition is

$$A \geq B \geq C \geq 0 \quad \text{and} \quad A + B + C = 180°.$$

Here

$$\{U = A = 85° \geq B = 60° \geq C = 35° \geq O, A + B + C = 180\}.$$

The membership for the triangle shown in Fig. 4.7, for, each triangle types are:

$$
\begin{aligned}
(1) \quad \mu_{\underset{\sim}{I}}(x) &= 1 - \frac{1}{60°} \min(A - B, B - C) \\
&= 1 - \frac{1}{60°} \min(85 - 60°, 60 - 35) \\
&= 1 - \frac{1}{60°} \min(25°, 25°) \\
&= 1 - \frac{1}{60°} \times 25° \\
\mu_{\underset{\sim}{I}}(x) &= 0.583,
\end{aligned}
$$

$$
\begin{aligned}
(2) \quad \mu_{\underset{\sim}{R}}(x) &= 1 - \frac{1}{90°}(A - 90°) \\
&= 1 - \frac{1}{90°}(85 - 90) \\
&= 1 - \frac{1}{90°} \times 5 \\
\mu_{\underset{\sim}{R}}(x) &= 0.944,
\end{aligned}
$$

$$
\begin{aligned}
(3) \quad \mu_{\underset{\sim}{O}}(x) &= \min\left[1 - \mu_{\underset{i\sim}{\pm}}(x), 1 = \mu_{\underset{\sim}{R}}(x)\right] \\
&= \min\{1 - 0.583, 1 - 0.944\} \\
&= \min\{0.417, 0.55\} \\
\mu_{\underset{\sim}{O}}(x) &= 0.055.
\end{aligned}
$$

Table 4.1. Pairwise preferences among five cars between 1000 people

	Number who preferred					Total	Percentage	Rank order
	Palio	Siena	Astra	Easter	Baleno			
Palio	–	515	545	523	671	2,254	22.5	2
Siena	481	–	475	845	580	2,381	23.8	1
Astra	469	624	–	141	536	1,770	17.7	4
Easter	457	530	470	–	649	2,114	21.1	3
Baleno	265	425	402	389	–	1,481	14.8	5
Total						10,000		

Hence there is highest membership for $\mu_R(x)$. Thus inference method can be used to calculate the membership values.

4.5.3 Rank Ordering

The polling concept is used to assign membership values by rank ordering process. Preferences are above for pairwise comparisons and from this the ordering of the membership is done.

Example 4.2. Suppose 1,000 people responds to a questionnaire about the pairwise preference among five cars, $x-$ {Palio, Siena, Astra, Easter, Baleno}. Define a fuzzy set as A on the universe of cars, "best cars".

Solution.
The pairwise comparison is made among 1,000 people and their views are summarized in Table 4.1.

From the table, it is clear that 515 preferred Siena compared to Palio, 545 Astra to Palio, etc. The table forms an antisymmetric matrix. There are about ten comparisons made which gives a ground total of 10,000. Based on preferences, the percentage is calculated. The ordering is then performed. It is found that siena is selected as the best car.

Figure 4.8 shows the membership function for this example.

4.5.4 Angular Fuzzy Sets

The angular fuzzy sets are different from the standard fuzzy sets in their coordinate description. These sets are defined on the universe of angles, hence are repeating shapes every 2Π cycles. Angular fuzzy sets are applied in quantitative description of linguistic variables known truth-values. When membership of value 1 is true and that of 0 is false, then in between '0' and '1' is partially true or partially false.

The linguistic values are formed to vary with θ, the angle defined on the unit circle and their membership values are on $\mu(\theta)$. The membership of this linguistic term can be obtained from

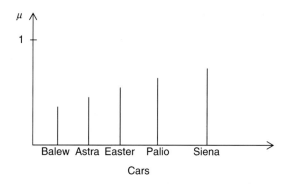

Fig. 4.8. Membership for best car

$$\mu_t(\theta) = t\tan\theta,$$

where t is the horizontal projection of the radial vector and is given as $\cos\theta$, i.e., $t = \cos\theta$. When the coordinates are in polar form, angular fuzzy sets can be used.

Example 4.3. Consider a motor, which is used in computer peripheral applications. From the membership functions based on its rotation using angular fuzzy sets.

Solution. The linguistic terms relating to the direction of motion of the motor is given as

Fully anticlockwise (FA)	–	$\theta = \Pi/2$
Partially anticlockwise (PA)	–	$\theta = \Pi/4$
No rotation (NR)	–	$\theta = 0$
Partially clockwise (PC)	–	$\theta = -\Pi/4$
Fully clockwise (FC)	–	$\theta = -\Pi/2$

The angular fuzzy set for this is shown in Fig. 4.9.
The membership function is shown in Fig 4.10.
The values for membership functions used in Fig. 4.9 is obtained as follows

$$\mu_t(Z) = Z\tan\theta,$$

where $Z = cos\ \theta$.
Therefore, the angular fuzzy membership values are shown in Table 4.2.
Hence, angular fuzzy sets can be used to obtain fuzzy membership values.

4.5.5 Neural Networks

Neural networks are used to simulate the working network of the neurons in the human brain. The concept of the human brain is used to perform computation on computers.

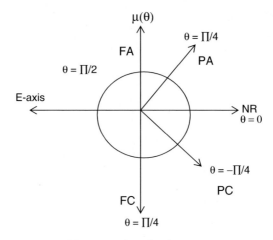

Fig. 4.9. Angular fuzzy set

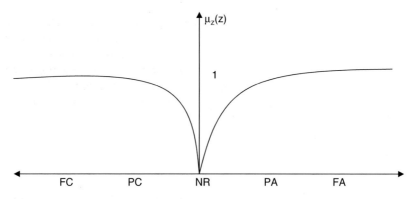

Fig. 4.10. Angular fuzzy membership function

Table 4.2. Angular fuzzy membership values

θ	Tan θ	$Z = \cos\theta$	$\mu_t(Z) = (Z\tan\theta)$
$\Pi/2(90°)$	$\propto$	0	1
$\Pi/4(45°)$	1	0.707	0.707
0	0	2	0

In this case, the fuzzy membership function may be created for fuzzy classes of an input data set. The procedure is, the number of input data values are selected. Then it is divided into training data set and testing data set. The training data set may be used to train the network.

The generations of membership function from neural network are shown in Fig. 4.11.

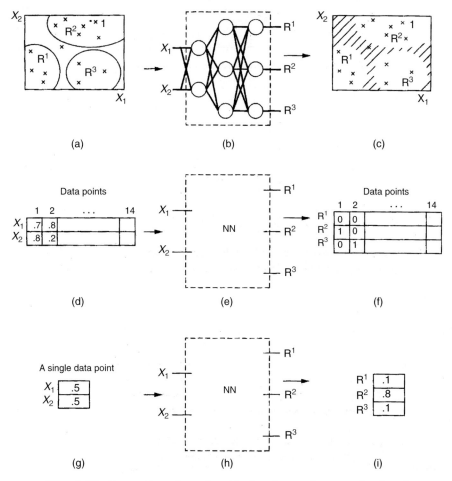

Fig. 4.11. Generation of membership functions using neural network

Figure 4.11a shows the training data set. This is passed through a neural network shown in Fig. 4.11b and this data points of Fig. 4.11a is divided into three regions as R^1, R^2, and R^3 as in Fig. 4.11c. Depending upon the data points, the regions are classified. If the data point is in region 1, then we assign full membership in regions 1 and zero membership in regions 2 and 3. Similarly if the data points are in regions 2 and 3, it will have full membership in regions 2 and 3 and zero membership in regions 1 and 3, and regions 1 and 2, respectively.

The neural network is then created, from which the training is done between corresponding membership values in different classes, to simulate the relationship between the coordinate locations and membership values. The neural network uses the set of data value and membership values to train itself as shown in Fig. 4.11d. This training process is continued until the neural network can simulate for the given entire set of input and output value.

After the net is trained, its performance can be checked by the testing data. After full training and testing process is completed, the neural network is ready and it can be used to determine the membership values of any input data in the different regions. These are all shown in Fig. 4.11g–i.

The complete mapping of the membership of different data points in different fuzzy classes can be determined by using neural network approach.

4.5.6 Genetic Algorithm

Genetic algorithm (GA) uses the concept of Darwin's theory of evolution. Darwin's theory is based on the rule, "survival of the fittest." Darwin also postulated that the new classes of living things came into existence through the process of reproduction, crossover, and mutation among existing organisms.

The steps involved in computing membership functions using GA are:

(1) For the given functional mapping of a system, some membership functions and their shapes are assumed for various fuzzy variables to be defined.
(2) These membership functions are then coded as bit stings.
(3) These bit strings are then concatenated (joined).
(4) Similar to activation function in neural networks, GA has a fitness function.
(5) This fitness function is used to evaluate the fitness of each set of membership functions.
(6) These membership functions are the parameters that define that functional mapping of the system.

Thus, GA can be used to determine the membership functions.

4.5.7 Inductive Reasoning

The membership can also be generated by the characteristics of inductive reasoning. The induction is performed by the entropy minimization principle, which clusters the parameters corresponding to the output classes. For inductive reasoning method, there should be a well-defined database for the input–output relationships. This method can be suited for complex systems where the data are abundant and static. When the data' are dynamic, this method is not suited, since the membership functions continually change with time.

There are three laws of induction (Christensen 1980).

(1) Given a set of irreducible outcomes of an experiment, the induced probabilities are those probabilities consistent with all available information that maximize the entropy of the set.

(2) The induced probability of a set of independent observations is proportional to the probability density of the induced probability of a single observation.

(3) The induced rule is that of rule consistent with all available information of which the entropy is minimum.

The third law stated here is the mostly used for membership function development.

The steps involved in generating membership functions using inductive reasoning are as follows:

(1) It is necessary to establish a fuzzy threshold between classes of data.
(2) First, determine the threshold line with an entropy minimization screening method.
(3) After this, start the segmentation process.
(4) The segmentation, process, first results into two classes.
(5) Further partitioning the first two classes one more time, there is three different classes.
(6) The partitioning is repeated with threshold value calculations, which lead us to partition the data set into a number of classes or fuzzy sets.
(7) Then based on shape, membership function is determined.

Thus the generation of membership function is based on partitioning or analog screening concept. This draws a threshold line between two classes of sample data. The main concept behind drawing the threshold line is to classify the samples when minimizing the entropy for optimum partitioning.

4.6 Solved Examples

Example 4.4. Using your own intuition and definitions of the universe of discourse, plot fuzzy membership functions for "weight of people."

Solution. The universe of discourse is the weight of people. Let the weights be in "kg" – kilogram.

Let the linguistic variables are:

Very light	–	$w \leq 30$
Light	–	$30 < w \leq 45$
Average	–	$45 < w \leq 60$
Heavy	–	$60 < w \leq 75$
Very heavy	–	$w > 75$

Representing this using triangular membership function, as shown in Fig. 4.12.

Example 4.5. Using your own intuition, plot the fuzzy membership function for the age of people.

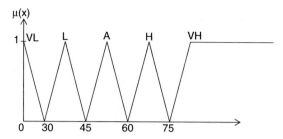

Fig. 4.12. Membership function of weight of people

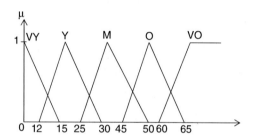

Fig. 4.13. Membership function for age of profile

Solution.
The linguistic variables are defined as, let A denotes age in years.
  (1) Very young (vy)   – $A < 15$
  (2) Young (y)   – $12 \leq A < 30$
  (3) Middle aged (m)   – $25 \leq A < 50$
  (4) Old (o)   – $45 \leq A < 65$
  (5) Very old (vo)   – $60 < A$
This is represented using triangular membership, as shown in Fig. 4.13.

Example 4.6. Using the inference approach, find the membership values for the triangular shapes $(\underset{\sim}{I}\ \underset{\sim}{R}\ \underset{\sim}{O})$ for a triangle with angles as $45°, 75°, 60°$.

Solution. Let U-universe of discourse is

$$\{u : x = 75° \geq y = 60° \geq z = 45°, x + y + z = 180°\}.$$

(1) Calculating membership of isosceles triangle

$$\mu_{\underset{\sim}{I}}(u) = 1 - \frac{1}{60°} \min(x - y, y - z)$$

$$= 1 - \frac{1}{60°} \min(15°, 15°)$$

$$= 1 - \frac{1}{60°} \times 15°,$$

$$\mu_{\underset{\sim}{I}}(x) = 1 - 0.25 = 0.75.$$

(2) Calculating membership of right triangle

$$\mu_{\underset{\sim}{R}}(u) = 1 - \frac{1}{90°}(A - 90°)$$

$$= 1 - \frac{1}{90°}(75 - 90)$$

$$= 1 - 0.166,$$

$$\mu_{\underset{\sim}{R}}(u) = 0.833.$$

(3) Calculating membership of other triangle

$$\mu_{\underset{\sim}{O}}(u) = \min\left[1 - \mu_{\underset{i\sim}{I}}(u), 1 - \mu_{\underset{\sim}{R}}(u)\right]$$

$$= \min\{1 - 0.75, 1 - 0.833\}$$

$$= \min\{0.25, 0.167\},$$

$$\mu_{\underset{\sim}{O}}(x) = 0.167.$$

Thus the membership values are calculated

Example 4.7. The energy E of a particle spinning in a magnetic field B is given by the equation

$$E = \mu\beta\sin\theta,$$

where μ is magnetic moment of spinning particle and θ is complement angle of magnetic moment with respect to the direction of the magnetic field.

Assuming the magnetic field B and magnetic moment μ to be constants, the linguistic terms for the complement angle of magnetic moment are given as:

High moment (H)	–	$\theta = \Pi/2$
Slighly high moment (SH)	–	$\theta = \Pi/4$
No moment (z)	–	$\theta = 0$
Slightly low moment (SL)	–	$\theta = -\Pi/4$
Low moment (L)	–	$\theta = -\Pi/2$

Find the membership values using the angular fuzzy set approach for these linguistic labels and plot these values versus θ.

Solution.
The linguistic variables are given by:

High moment (H)	–	$\theta = \Pi/2$
Slighly high moment (SH)	–	$\theta = \Pi/4$
No moment (z)	–	$\theta = 0$
Slightly low moment (SL)	–	$\theta = -\Pi/4$
Low moment (L)	–	$\theta = -\Pi/2$

The angular fuzzy set is shown in Fig. 4.14.
Calculating the angular fuzzy membership values as shown in Table 4.3.
The plot for this calculated membership value is shown in Fig. 4.15.

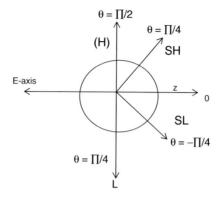

Fig. 4.14. Angular fuzzy set

Table 4.3. Angular fuzzy membership values

θ	Tan θ	$Z = \cos\theta$	$\mu_t = (z \ \tan\theta)$
$\Pi/2$	$\propto$	0	1
$\Pi/4$	1	0.707	0.707
0	0	1	0
$-\Pi/4$	-1	0.707	$+0.707$
$-\Pi/2$	$\propto$	0	1

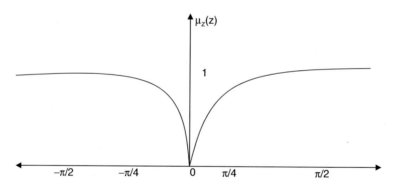

Fig. 4.15. Plot of membership function

Example 4.8. Use Matlab command line commands to display the Gaussian membership function. Given $x = 0\text{--}10$ with increment of 0.1 and Gaussian function is defined between 0.5 and -5.

Solution.
Step 1:First enter the x value
   x = (0:0.1:10)';
   Step 2:enter gaussmembership function
   >> y1 = gaussmf(x, [0.5 5]);

Step 3:plot the curve
>> plot(x, [y1])

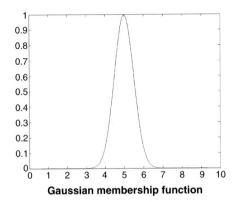

Gaussian membership function

Example 4.9. Use Matlab command line commands to display the triangu-lar membership function. Given $x = 0$–10 with increment of 0.2 triangular membership function is defined between [3 4 5]

Solution.
Step 1:First enter the x value
>> x = (0:0.2:10)';
Step 2:enter triangular membership function
>> y1 = trimf(x, [3 4 5]);
Step 3:plot the curve
>> plot(x,y1)

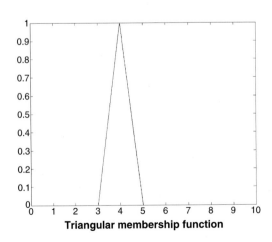

Triangular membership function

Example 4.10. Illustrate different types of generalized bell membership func-tions using Matlab program

Solution.

The Matlab program for illustrating bell membership function is given by:

Program

```
% Illustration of different generalized bell MFs
x = (-10:0.4:10)';
b = 2;c = 0;
mf1 = gbell_mf(x, [2, b, c]);
mf2 = gbell_mf(x, [4, b, c]);
mf3 = gbell_mf(x, [6, b, c]);
mf = [mf1 mf2 mf3];
subplot(221); plot(x, mf); title('(a) Changing "a"');
axis([-inf inf 0 1.2]);
a = 5;c = 0;
mf1 = gbell_mf(x, [a, 1, c]);
mf2 = gbell_mf(x, [a, 2, c]);
mf3 = gbell_mf(x, [a, 4, c]);
mf = [mf1 mf2 mf3];
subplot(222); plot(x, mf); title('(b) Changing "b"');
axis([-inf inf 0 1.2]);
a = 5;b = 2;
mf1 = gbell_mf(x, [a, b, -5]);
mf2 = gbell_mf(x, [a, b, 0]);
mf3 = gbell_mf(x, [a, b, 5]);
mf = [mf1 mf2 mf3];
subplot(223); plot(x, mf); title('(c) Changing "c"');
axis([-inf inf 0 1.2]);
c = 0;
mf1 = gbell_mf(x, [4, 4, c]);
mf2 = gbell_mf(x, [6, 6, c]);
mf3 = gbell_mf(x, [8, 8, c]);
mf = [mf1 mf2 mf3];
subplot(224); plot(x, mf); title('(d) Changing "a" and "b"');
axis([-inf inf 0 1.2]);
```

Output

The output membership functions for different values of a, b and c are shown in Fig. 4.16.

Summary

This chapter has described the different methods of obtaining the membership functions. The entire fuzzy system operation is based on the formation of the membership functions. The sense of reasoning is very important in forming

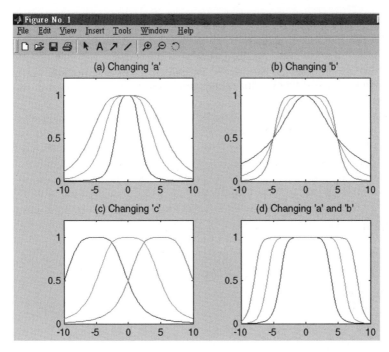

Fig. 4.16. Bell membership functions

the membership functions. The inference and the angular fuzzy sets are based upon the angular features. In the case of neural networks and reasoning methods the memberships are tuned in a cyclic fashion and are associated with the rule structure. In genetic algorithms, improvements have been made to achieve the optimum solution. Thus by using any one of the method discussed earlier, the membership function may be formed.

Review Questions

1. State the features of membership functions.
2. Define normal and subnormal fuzzy set.
3. What is a convex fuzzy set?
4. State the properties of a convex fuzzy set.
5. How is the crossover point and the height defined based on the membership function?
6. Define fuzzy number.
7. Compare normal and convex fuzzy set.
8. Define fuzzification.
9. What are the various methods employed for the membership value assignment?

10. Justify intuition is based on human reasoning. Give some suitable examples.
11. Discuss in detail on the inference method adopted for assigning membership values. Give details on the concepts of triangle used.
12. How is the polling concept adopted in rank ordering method to define the membership values?
13. Give details about the method of assigning membership values using angular fuzzy set with example.
14. Explain the method of generating membership function by means of neural networks and genetic algorithm.
15. How is membership value assigned based on inductive reasoning?

Exercise Problems

1. Using your own intuition, develop fuzzy membership functions for the fuzzy number 3, using the following shapes: (a) right angle triangle, (b) quadrilateral, (c) Gaussian function, (d) trapezoid, and (e) isosceles triangle.
2. Using intuition, assign the membership functions for
   (a) population of people, (b) employment strategy, and (c) usage of library.
3. Using the inference method, find the membership values of the triangular shapes for each of the following triangles: (a) $60°$, $40°$, $80°$, (b) $45°$, $65°$, $70°$, and (c) $75°$, $55°$, $50°$.
4. The following data were determined by the pairwise comparison of work preferences of 100 people. When it was compared with Software (S), 69 of persons polled preferred Hardware (H), 45 of them preferred Educational (E), 55 of them preferred Business (B) and 25 preferred Textile (T). When it was compared with hardware (H), the preferences was 58-S, 45-E, 60-B, 30-T. When it was compared with educational, 39-S, 56-H, 34-B, 25-T. When it was compared business, the preferences was 52-S, 49-H, 38-E, 20-T. When it was compared with textile, the preferences was 69-S, 65-H, 44-E, 40-B. Using rank ordering, plot the membership function for the "most preferred work."
5. Using your own intuition, develop fuzzy membership functions on the real line for the fuzzy number 4, using the following function shapes:
   (1) Symmetric triangle
   (2) Trapezoid
   (3) Gaussian function
6. Using your own intuition, develop fuzzy number "approximately 4 or approximately 8" using the following function shapes:
   (1) Symmetric triangle
   (2) Trapezoids
   (3) Gaussian functions.

7. Using your own intuition and your own definition of the universe of discourse plot fuzzy membership functions to the following variables:
   (1) Height of liquid in a tank
       (a) Very full
       (b) Full
       (c) Medium
       (d) Small
       (e) Very small
   (2) Race of people
       (a) Very white
       (b) White
       (c) Moderate
       (d) Black
       (e) Very black
   (3) Age of people
       (a) Very young
       (b) Young
       (c) Middle ages
       (d) Old
       (e) Very old
8. Using the Inference approach outlined in this chapter find the membership values for each of the triangular shapes $(\underset{\sim}{I}, \underset{\sim}{R}, \underset{\sim}{IR}, \underset{\sim}{E}, \underset{\sim}{R})$ for each of the following
   (1) $80°, 75°, 25°$,
   (2) $60°, 75°, 45°$,
   (3) $50°, 75°, 55°$, and
   (4) $45°, 45°, 90°$.
9. Develop membership function for trapezoidal similar to algorithm developed for triangle and the function should have two independent variables hence it can be passed. For the shown in table, show the first iteration in trying to compute the membership values for input variables x_1, x_2, and x_3 in the output regions R^1 and R^2

x_1	x_2	x_3	R^1	R^2
1.0	0.5	2.3	1.0	0.0

   (a) Use $3 \times 3 \times 1$ neural network,
   (b) Use $3 \times 3 \times 2$ neural network.
10. For data shown in the following table (Table A) shows the first two iteration using a genetic algorithm in trying to find the optimum membership function (right triangular function S) for the input variable x and output variable y in the rule table out.
11. The following raw data were determined in a pairwise comparison of new scooter in a poll 100 people. When it was compare with Splender (S), 79 of house preferred TVS Suzuki (T) 59, preferred Hero Honda (H) and 88

Table A. data

x	0	0.3	0.6	1.0
y	1	0.74	0.53	0.35

Table B. rules

x	L	S	Z – Zero
y	Z	S	L – Large
			S – Small

preferred Enfield (E), and 67 preferred infinity (I) when (T) was compared the preferences when (T) was compared, the preferences were 21-S, 23-H, 37-H, and 45-I when H1 was compared the preferences were 15-S, 77-T, 35-E, 48-I finally when an infinity was compared the preferences were 33-S, 55-T, 52-H, and 49-E. Using rank ordering, plot the membership function for "most preferred bike."

12. The energy E of a particle spinning is a magnetic field B is given by the equation

$$E = \mu B \, \sin \theta,$$

where μ is complement angle of magnetic moment with respect to direction of the magnetic field.

Assuming the magnetic field B and magnetic moment μ to be constant, we propose/linguistic terms for the complement angle of magnetic moment as follows:

High moment (H)	$\theta = \Pi/4$
Slightly high moment (SH)	$\theta = 3\Pi/4$
No moment	$\theta = 0$
Slightly low moment (SL)	$\theta = 3\Pi/4$
Low moment (L)	$\theta = \Pi/4$

Find the membership values using the angular fuzzy set approach for these linguistic labels for the complement angles and plot these values versus θ.

13. Use Matlab command line commands to display the triangular membership function. Given $x = 0$–20 with increment of 0.4 triangular membership function is defined between [6 7 8].

5

Defuzzification

5.1 Introduction

Defuzzification means the fuzzy to crisp conversions. The fuzzy results generated cannot be used as such to the applications, hence it is necessary to convert the fuzzy quantities into crisp quantities for further processing. This can be achieved by using defuzzification process. The defuzzification has the capability to reduce a fuzzy to a crisp single-valued quantity or as a set, or converting to the form in which fuzzy quantity is present. Defuzzification can also be called as "rounding off" method. Defuzzification reduces the collection of membership function values in to a single sealer quantity. In this chapter we will discuss on the various methods of obtaining the defuzzified values.

5.2 Lambda Cuts for Fuzzy Sets

Consider a fuzzy set A, then the lambda cut set can be denoted by A_λ, where λ ranges between 0 and 1 ($0 \leq \lambda \leq 1$).

The set A_λ is going to be a crisp set. This crisp set is called the lambda cut set of the fuzzy set A, where

$$A_\lambda = \left\{ x / \mu_A(x) \geq \lambda \right\},$$

i.e., the value of lambda cut set is x, when the membership value corresponding to x is greater that or equal to the specified λ. This lambda cut set can also be called as alpha cut set. The λ cut set A_λ does not have title underscore, because it is derived from parent fuzzy set A. Since the lambda λ ranges in the interval [0, 1], the fuzzy set A can be transformed to infinite number of λ cut sets.

Properties of Lambda Cut Sets:

There are four properties of the lambda cut sets, they are:

(1) $\left(\underset{\sim}{A} \cup \underset{\sim}{B}\right)_\lambda = A_\lambda \cup B_\lambda$

(2) $\left(\underset{\sim}{A} \cap \underset{\sim}{B}\right)_\lambda = A_\lambda \cap B_\lambda$

(3) $\left(\overline{\underset{\sim}{A}}\right)_\lambda \neq \left(\overline{A_\lambda}\right)$ except for a value of $\lambda = 0.5$

(4) For any $\lambda \leq \alpha$, where α varies between 0 and 1, it is true that, $A_\alpha \subseteq A_\lambda$, where the value of A_0 will be the universe defined.

From the properties it is understood that the standard set of operations or fuzzy sets is similar to the standard set operations on lambda cut sets.

5.3 Lambda Cuts for Fuzzy Relations

The lambda cut procedure for relations is similar to that for the lambda cut sets. Considering a fuzzy relation $\underset{\sim}{R}$, in which some of the relational matrix represents a fuzzy set. A fuzzy relation can be converted into a crisp relation by depending the lambda cut relation of the fuzzy relation as:

$$R_\lambda = \{x, y / \mu_R(x, y) \geq \lambda\}.$$

Properties of Lambda Cut Relations:

Lambda cut relations satisfy some of the properties similar to lambda cut sets.

(1) $\left(\underset{\sim}{R} \cup \underset{\sim}{S}\right)_\lambda = R_\lambda \cup S_\lambda$.

(2) $\left(\underset{\sim}{R} \cap \underset{\sim}{S}\right)_\lambda = R_\lambda \cap S_\lambda$.

(3) $\left(\overline{\underset{\sim}{R}}\right)_\lambda \neq \left(\overline{R_\lambda}\right)$.

(4) For $\lambda \leq \alpha$, where α between 0 and 1, then $R_\alpha \subseteq R_\lambda$.

5.4 Defuzzification Methods

Apart from the lambda cut sets and relations which convert fuzzy sets or relations into crisp sets or relations, there are other various defuzzification methods employed to convert the fuzzy quantities into crisp quantities. The output of an entire fuzzy process can be union of two or more fuzzy membership functions. To explain this in detail, consider a fuzzy output, which is formed by two parts, one part being triangular shape (Fig. 5.1a) and other part being trapezoidal (Fig. 5.1b). The union of these two forms (Fig. 5.1c) the outer envelop of the two shapes.

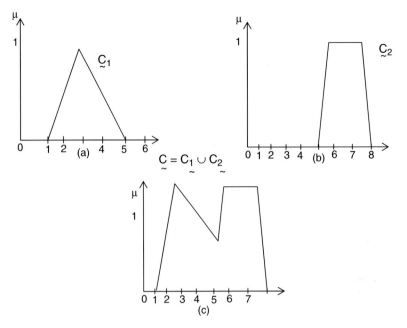

Fig. 5.1. Typical fuzzy output

Generally this can be given as:

$$C_n = \sum_{r}^{n} C_i = \underset{\sim}{C}.$$

There are seven methods used for defuzzifying the fuzzy output functions. They are:

(1) Max-membership principle,
(2) Centroid method,
(3) Weighted average method,
(4) Mean–max membership,
(5) Centre of sums,
(6) Centre of largest area, and
(7) First of maxima or last of maxima

(1) Max-membership-principle
This method is given by the expression,

$$\mu_{\underset{\sim}{C}}(z^*) \geq \mu_{\underset{\sim}{C}}(z) \quad \text{for all} \quad z \in \mathbf{z}.$$

This method is also referred as height method. This is shown in Fig. 5.2.

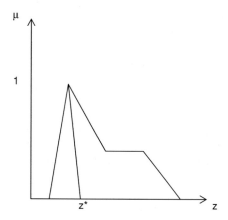

Fig. 5.2. Max-membership method

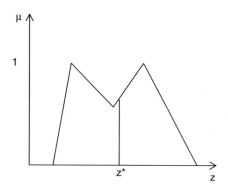

Fig. 5.3. Centroid method

(2) Centroid method

This is the most widely used method. This can be called as center of gravity or center of area method. It can be defined by the algebraic expression

$$z^* = \int \frac{\mu_{\underset{\sim}{C}}(z)\, z\, \mathrm{d}z}{\mu_{\underset{\sim}{C}}(z)\, \mathrm{d}z},$$

$\int$ is used for algebraic integration. Figure 5.3 represents this method graphically.

(3) Weighted average method

This method cannot be used for asymmetrical output membership functions, can be used only for symmetrical output membership functions. Weighting each membership function in the obtained output by its largest membership value forms this method. The evaluation expression for this method is

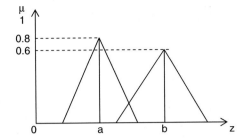

Fig. 5.4. Weighted average method

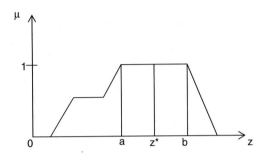

Fig. 5.5. Mean–max-membership

$$z^* = \frac{\sum \mu_C(\bar{z}) \, \bar{z}}{\sum \mu_C(\bar{z})},$$

Σ is used for algebraic sum.

From Fig. 5.4

$$z^* = \frac{a\,(0.8) + b\,(0.6)}{0.8 + 0.6}.$$

(4) Mean–max-membership

This method is related to max-membership principle, but the present of the maximum membership need not be unique, i.e., the maximum membership need not be a single point, it can be a range. This method is also called as middle of maxima method the expression is given as

$$z^* = \frac{a + b}{2},$$

where a $\times$ b are the end point of the maximum membership range as shown in Fig. 5.5.

(5) Centre of sums

It involves the algebraic sum of individual output fuzzy sets, say c_1 and c_2 instead of union. In this method, it is noted that the intersecting areas are

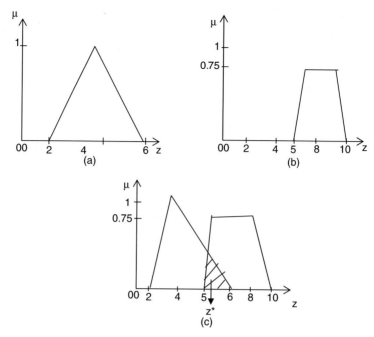

Fig. 5.6. (a) First membership, (b) second membership, and (c) defuzzification step

added twice. This method is similar to the weighted average method, but in center of sums, the weights are the areas of the respective membership functions whereas in the weighted average method, the weights are individual membership values.

The defuzzified value z^* is given as

$$z^* = \frac{\int_2 z \sum_{k=1}^{n} \mu_{C_k} (z)\, dz}{\int_2 z \sum_{k=1}^{n} \mu_{C_k} (z)\, dz}.$$

Figure 5.6 represents the center of sums method.

(6) Center of largest area

If the fuzzy set has two convex subregions, then the entire of gravity of the convex subregion with the largest area can be used to calculate the defuzzification value. The equation is given as

$$z^* = \frac{\int \mu_{c_m} (z)\, z dz}{\int \mu_{c_m} (z)\, dz},$$

where c_m is the convex region with largest area. The value z^* is same as the value z^* obtained by centroid method. This can be done even for non-convex

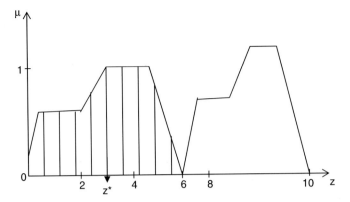

Fig. 5.7. Center of largest area

regions. Figure 5.7 represents the center of largest area method.

(7) First of maxima or last of maxima

Here, the compute output of all individual output fuzzy sets c_k is used to determine the smallest value, with maximized membership degree in c_k.

The evaluation expressions are

Let largest height in the union is represents by $hgt(c_k)$, then it is found by:

$$hgt\left(c_k\right) = \sup_{z \in z} \mu\, c_k\left(z\right).$$

First of maxima is found by

$$z^* = \inf_{z \in z}\left\{z \in z/\mu\, c_k\left(z\right) = hgt\left(c_k\right)\right\}.$$

Fast of maxima is found by,

$$z^* = \sup_{z \in z}\left\{z \in z/\mu\, c_k\left(z\right) = hgt\left(c_k\right)\right\}.$$

The inf denotes infirm (greatest lower bound) and the sup denotes supremum (least upper bound). This method is shown in Fig. 5.8.

5.5 Solved Examples

Example 5.1. Two fuzzy sets P and Q are defined on x as follows:

$\mu(x_1)$	x_1	x_2	x_3	x_4	x_5
P	0.1	0.2	0.7	0.5	0.4
Q	0.9	0.6	0.3	0.2	0.8

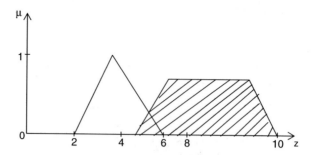

Fig. 5.8. First of or last of maxima

Find the following λ cut sets

(a) $\left(\overline{\underset{\sim}{P}}\right)_{0.2}$ (b) $(\underset{\sim}{Q})_{0.3}$ (c) $\left(\underset{\sim}{P} \cup \underset{\sim}{Q}\right)_{0.5}$ (d) $\left(\underset{\sim}{P} \cap \underset{\sim}{Q}\right)_{0.4}$

(e) $\left(\underset{\sim}{Q} \cup \overline{\underset{\sim}{P}}\right)_{0.8}$ (f) $\left(\underset{\sim}{P} \cup \overline{\underset{\sim}{P}}\right)_{0.2}$.

Solution. Given

$$P = \left\{ \frac{0.1}{x_1} + \frac{0.2}{x_2} + \frac{0.7}{x_3} + \frac{0.5}{x_4} + \frac{0.4}{x_5} \right\},$$

$$Q = \left\{ \frac{0.9}{x_1} + \frac{0.6}{x_2} + \frac{0.3}{x_3} + \frac{0.2}{x_4} + \frac{0.8}{x_5} \right\}.$$

Finding

$$\overline{P} = \left\{ \frac{0.9}{x_1} + \frac{0.8}{x_2} + \frac{0.3}{x_3} + \frac{0.5}{x_4} + \frac{0.6}{x_5} \right\},$$

$$\overline{Q} = \left\{ \frac{0.1}{x_1} + \frac{0.4}{x_2} + \frac{0.7}{x_3} + \frac{0.8}{x_4} + \frac{0.2}{x_5} \right\}.$$

(a) $\left(\overline{\underset{\sim}{P}}\right)_{0.2} = \left\{ \frac{1}{x_1} + \frac{1}{x_2} + \frac{1}{x_3} + \frac{1}{x_4} + \frac{1}{x_5} \right\},$

(b) $\left(\overline{\underset{\sim}{Q}}\right)_{0.3} = \left\{ \frac{0}{x_1} + \frac{1}{x_2} + \frac{1}{x_3} + \frac{1}{x_4} + \frac{0}{x_5} \right\},$

(c) $\left(\underset{\sim}{P} \cup \underset{\sim}{Q}\right) = \left\{ \frac{0.9}{x_1} + \frac{0.6}{x_2} + \frac{0.7}{x_3} + \frac{0.5}{x_4} + \frac{0.8}{x_5} \right\},$

$\left(\underset{\sim}{P} \cup \underset{\sim}{Q}\right)_{0.6} = \left\{ \frac{1}{x_1} + \frac{1}{x_2} + \frac{1}{x_3} + \frac{0}{x_4} + \frac{1}{x_5} \right\},$

(d) $\left(\underset{\sim}{P} \cup \overline{\underset{\sim}{P}}\right) = \left\{ \frac{0.9}{x_1} + \frac{0.8}{x_2} + \frac{0.7}{x_3} + \frac{0.5}{x_4} + \frac{0.6}{x_5} \right\},$

$\left(\underset{\sim}{P} \cup \overline{\underset{\sim}{P}}\right)_{0.8} = \left\{ \frac{1}{x_1} + \frac{1}{x_2} + \frac{0}{x_3} + \frac{0}{x_4} + \frac{0}{x_5} \right\}$

(e) $\left(\underset{\sim}{P} \cap \underset{\sim}{Q} \right) = \left\{ \dfrac{0.9}{x_1} + \dfrac{0.8}{x_2} + \dfrac{0.7}{x_3} + \dfrac{0.5}{x_4} + \dfrac{0.6}{x_5} \right\}$,

$\left(\underset{\sim}{P} \cap \underset{\sim}{Q} \right)_{0.4} = \left\{ \dfrac{0}{x_1} + \dfrac{0}{x_2} + \dfrac{0}{x_3} + \dfrac{0}{x_4} + \dfrac{1}{x_5} \right\}$,

(f) $\left(\underset{\sim}{P} \cap \overline{\underset{\sim}{P}} \right) = \left\{ \dfrac{0.1}{x_1} + \dfrac{0.2}{x_2} + \dfrac{0.3}{x_3} + \dfrac{0.5}{x_4} + \dfrac{0.4}{x_5} \right\}$,

$\left(\underset{\sim}{P} \cap \overline{\underset{\sim}{P}} \right)_{0.8} = \left\{ \dfrac{0}{x_1} + \dfrac{1}{x_2} + \dfrac{1}{x_3} + \dfrac{1}{x_4} + \dfrac{1}{x_5} \right\}$.

Example 5.2. Given three fuzzy sets:

$$\underset{\sim}{A} = \left\{ \dfrac{0.9}{x_1} + \dfrac{0.5}{x_2} + \dfrac{0.2}{x_3} + \dfrac{0.3}{x_4} \right\},$$

$$\underset{\sim}{B} = \left\{ \dfrac{0.2}{x_1} + \dfrac{1.0}{x_2} + \dfrac{0.8}{x_3} + \dfrac{0.4}{x_4} \right\},$$

$$\underset{\sim}{C} = \left\{ \dfrac{0.1}{x_1} + \dfrac{0.7}{x_2} + \dfrac{0.5}{x_3} + \dfrac{0.6}{x_4} \right\}.$$

Find $A_{0.6}$, $B_{1.0}$, $C_{0.3}$, $\overline{\underset{\sim}{A}}_{0.2}$, $\overline{\underset{\sim}{B}}_{0.8}$, $\overline{\underset{\sim}{C}}_{0.5}$.

Solution.

(a) $A_{0.6} = \left\{ \dfrac{1}{x_1} + \dfrac{0}{x_2} + \dfrac{0}{x_3} + \dfrac{0}{x_4} \right\}$,

(b) $B_{1.0} = \left\{ \dfrac{0}{x_1} + \dfrac{1}{x_2} + \dfrac{0}{x_3} + \dfrac{0}{x_4} \right\}$,

(c) $C_{0.3} = \left\{ \dfrac{0}{x_1} + \dfrac{1}{x_2} + \dfrac{1}{x_3} + \dfrac{1}{x_4} \right\}$,

(d) $\overline{\underset{\sim}{A}} = \left\{ \dfrac{0.1}{x_1} + \dfrac{0.5}{x_2} + \dfrac{0.8}{x_3} + \dfrac{0.7}{x_4} \right\}$,

$\overline{\underset{\sim}{A}}_{0.2} = \left\{ \dfrac{0}{x_1} + \dfrac{1}{x_2} + \dfrac{1}{x_3} + \dfrac{1}{x_4} \right\}$,

(e) $\overline{\underset{\sim}{B}} = \left\{ \dfrac{0.8}{x_1} + \dfrac{0.0}{x_2} + \dfrac{0.2}{x_3} + \dfrac{0.6}{x_4} \right\}$,

$\overline{\underset{\sim}{B}}_{0.8} = \left\{ \dfrac{1}{x_1} + \dfrac{0}{x_2} + \dfrac{0}{x_3} + \dfrac{0.9}{x_4} \right\}$,

(f) $\overline{\underset{\sim}{C}} = \left\{ \dfrac{0.9}{x_1} + \dfrac{0.3}{x_2} + \dfrac{0.5}{x_3} + \dfrac{0.4}{x_4} \right\}$,

$\overline{\underset{\sim}{C}}_{0.5} = \left\{ \dfrac{1}{x_1} + \dfrac{0}{x_2} + \dfrac{1}{x_3} + \dfrac{0}{x_4} \right\}$.

Example 5.3. The fuzzy sets $\underset{\sim}{A}$ and $\underset{\sim}{B}$ are defined as universe, $x = \{0, 1, 2, 3\}$, with the following membership fractions:

$$\mu_A(x) = \frac{2}{x+3},$$

$$\mu_B(x) = \frac{4x}{x+5}.$$

Define the intervals along x-axis corresponding to the λ cut sets for each fuzzy set $\underset{\sim}{A}$ and $\underset{\sim}{B}$ for following values of λ. $\lambda = 0.2, 0.5, 0.6$.

Solution.

$$x = \{0, 1, 2, 3\}$$

X	0	1	2	3
$\mu_{A\sim}(x) = 2/x + 3$	2/3	2/4	2/5	2/6
$\mu_{B\sim}(x) = 4x/2(x+5)$	0	4/12	8/14	12/16

Therefore the values are:

X	0	1	2	3
$\mu_{A\sim}(x)$	0.67	0.5	0.4	0.33
$\mu_{B\sim}(x)$	0	0.33	0.57	0.75

(a) When $\lambda = 0.2$
   $A_{0.2} = \{0, 1, 2, 3\}$,
   $A_{0.2} = \{1, 2, 3\}$.
(b) When $\lambda = 0.5$
   $A_{0.5} = \{0, 1\}$,
   $B_{0.5} = \{2, 3\}$.
(c) When $\lambda = 0.6$
   $A_{0.6} = \{0\}$,
   $A_{0.6} = \{3\}$.

Example 5.4. For the fuzzy relation

$$\underset{\sim}{R} = \begin{bmatrix} 1 & 0.2 & 0.3 \\ 0.5 & 0.9 & 0.6 \\ 0.4 & 0.8 & 0.7 \end{bmatrix},$$

find the λ cut relations for the following values of $\lambda = 0^+, 0.2, 0.9, 0.5$.

Solution. Given,

$$\underset{\sim}{R} = \begin{bmatrix} 1 & 0.2 & 0.3 \\ 0.5 & 0.9 & 0.6 \\ 0.4 & 0.8 & 0.7 \end{bmatrix}.$$

(a) $\lambda = 0^+$,

$$R_{0+} = \begin{bmatrix} 1 & 1 & 1 \\ 1 & 1 & 1 \\ 1 & 1 & 1 \end{bmatrix}.$$

(b) $\lambda = 0.2$,

$$R_{0.2} = \begin{bmatrix} 1 & 1 & 1 \\ 1 & 1 & 1 \\ 1 & 1 & 1 \end{bmatrix}.$$

(c) $\lambda = 0.9$,

$$R_{0.9} = \begin{bmatrix} 1 & 0 & 0 \\ 0 & 1 & 0 \\ 0 & 0 & 0 \end{bmatrix}.$$

(d) $\lambda = 0.5$,

$$R_{0.5} = \begin{bmatrix} 1 & 0 & 0 \\ 1 & 1 & 1 \\ 0 & 1 & 1 \end{bmatrix}.$$

Example 5.5. For the given fuzzy relation

$$R = \begin{bmatrix} 0.2 & 0.5 & 0.47 & 1 & 0.9 \\ 0.3 & 0.5 & 0.6 & 1 & 0.8 \\ 0.4 & 0.6 & 0.4 & 0.5 & 0.3 \\ 0.9 & 1 & 0.3 & 0.3 & 0.2 \end{bmatrix},$$

find the cut λ cut relation for the following values of $\lambda = 0.4, 0.7, 0.8$.

Solution. (a) $\lambda = 0.4$,

$$R_{0.4} = \begin{bmatrix} 0 & 1 & 1 & 1 & 1 \\ 0 & 1 & 1 & 1 & 1 \\ 1 & 1 & 1 & 1 & 0 \\ 1 & 1 & 0 & 0 & 0 \end{bmatrix}.$$

(b) $\lambda = 0.7$,

$$R_{0.7} = \begin{bmatrix} 0 & 0 & 0 & 1 & 1 \\ 0 & 0 & 0 & 1 & 1 \\ 0 & 0 & 0 & 0 & 0 \\ 1 & 1 & 0 & 0 & 0 \end{bmatrix}.$$

(c) $\lambda = 0.8$,

$$R_{0.8} = \begin{bmatrix} 0 & 0 & 0 & 1 & 1 \\ 0 & 0 & 0 & 1 & 1 \\ 0 & 0 & 0 & 0 & 0 \\ 1 & 1 & 0 & 0 & 0 \end{bmatrix}.$$

Example 5.6. For the given membership function as shown in Fig. 5.9 determines the defuzzified output value by seven methods.

Solution. (a) *Centroid method*
$A_{11}(0,0), (2 - 0.7)$
The straight line may be:

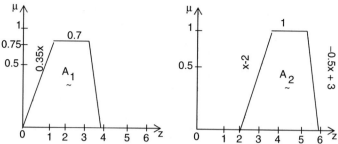

Fig. 5.9. Membership function

$$y - 0 = \frac{0.7}{2}(x - 0),$$

$$y = 0.35x.$$

$$A_{12} : y = 0.7.$$

$$A_{13} : \text{not needed.}$$

$$A_{21} : (2, 0)(3, 1),$$

$$y - 0 = \frac{1 - 0}{3 - 2}(x - 2),$$

$$y = x - 2.$$

$$A_{22} : y = 1.$$

$$A_{23} : (4, 1)(6, 0),$$

$$y - 1 = \frac{0 - 1}{6 - 4}(x - 4),$$

$$y = \frac{-1}{2}(x - 4) + 1 = -0.5x + 3.$$

Solving A_{12} and A_{21},

$$Y = 0.7 \quad y = x - 2,$$

$$x - 2 = 0.7,$$

$$x = 2.7,$$

$$y = 0.7.$$

$$\text{Numerator} = \int_0^2 0.35\, z^2\, \mathrm{d}z + \int_2^{2.7} 0.7K \mathrm{d}z + \int_{2.7}^3 \left(z^2 - 2z\right) \mathrm{d}z$$

$$+ \int_3^4 z\, \mathrm{d}z + \int_4^6 \left(-0.5z^2 + 3z\right) \mathrm{d}z$$

$$= 10.98.$$

$$\text{Denominator} = \int_0^2 0.35\, z^2\, \mathrm{d}z + \int_2^{2.7} 0.7K\, \mathrm{d}z + \int_{2.7}^3 \left(z^2 - 2\right) \mathrm{d}z$$

$$+ \int_3^4 \mathrm{d}z + \int_4^6 \left(-0.5z^2 + 3z\right) \mathrm{d}z$$

$$= 3.445.$$

$$z^* = \frac{\text{Numerator}}{\text{Demoninator}} = \frac{10.98}{3.445} = 3.187.$$

(b) Weighted average method

$$z^* = \frac{2 \times 0.7 + 4 \times 1}{1 + 0.7} = 3.176.$$

(c) Mean–max method

$$z^* = \frac{2.5 + 3.5}{2} = 3.$$

(d) Center of saws method

$$z^* = \frac{\int_0^6 \left(\frac{1}{2} \times 0.7 \times (3 + 2) \times 2 + \frac{1}{2} \times 1 \times (2 + 4) \times 4\right)}{\int_0^6 \left(\frac{1}{2} \times 0.7 \times (3 + 2) + \frac{1}{2} \times 1 \times (2 + 4) \times 4\right)}$$

$$= \frac{\int_0^b (3.5 + 12)\, dz}{\int_0^b (1.75 + 3)\, dz} = 2.84.$$

(e) First of maximum
$z^* = 3.$
(f) Last of maxima
$z^* = 4.$
(g) Center of largest area

$$\text{Area of } I = \frac{1}{2} \times 0.7 \times (2.7 + 0.7) = 1.19,$$

$$\text{Area of } II = \frac{1}{2} \times 1 \times (2 + 3) \times \frac{1}{2} \times 0.7 \times = 2.255.$$

Area of II is larger, So,

$$z^* = \frac{\int_{2.7}^3 \frac{1}{2} \times 0.3 \times 0.3 \times 2.85\, dz + \int_3^4 1 \times 1 \times 3.5\, dz + \int_4^6 \frac{1}{2} \times 2 \times 1\, dz}{\int_{2.7}^3 \frac{1}{2} \times 0.3 \times 0.3\, dz + \int_4^3 1 \times 1\, dz + \int_4^6 \frac{1}{2} \times 2 \times 1\, dz}$$

$$= \frac{\int_{2.7}^3 0.12825\, dz + \int_3^4 3.5\, dz + \int_4^6 5\, dz}{\int_{2.7}^3 0.045\, dz + \int_3^4 dz + \int_4^6 dz}$$

$$z^* = 4.49.$$

Example 5.7. Using Matlab program find the crisp lambda cut set relations for $\lambda = 0.2$, the fuzzy matrix is given by

$$\underset{\sim}{R} = \begin{bmatrix} 0.2 & 0.7 & 0.8 & 1 \\ 1 & 0.9 & 0.5 & 0.1 \\ 0 & 0.8 & 1 & 0.6 \\ 0. & 0.4 & 1 & 0.3 \end{bmatrix}.$$

Solution. The Matlab program is

Program

```
clear all
% Enter the matrix value
R=input('Enter the matrix value')
% Enter the lambda value
lambda=input('enter the lambda value')
[m,n]=size(R);
    for i=1:m
        for j=1:n
                if(R(i,j)<lambda)
                b(i,j)=0;
                else
                b(i,j)=1;
                end
        end
    end
% output value
display('the crisp value is')
display(b)
```

Output

Enter the matrix value
    [0.2 0.7 0.8 1;1 0.9 0.5 0.1;0 0.8 1 0.6;0.2 0.4 1 0.3]
    R =

0.2000	0.7000	0.8000	1.0000
1.0000	0.9000	0.5000	0.1000
0	0.8000	1.0000	0.6000
0.2000	0.4000	1.0000	0.3000

Enter the lambda value 0.2
    lambda = 0.2000
    The crisp value is
    b =

1	1	1	1
1	1	1	0
0	1	1	1
1	1	1	1

Summary

Defuzzification is thus a natural and necessary process. Because the output to any practical system cannot be given using the linguistic variables like "moderately high," "medium," "very positive," etc., it has to be given only in crisp quantities. These crisp quantities are thus obtained from the fuzzy quantities using the various defuzzification methods discussed in this chapter.

Review Questions

1. Define defuzzification process.
2. What is the necessity to convert the fuzzy quantities into crisp quantities?
3. State the method lambda cuts employed for the conversion of the fuzzy set into crisp.
4. Discuss in detail on the special properties of lambda cut sets.
5. How is lambda cut method employed for a fuzzy relation?
6. List some of the methods to perform defuzzification process.
7. How does the max-membership method convert the fuzzy quantity to crisp quantity?
8. Centroid method is very efficient method for defuzzification, Justify. Give suitable example.
9. In what way does the weighted average method perform the defuzzification process?
10. Explain about the mean–max-membership method for converting the fuzzy quantity to crisp quantity. Give some details on the accuracy of the output obtained.
11. Compare the methods center of sums and center of largest area with necessary examples.
12. What is difference between first and last of maxima? Explain the process of conversion in each case with example.
13. Compare and contrast the methods employed for defuzzification process on the basis of accuracy and time consumption.
14. What are the four important criteria on which the defuzzification method is defined?

Exercise Problems

1. Determine crisp λ cut relation for $\lambda = 0.2$; for $j = 0, 1, \ldots, 10$ for the following fuzzy relation matrix $\underset{\sim}{R}$:

$$
\underset{\sim}{R} = \begin{bmatrix}
0.3 & 0.8 & 0.7 & 0.9 \\
1 & 0.7 & 0.6 & 0.2 \\
0.1 & 0.7 & 1 & 0.9 \\
0.5 & 0.6 & 0.2 & 0.5
\end{bmatrix}.
$$

2. The fuzzy set $\underset{\sim}{A}$, $\underset{\sim}{B}$, $\underset{\sim}{C}$ are all defined on the universe $X = [0, 5]$ with the following membership functions:

$$
\mu_{\underset{\sim}{A}}(x) = \frac{1}{1 + 5(x-5)^2},
$$

$$
\mu_{\underset{\sim}{B}}(x) = 2^{-x},
$$

$$
\mu_{\underset{\sim}{C}}(x) = \frac{2x}{x+5}.
$$

(a) Sketch the membership functions

(b) Define the intervals along x-axis corresponding to the λ cut sets for each of the fuzzy sets A, B, C for the following values of λ.

$\lambda = 0.2, \lambda = 0.4, \lambda = 0.7, \lambda = 0.9.$

3. Two fuzzy sets $\underset{\sim}{A}$ and $\underset{\sim}{B}$ both defined on x are as follows:

$$\underset{\sim}{A} = \left\{ \frac{0.1}{x_1} + \frac{0.6}{x_2} + \frac{0.4}{x_3} + \frac{0.7}{x_4} + \frac{0.5}{x_5} + \frac{0.2}{x_6} \right\},$$

$$\underset{\sim}{B} = \left\{ \frac{0.8}{x_1} + \frac{0.6}{x_2} + \frac{0.3}{x_3} + \frac{0.2}{x_4} + \frac{0.6}{x_5} + \frac{0}{x_6} \right\}.$$

Find

(a) $\left(\underset{\sim}{\bar{A}} \right)_{0.5}$

(b) $\left(\underset{\sim}{B} \right)_{0.3}$

(c) $\left(\underset{\sim}{A} \cup \underset{\sim}{\bar{A}} \right)_{0.5}$

(d) $\left(\underset{\sim}{A} \cup \underset{\sim}{B} \right)_{0.4}$

(e) $\left(\overline{\underset{\sim}{A} \cap \underset{\sim}{B}} \right)_{0.6}$

(f) $\left(\underset{\sim}{A} \cap \underset{\sim}{B} \right)_{0.64}$

4. For fuzzy relation R find λ cut relations for the following values of λ

$$R = \begin{bmatrix} 0.4 & 0.3 & 0.7 & 0.5 \\ 0.6 & 0.2 & 0.1 & 1 \\ 0.9 & 0.8 & 0.5 & 0.6 \\ 0.7 & 0.4 & 0.3 & 0.2 \end{bmatrix}.$$

(a) $\lambda = 0^+$ (c) $\lambda = 0.4$ (e) $\lambda = 0.3$
(b) $\lambda = 0.2$ (d) $\lambda = 0.7$ (f) $\lambda = 0.6$

5. Show that any λ cut relation of fuzzy tolerance relation results in a crisp tolerance relation.

Show that any λ cut relation of a fuzzy equivalence relation results in a crisp equivalence relation.

6. For fuzzy relation $\underset{\sim}{A}$ and $\underset{\sim}{E}$ determine λ cut relations for the following values of λ:

   (a) $\lambda = 0^+$ (b) $\lambda = 0.5$ (c) $\lambda = 0.9$

$$\underset{\sim}{A} = \begin{bmatrix} 0.8 & 1 & 0.5 & 0.3 & 0.1 & 0 \\ 0.2 & 0.3 & 0.5 & 0.7 & 0.1 & 0.2 \\ 0.1 & 0.2 & 0.4 & 0.8 & 0.7 & 0.1 \\ 0.2 & 0.1 & 0.4 & 0.7 & 0.7 & 0.3 \end{bmatrix},$$

$$\underset{\sim}{E} = \begin{bmatrix} 0.8 & 0.7 & 0.4 & 0.1 & 0 \\ 0.6 & 0.5 & 0.3 & 0.2 & 0.1 \\ 0.9 & 0.6 & 0.7 & 0.4 & 0.3 \\ 0.2 & 0.4 & 0.5 & 0.9 & 0.6 \\ 0.1 & 0.4 & 0.3 & 0.6 & 0.9 \\ 0.1 & 0 & 1 & 0.8 & 0.7 \end{bmatrix}.$$

7. Determine the λ cut sets for the six set operation for two fuzzy set $\underset{\sim}{R}$ and $\underset{\sim}{S}$ using $\lambda = 0.2$ and 0.8:

$$\underset{\sim}{A} = \left\{ \frac{0.1}{20} + \frac{0.6}{40} + \frac{0.4}{60} + \frac{0.3}{80} + \frac{0.9}{100} \right\},$$

$$\underset{\sim}{B} = \left\{ \frac{0.3}{20} + \frac{0.4}{40} + \frac{0.7}{60} \frac{0.4}{80} \frac{0.2}{100} \right\}.$$

For the fuzzy sets operation:

   (a) $\underset{\sim}{A} \cup \underset{\sim}{B}$ (b) $\underset{\sim}{A} \cap \underset{\sim}{B}$ (c) $\bar{\underset{\sim}{A}}$ (d) $\underset{\sim}{A} / \underset{\sim}{B}$ (e) $\overline{\underset{\sim}{A} \cup \underset{\sim}{B}}$ (f) $\overline{\underset{\sim}{A} \cap \underset{\sim}{B}}$

8. By using centroid method of defuzzification convert fuzzy value z to precise value Z^* for the following graph.

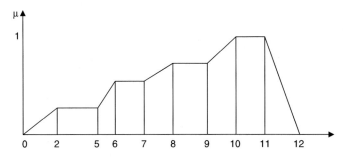

9. Find the defuzzified value by weighted average method shown in figure.

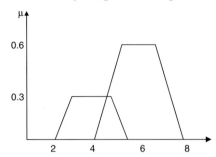

10. Find the defuzzified values using (a) center of sums methods and (b) center of largest area for the figure shown.

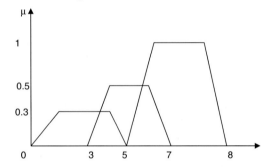

11. Find the defuzzified values for the figure shown above using first of maxima and last of maxima.

12. Two companies bid for a contract. The fuzzy set of two companies B_1 and B_2 is shown in the following figure. Find the defuzzified value z^* using different methods.

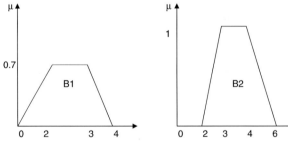

13. Using Matlab program find the crisp lambda cut set relations for $\lambda = 0.4$, the fuzzy matrix is given by:

$$\underset{\approx}{R} = \begin{bmatrix} 0.3 & 0.2 & 0.8 & 0 \\ 1 & 0.1 & 0.5 & 0.1 \\ 0 & 0.8 & 1 & 0.5 \\ 0.7 & 0.6 & 1 & 0.3 \end{bmatrix}.$$

6

Fuzzy Rule-Based System

6.1 Introduction

Rules form the basis for the fuzzy logic to obtain the fuzzy output. The rule-based system is different from the expert system in the manner that the rules comprising the rule-based system originates from sources other than that of human experts and hence are different from expert systems. The rule-based form uses linguistic variables as its antecedents and consequents. The antecedents express an inference or the inequality, which should be satisfied. The consequents are those, which we can infer, and is the output if the antecedent inequality is satisfied. The fuzzy rule-based system uses IF–THEN rule-based system, given by, IF antecedent, THEN consequent. The formation of the fuzzy rules is discussed in this chapter.

6.2 Formation of Rules

The formation of rules is in general the canonical rule formation. For any linguistic variable, there are three general forms in which the canonical rules can be formed. They are:

(1) Assignment statements
(2) Conditional statements
(3) Unconditional statements

(1) Assignment statements

These statements are those in which the variable is assignment with the value. The variable and the value assigned are combined by the assignment operator "=." The assignment statements are necessary in forming fuzzy rules. The value to be assigned may be a linguistic term.

The examples of this type of statements are:

$$y = \text{low},$$
$$\text{Sky color} = \text{blue},$$
$$\text{Climate} = \text{hot}$$
$$a = 5$$
$$p = q + r$$
$$\text{Temperature} = \text{high}$$

The assignment statement is found to restrict the value of a variable to a specific equality.

(2) Conditional statements

In this statements, some specific conditions are mentioned, if the conditions are satisfied then it enters the following statements, called as restrictions.

If $x = y$ Then both are equal,

If Mark > 50 Then pass,

If Speed $> 1,500$ Then stop.

These statements can be said as fuzzy conditional statements, such as
If condition C' Then restriction F'

(3) Unconditional statements

There is no specific condition that has to be satisfied in this form of statements. Some of the unconditional statements are:

Go to F/o

Push the value

Stop

The control may be transferred without any appropriate conditions. The unconditional restrictions in the fuzzy form can be:

R^1 :   Output is B^1
                   AND
R^2 :   Output is B^2
                   AND
         ...,   etc.

where B^1 and B^2 are Fuzzy consequents.

Both conditional and unconditional statements place restrictions on the consequent of the rule-based process because of certain conditions. The fuzzy

Table 6.1. Canonical form – Fuzzy rule-based system

Rule 1:	IF condition C^1 THEN restriction R^1
Rule 2:	IF condition C^2 THEN restriction R^2
$\ldots$	
Rule n:	IF condition C^n THEN restriction R^n

sets and relations model the restrictions. The linguistic connections like "and," "or," "else" connects the conditional, unconditional, and restriction statements. the consequent of rules or output is denoted by the restrictions R^1, $R^2, \ldots, R^n$.

The rule-based system with a set of conditional rules (canonical form of rules) is shown in Table 6.1.

6.3 Decomposition of Rules

There might be a compound rule structure involved in many applications. An example for a compound rule structure is

IF $x = y$ THEN both are equal
ELSE
      IF $x \neq y$
THEN
      IF $x > y$ THEN X is highest
ELSE
      IF $y > x$ THEN Y is highest
ELSE

IF x and y are equal to zero THEN no output is obtained.
By the properties and operations defined on fuzzy sets in Chap. 2, any compound rule structure can be decomposed and reduced to number of simple canonical rules. There are various methods for decomposition of rules. They are:

(1) Multiple conjunction antecedents

This uses fuzzy intersection operation. Since it involves linguistic "AND" connective

IF x is $\underset{\sim}{P^1}$ AND $\underset{\sim}{P^2} \cdots$ AND $\underset{\sim}{P^n}$ THEN y is Q^r,

where

$$\underset{\sim}{P^r} = \underset{\sim}{P^1} \text{ AND } \underset{\sim}{P^2} \cdots \text{ OR } \underset{\sim}{P^n}.$$

The membership for this can be

$$\mu_{\underset{\sim}{P^r}}(x) = \min\left[\mu_{\underset{\sim}{P^1}}(x), \mu_{\underset{\sim}{P^2}}(x), \ldots, \mu_{\underset{\sim}{P^n}}(x)\right].$$

Hence the rule can be
IF x is $\underset{\sim}{P^r}$ THEN $\underset{\sim}{Q^r}$.

(2) Multiple disjunctive antecedents

This uses fuzzy union operations. It involves linguistic "OR" connections
IF x is $\underset{\sim}{P^1}$ OR $\underset{\sim}{P^2}\cdots$ OR $\underset{\sim}{P^n}$ THEN y is Q^r,
where

$$\underset{\sim}{P^r} = \underset{\sim}{P^1} \ \text{OR} \ \underset{\sim}{P^2} \cdots \ \text{OR} \ \underset{\sim}{P^n}$$
$$= \underset{\sim}{P^1} \cup \underset{\sim}{P^2} \cdots \cup \underset{\sim}{P^n}.$$

The membership for this can be

$$\mu_{\underset{\sim}{P^r}}(x) = \max\left[\mu_{\underset{\sim}{P^1}}(x), \mu_{\underset{\sim}{P^2}}(x), \ldots, \mu_{\underset{\sim}{P^n}}(x)\right].$$

Hence the rule can be
IF x is $\underset{\sim}{P^r}$ THEN y is $\underset{\sim}{Q^r}$.

(3) Conditional statements with ELSE

(a) IF $\underset{\sim}{P^1}$ THEN $\left(\underset{\sim}{Q^1} \text{ ELSE } \underset{\sim}{Q^2}\right)$.

Considering this as one compound statement, splitting this into two canonical form rules, we get
IF $\underset{\sim}{P^1}$ THEN Q^1 OR IF NOT $\underset{\sim}{P^1}$ THEN Q^2.

(b) IF $\underset{\sim}{P^1}$ THEN $\left(\underset{\sim}{Q^1} \text{ ELSE } \underset{\sim}{P^2} \text{ THEN } \left(\underset{\sim}{Q^2}\right)\right)$.

The decomposition for this can be of the form

$$\text{IF } \underset{\sim}{P^1} \text{ THEN } Q^1 \text{ OR}$$

$$\text{IF NOT } \underset{\sim}{P^1} \text{ AND } \underset{\sim}{P^2} \text{ THEN } Q^2.$$

(4) Nested IF–THEN rules

IF $\underset{\sim}{P^1}$ THEN $\left(\text{IF } \underset{\sim}{P^2} \text{ THEN } (Q^2)\right)$.
This can be decomposed into
IF $\underset{\sim}{P^1}$ AND $\underset{\sim}{P^2}$ THEN Q^1.

Thus the compound rules are decomposed into single canonical rules. Then this rules may be reduced to a series of relations.

6.4 Aggregation of Fuzzy Rules

The fuzzy rule-based system may involve more than one rule. The process of obtaining the overall conclusion from the individually mentioned consequents contributed by each rule in the fuzzy rule this is known as aggregation of rule. There are two methods for determining the aggregation of rules:

(1) Conjunctive system of rules

The rules that are connected by "AND" connectives satisfy the connective system of rules. In this case, the aggregated output may be found by the fuzzy intersection of all individual rule consequents,

$y = y^1$ AND y^2 and $\cdots$ AND y^r
(or) $y = y^1 \wedge y^2 \wedge \cdots \wedge y^r$.
Then the membership friction is defined as
$\mu_y(y) = \min\left(\mu_{y^1}(y), \mu_{y^2}(y), \ldots, \mu_{y^n}(y),\right)$ for $y \in y$.

(2) Disjunctive system of rules

The rules that are connected by "OR" connectives satisfies the disjunctive system of rules. In this case, the aggregated output may be found by the fuzzy union of all individual rule consequents

$y = y^1$ OR y^2 OR $\cdots$ OR y^r
(or)
$y = y^1 \cup y^2 \cup \cdots \cup y^r$.
Then the membership function is defined as
$\mu_y(y) = \min\left(\mu_{y^1}(y), \mu_{y^2}(y), \ldots, \mu_{y^n}(y),\right)$ for $y \in y$.

6.5 Properties of Set of Rules

The properties for the sets of rules are

- Completeness,
- Consistency,
- Continuity, and
- Interaction.

(a) Completeness

A set of IF–THEN rules is complete if any combination of input values result in an appropriate output value.

(b) Consistency

A set of IF–THEN rules is inconsistent if there are two rules with the same rules-antecedent but different rule-consequents.

(c) Continuity

A set of IF–THEN rules is continuous if it does not have neighboring rules with output fuzzy sets that have empty intersection.

(d) Interaction

In the interaction property, suppose that is a rule, "IF x is A THEN y is B," this meaning is represented by a fuzzy relation R^2, then the composition of A and R does not deliver B

$$\underset{\sim}{A} \circ \underset{\sim}{R} \neq \underset{\sim}{B}.$$

These are the properties of the fuzzy set of rules

6.6 Fuzzy Inference System

Fuzzy inference systems (FISs) are also known as fuzzy rule-based systems, fuzzy model, fuzzy expert system, and fuzzy associative memory. This is a major unit of a fuzzy logic system. The decision-making is an important part in the entire system. The FIS formulates suitable rules and based upon the rules the decision is made. This is mainly based on the concepts of the fuzzy set theory, fuzzy IF–THEN rules, and fuzzy reasoning. FIS uses "IF...THEN..." statements, and the connectors present in the rule statement are "OR" or "AND" to make the necessary decision rules. The basic FIS can take either fuzzy inputs or crisp inputs, but the outputs it produces are almost always fuzzy sets. When the FIS is used as a controller, it is necessary to have a crisp output. Therefore in this case defuzzification method is adopted to best extract a crisp value that best represents a fuzzy set. The whole FIS is discussed in detail in the following subsections.

6.6.1 Construction and Working of Inference System

Fuzzy inference system consists of a fuzzification interface, a rule base, a database, a decision-making unit, and finally a defuzzification interface. A FIS with five functional block described in Fig. 6.1. The function of each block is as follows:

- a *rule base* containing a number of fuzzy IF–THEN rules;
- a *database* which defines the membership functions of the fuzzy sets used in the fuzzy rules;
- a *decision-making unit* which performs the inference operations on the rules;
- a *fuzzification interface* which transforms the crisp inputs into degrees of match with linguistic values; and
- a *defuzzification interface* which transforms the fuzzy results of the inference into a crisp output.

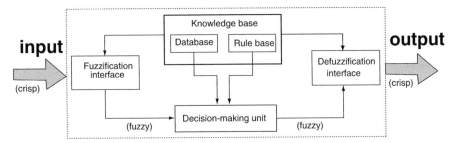

Fig. 6.1. Fuzzy inference system

The working of FIS is as follows. The crisp input is converted in to fuzzy by using fuzzification method. After fuzzification the rule base is formed. The rule base and the database are jointly referred to as the *knowledge base.* Defuzzification is used to convert fuzzy value to the real world value which is the output.

The steps of *fuzzy reasoning* (inference operations upon fuzzy IF–THEN rules) performed by FISs are:

1. Compare the input variables with the membership functions on the antecedent part to obtain the membership values of each linguistic label. (this step is often called *fuzzification.*)
2. Combine (through a specific *t*-norm operator, usually multiplication or min) the membership values on the premise part to get *firing strength* (*weight*) of each rule.
3. Generate the qualified consequents (either fuzzy or crisp) or each rule depending on the firing strength.
4. Aggregate the qualified consequents to produce a crisp output. (This step is called *defuzzification.*)

6.6.2 Fuzzy Inference Methods

The most important two types of fuzzy inference method are Mamdani's fuzzy inference method, which is the most commonly seen inference method. This method was introduced by Mamdani and Assilian (1975). Another well-known inference method is the so-called Sugeno or Takagi–Sugeno–Kang method of fuzzy inference process. This method was introduced by Sugeno (1985). This method is also called as TS method. The main difference between the two methods lies in the consequent of fuzzy rules. Mamdani fuzzy systems use fuzzy sets as rule consequent whereas TS fuzzy systems employ linear functions of input variables as rule consequent. All the existing results on fuzzy systems as universal approximators deal with Mamdani fuzzy systems only and no result is available for TS fuzzy systems with linear rule consequent.

6.6.3 Mamdani's Fuzzy Inference Method

Mamdani's fuzzy inference method is the most commonly seen fuzzy methodology. Mamdani's method was among the first control systems built using fuzzy set theory. It was proposed by Mamdani (1975) as an attempt to control a steam engine and boiler combination by synthesizing a set of linguistic control rules obtained from experienced human operators. Mamdani's effort was based on Zadeh's (1973) paper on fuzzy algorithms for complex systems and decision processes.

Mamdani type inference, as defined it for the Fuzzy Logic Toolbox, expects the output membership functions to be fuzzy sets. After the aggregation process, there is a fuzzy set for each output variable that needs defuzzification. It is possible, and in many cases much more efficient, to use a single spike as the output membership function rather than a distributed fuzzy set. This is sometimes known as a *singleton* output membership function, and it can be thought of as a pre-defuzzified fuzzy set. It enhances the efficiency of the defuzzification process because it greatly simplifies the computation required by the more general Mamdani method, which finds the centroid of a two-dimensional function. Rather than integrating across the two-dimensional function to find the centroid, the weighted average of a few data points. Sugeno type systems support this type of model. In general, Sugeno type systems can be used to model any inference system in which the output membership functions are either linear or constant.

An example of a Mamdani inference system is shown in Fig. 6.2. To compute the output of this FIS given the inputs, six steps has to be followed:

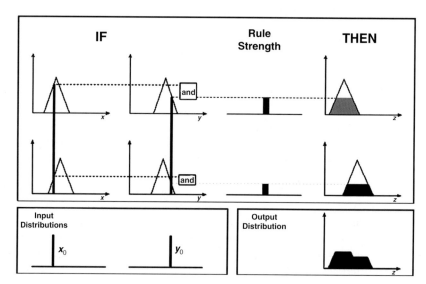

Fig. 6.2. A two input, two rule Mamdani FIS with crisp inputs

1. Determining a set of fuzzy rules
2. Fuzzifying the inputs using the input membership functions
3. Combining the fuzzified inputs according to the fuzzy rules to establish a rule strength
4. Finding the consequence of the rule by combining the rule strength and the output membership function
5. Combining the consequences to get an output distribution
6. Defuzzifying the output distribution (this step is only if a crisp output (class) is needed).

The following is a more detailed description of this process

Creating Fuzzy Rules

Fuzzy rules are a collection of linguistic statements that describe how the FIS should make a decision regarding classifying an input or controling an output. Fuzzy rules are always written in the following form:

if (input 1 is membership function 1) **and/or** *(input 2 is membership function 2)* **and/or**... **then** *(output$_n$ is output membership function$_n$).*

For example:

if temperature is high **and** *humidity is high* **then** *room is hot.*

There would have to be membership functions that define high temperature (input 1), high humidity (input 2), and a hot room (output 1). This process of taking an input such as temperature and processing it through a membership function to determine "high" temperature is called fuzzification and is discussed in section, "Fuzzification." Also, "AND"/"OR" in the fuzzy rule should be defined. This is called fuzzy combination and is discussed in following section.

Fuzzification

The purpose of fuzzification is to map the inputs from a set of sensors (or features of those sensors such as amplitude or spectrum) to values from 0 to 1 using a set of input membership functions. In the example shown in Fig. 6.2, there are two inputs, x_0 and y_0 shown at the lower left corner. These inputs are mapped into fuzzy numbers by drawing a line up from the inputs to the input membership functions above and marking the intersection point.

These input membership functions, as discussed previously, can represent fuzzy concepts such as "large" or "small," "old" or "young," "hot" or "cold," etc. For example, x_0 could be the EMG energy coming from the front of the forearm and y_0 could be the EMG energy coming from the back of the forearm. The membership functions could then represent large amounts of tension coming from a muscle or small amounts of tension. When choosing the input membership functions, the definition of large and small may be different for each input.

Consequence

The consequence of a fuzzy rule is computed using two steps:

1. Computing the rule strength by combining the fuzzified inputs using the fuzzy combination process discussed in previous section. This is shown in Fig. 6.2. In this example, the fuzzy "AND" is used to combine the membership functions to compute the rule strength.
2. Clipping the output membership function at the rule strength.

Combining Outputs into an Output Distribution

The outputs of all of the fuzzy rules must now be combined to obtain one fuzzy output distribution. This is usually, but not always, done by using the fuzzy "OR." Figure 6.2 shows an example of this. The output membership functions on the right-hand side of the figure are combined using the fuzzy OR to obtain the output distribution shown on the lower right corner of the Fig. 6.2.

Defuzzification of Output Distribution

In many instances, it is desired to come up with a single crisp output from an FIS. For example, if one was trying to classify a letter drawn by hand on a drawing tablet, ultimately the FIS would have to come up with a crisp number to tell the computer which letter was drawn. This crisp number is obtained in a process known as defuzzification. There are two common techniques for defuzzifying:

1. Center of mass. This technique takes the output distribution and finds its center of mass to come up with one crisp number. This is computed as follows:

$$z = \frac{\sum_{j=1}^{q} Z_j u_c(Z_j)}{\sum_{j=1}^{q} u_c(Z_j)},$$

where z is the center of mass and u_c is the membership in class c at value z_j. An example outcome of this computation is shown in Fig. 6.3.
2. Mean of maximum. This technique takes the output distribution and finds its mean of maxima to come up with one crisp number. This is computed as follows:

$$z = \sum_{j=1}^{l} \frac{z_j}{l},$$

where z is the mean of maximum, z_j is the point at which the membership function is maximum, and l is the number of times the output distribution reaches the maximum level. An example outcome of this computation is shown in Fig. 6.4.

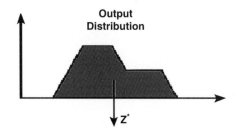

Fig. 6.3. Defuzzification using the center of mass

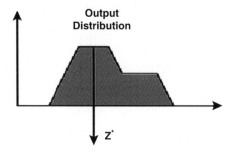

Fig. 6.4. Defuzzification using the mean of maximum

Fuzzy Inputs

In summary, Fig. 6.5 shows a two input Mamdani FIS with two rules. It fuzzi-
fies the two inputs by finding the intersection of the crisp input value with
the input membership function. It uses the minimum operator to compute the
fuzzy AND for combining the two fuzzified inputs to obtain a rule strength.
It clips the output membership function at the rule strength. Finally, it uses
the maximum operator to compute the fuzzy OR for combining the outputs
of the two rules.

6.6.4 Takagi–Sugeno Fuzzy Method (TS Method)

In this section, the basic of Sugeno fuzzy model which is implemented into
the neural-fuzzy system. The Sugeno fuzzy model was proposed by Takagi,
Sugeno, and Kang in an effort to formalize a system approach to generating
fuzzy rules from an input–output data set. Sugeno fuzzy model is also know
as Sugeno–Takagi model. A typical fuzzy rule in a Sugeno fuzzy model has
the format

$$\text{IF } x \text{ is } A \text{ and } y \text{ is } B \text{ THEN } z = f(x, y),$$

where AB are fuzzy sets in the antecedent; $Z = f(x, y)$ is a crisp function in
the consequent. Usually $f(x, y)$ is a polynomial in the input variables x and y,

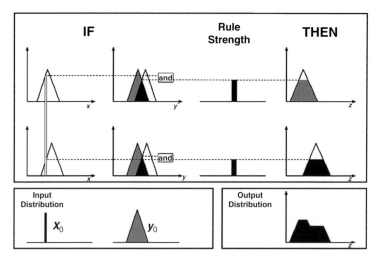

Fig. 6.5. A two input, two rule Mamdani FIS with a fuzzy input

but it can be any other functions that can appropriately describe the output of the output of the system within the fuzzy region specified by the antecedent of the rule. When $f(x, y)$ is a first-order polynomial, we have the *first-order* Sugeno fuzzy model. When f is a constant, we then have the *zero-order* Sugeno fuzzy model, which can be viewed either as a special case of the Mamdani FIS where each rule's consequent is specified by a fuzzy singleton, or a special case of Tsukamoto's fuzzy model where each rule's consequent is specified by a membership function of a step function centered at the constant. Moreover, a zero-order Sugeno fuzzy model is functionally equivalent to a radial basis function network under certain minor constraints.

The first two parts of the fuzzy inference process, fuzzifying the inputs and applying the fuzzy operator, are exactly the same. The main difference between Mamdani and Sugeno is that the Sugeno output membership functions are either linear or constant. A typical rule in a Sugeno fuzzy model has the form

IF Input 1 = x AND Input 2 = y, THEN Output is $z = ax + by + c$.

For a zero-order Sugeno model, the output level z is a constant ($a = b = 0$). The output level z_i of each rule is weighted by the firing strength w_i of the rule. For example, for an AND rule with Input 1 = x and Input 2 = y, the firing strength is

$$w_i = \text{AndMethod}(F_1(x), F_2(y)),$$

where $F_{1,2}(\cdot)$ are the membership functions for Inputs 1 and 2. The final output of the system is the weighted average of all rule outputs, computed as

$$\text{Final output} = \frac{\sum_{i=1}^{N} w_i z_i}{\sum_{i=1}^{N} w_i}.$$

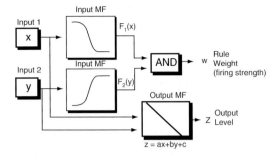

Fig. 6.6. Sugeno rule

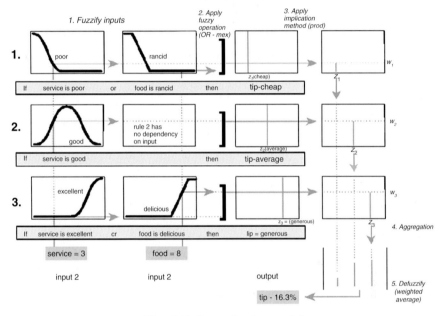

Fig. 6.7. Fuzzy tipping model

A Sugeno rule operates as shown in Fig. 6.6.

Figure 6.7 shows the fuzzy tipping model developed in previous sections of this manual adapted for use as a Sugeno system. Fortunately, it is frequently the case that singleton output functions are completely sufficient for the needs of a given problem. As an example, the system tippersg.fis is the Sugeno type representation of the now-familiar tipping model (Fig. 6.7).

- a = readfis('tippersg');
  gensurf(a)

The above command gives the surface view of fuzzy tipping model as shown in Fig. 6.8. The easiest way to visualize first-order Sugeno systems is to think of each rule as defining the location of a "moving singleton." That is,

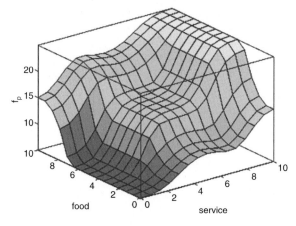

Fig. 6.8. Surface view

the singleton output spikes can move around in a linear fashion in the output space, depending on what the input is. This also tends to make the system notation very compact and efficient. Higher-order Sugeno fuzzy models are possible, but they introduce significant complexity with little obvious merit. Sugeno fuzzy models whose output membership functions are greater than first-order are not supported by the Fuzzy Logic Toolbox.

Because of the linear dependence of each rule on the input variables of a system, the Sugeno method is ideal for acting as an interpolating supervisor of multiple linear controllers that are to be applied, respectively, to different operating conditions of a dynamic nonlinear system. For example, the performance of an aircraft may change dramatically with altitude and Mach number. Linear controllers, though easy to compute and well suited to any given flight condition, must be updated regularly and smoothly to keep up with the changing state of the flight vehicle. A Sugeno FIS is extremely well suited to the task of smoothly interpolating the linear gains that would be applied across the input space; it is a natural and efficient gain scheduler. Similarly, a Sugeno system is suited for modeling nonlinear systems by interpolating between multiple linear models.

Because it is a more compact and computationally efficient representation than a Mamdani system, the Sugeno system lends itself to the use of adaptive techniques for constructing fuzzy models. These adaptive techniques can be used to customize the membership functions so that the fuzzy system best models the data.

6.6.5 Comparison Between Sugeno and Mamdani Method

The main difference between Mamdani and Sugeno is that the Sugeno output membership functions are either linear or constant. Also the difference lies in the consequents of their fuzzy rules, and thus their aggregation and defuzzification procedures differ suitably. The number of the input fuzzy sets and fuzzy rules needed by the Sugeno fuzzy systems depend on the number and locations of the extrema of the function to be approximated. In Sugeno method a large number of fuzzy rules must be employed to approximate periodic or highly oscillatory functions. The minimal configuration of the TS fuzzy systems can be reduced and becomes smaller than that of the Mamdani fuzzy systems if nontrapezoidal or nontriangular input fuzzy sets are used. Sugeno controllers usually have far more adjustable parameters in the rule consequent and the number of the parameters grows exponentially with the increase of the number of input variables. Far fewer mathematical results exist for TS fuzzy controllers than do for Mamdani fuzzy controllers, notably those on TS fuzzy control system stability. Mamdani is easy to form compared to Sugeno method.

6.6.6 Advantages of Sugeno and Mamdani Method

Advantages of the Sugeno Method

- It is computationally efficient.
- It works well with linear techniques (e.g., PID control).
- It works well with optimization and adaptive techniques.
- It has guaranteed continuity of the output surface.
- It is well suited to mathematical analysis.

Advantages of the Mamdani Method

- It is intuitive.
- It has widespread acceptance.
- It is well suited to human input.

Fuzzy inference system is the most important modeling tool based on fuzzy set theory. The FISs are built by domain experts and are used in automatic control, decision analysis, and various other expert systems.

6.7 Solved Examples

Example 6.1. Temperature control of the reactor where the error and change in error is given to the controller. Here the temperature of the reactor is controlled by the temperature bath around the reactor thus the temperature is controlled by controlling the flow of the coolant into the reactor. Form the membership function and the rule base using FIS editor.

Solution. In the FIS editor (choose either Mamdani or Sugeno model), we choose triangular membership function, and we can set the linguistic variables. The connective rules are formed and based on these rules the fuzzy associative memory table is formed.

Membership Function for error is

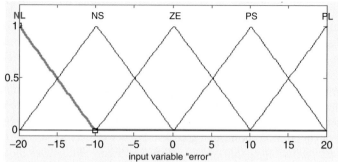

Membership Function for change in error is

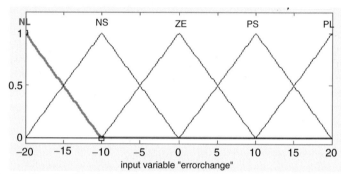

Membership Function for valve position is

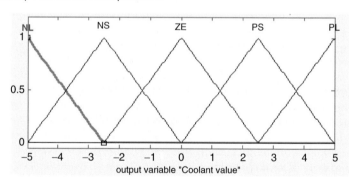

Rule base for the above is

e	δc				
	NL	NS	NE	PS	PL
PL	ZE	PS	PL	PL	PL
PS	NS	ZE	PS	PL	PL
ZE	NL	NS	ZE	PS	PL
NS	NL	NL	NS	ZE	PS
NL	NL	NL	NL	NS	ZE

Example 6.2. Consider the water tank with following rules

1. IF (level is okay) THEN (valve is no_change) (1)
2. IF (level is low) THEN (valve is open_fast) (1)
3. IF (level is high) THEN (valve is close_fast) (1)

Using Mamdani method and max–min method for fuzzification and method of centroid for defuzzification method construct a FIS. Before editing that rules, membership functions must be defined with membership function editor.

Solution. The following step should be followed for constructing the FIS.

Step 1: Open the FIS editor and edit the membership function for the input and output as shown in the figure. Select the Mamdani or Sugeno method, fuzzification method, and defuzzification method.

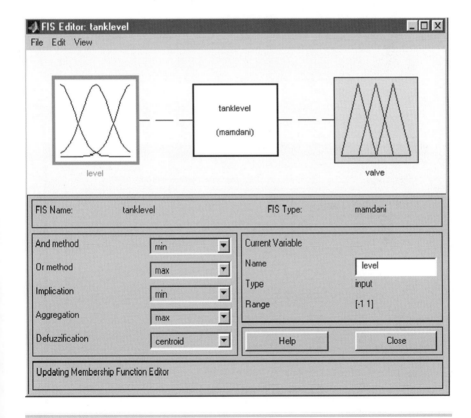

Step 2: Click the input variable and edit the membership function as shown below.

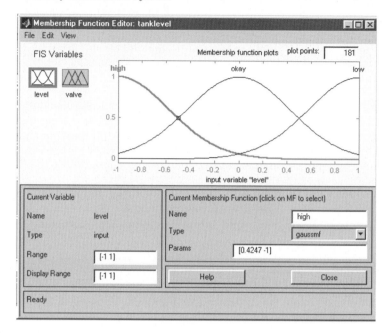

Step 3: Click the output variable and edit the member ship function as
shown below.

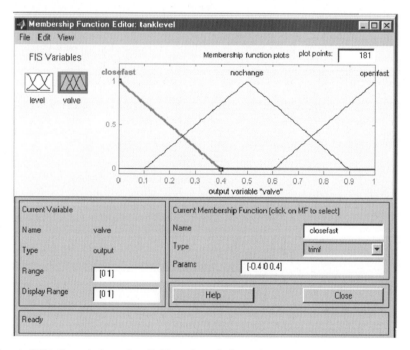

Step 4: Edit the rule base by clicking the rule base from view menu.

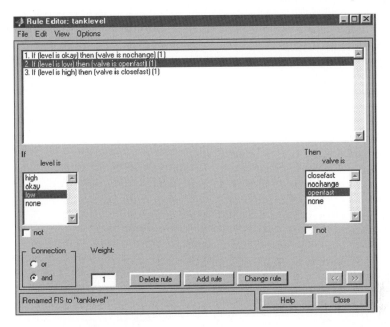

Step 5: To view the rule viewer go to the view and click rule viewer.

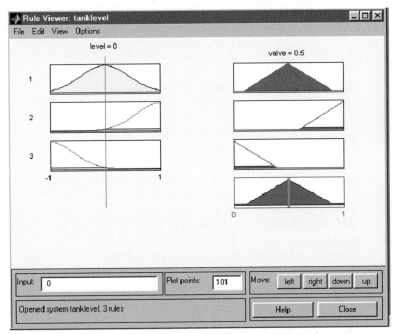

Step 6: To view the surface viewer go to the view and click surface viewer.

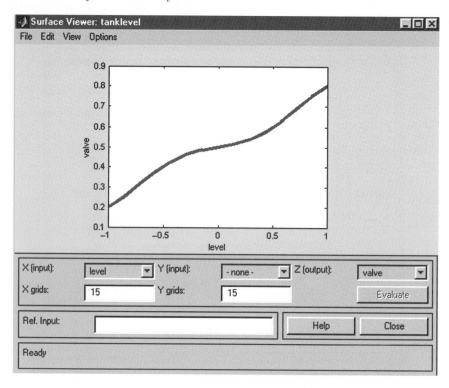

Step 7: Save the file to the workspace as tanklevel and go to the command window and enter the command tanklevel.

>> tanklevel
   tanklevel =
   name: 'tanklevel'
   type: 'mamdani'
   andMethod: 'min'
   orMethod: 'max'
   defuzzMethod: 'centroid'
   impMethod: 'min'
   aggMethod: 'max'
   input: [1x1 struct]
   output: [1x1 struct]
   rule: [1x3 struct]
Thus an FIS editor is formed for a tanklevel controller.

Example 6.3. Let $y = -2x + x^2$.

(a) Form a fuzzy system, which approximates function f, when $x \in [-10, 10]$. Repeat the same by adding random, normally distributed noise with zero mean and unit variance.

(b) Simulate the output when the input is sin(t). Observe what happens to the signal shape at the output.

Solution.

Method 1
This is achieved by writing a Matlab program
Program

```
% Generate input-output data and plot it
x=[-10:.5:10]'; y=-2*x-x.*x;
% Plot of parabola
plot(x,y)
grid
xlabel('x');ylabel('output');title('Nonlinear characteristics')
% Store data in appropriate form for genfis1 and anfis and plot it
data=[x y];
trndata=data(1:2:size(x),:);
chkdata=data(2:2:size(x),:);
% Plot of training and checking data generated from parabolic equation
plot(trndata(:,1),trndata(:,2),'o',chkdata(:,1),chkdata(:,2),'x')
xlabel('x');ylabel('output');title('Measurement data'); grid
%Initialize the fuzzy system with command genfis1. Use 5 bellshaped
    membership functions.
nu=5; mftype='gbellmf'; fismat=genfis1(trndata, nu, mftype);
%The initial membership functions produced by genfis1 are plotted
plotmf(fismat,'input',1)
xlabel('x');ylabel('output');title('Initial membership functions');
grid
% Apply anfis-command to find the best FIS system - max number of
iterations = 100
numep=100;
[parab, trnerr,ss,parabcheck,chkerr]=anfis(trndata,fismat,numep,[],chkdata);
%Evaluate the output of FIS system using input x
anfi=evalfis(x,parab);
% Plot of trained fuzzy system using trained data
plot(trndata(:,1),trndata(:,2),'o',chkdata(:,1),chkdata(:,2),'x',x,anfi,'-')
grid
xlabel('x');ylabel('output');title('Goodness of fit')
```

Output
iterations = 100
ANFIS info:

> Number of nodes: 24
> Number of linear parameters: 10
> Number of nonlinear parameters: 15

Total number of parameters: 25
Number of training data pairs: 21
Number of checking data pairs: 20
Number of fuzzy rules: 5

Start training ANFIS.

1	1.11035	1.11725
2	1.1055	1.11205
3	1.10065	1.10685
4	1.09578	1.10165
5	1.09091	1.09644

Step size increases to 0.011000 after epoch 5.

6	1.08602	1.09123
7	1.08063	1.08548
8	1.07523	1.07974
9	1.06982	1.07399

Step size increases to 0.012100 after epoch 9.

10	1.0644	1.06823
11	1.05842	1.06189
12	1.05242	1.05555
13	1.04642	1.0492

Step size increases to 0.013310 after epoch 13.

.
.
.
.
.

.

86	0.021108	0.0264632
87	0.0107584	0.0173793
88	0.0173351	0.0221364
89	0.00977897	0.0162064
90	0.0165189	0.0217102
91	0.00934733	0.0162595

Step size decreases to 0.066602 after epoch 91.

92	0.0160967	0.0218316
93	0.00734716	0.0157201
94	0.0155897	0.0219794
95	0.00732559	0.016182

Step size decreases to 0.059942 after epoch 95.

96	0.015227	0.0222669
97	0.0061041	0.0163292
98	0.0147314	0.022553
99	0.00645142	0.0170149

Step size decreases to 0.053948 after epoch 99.

| 100 | 0.0144049 | 0.0229312 |

Designated epoch number reached –> ANFIS training completed at epoch 100.

The respective plots obtained are:

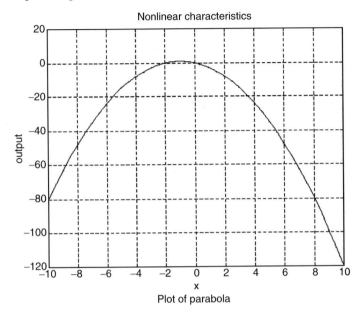

Training data (o) and checking data (x) generated from the parabolic equation

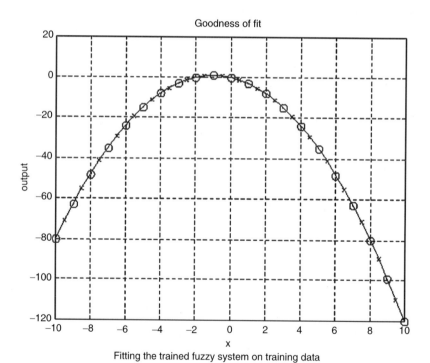

Initial membership functions

Initial fuzzy system (fismat) for anfis

Goodness of fit

Fitting the trained fuzzy system on training data

Method 2

Using a fuzzy logic toolbox graphical user interface (GUI) can perform the same problem.

Open fuzzy toolbox GUI, choose new sugeno system

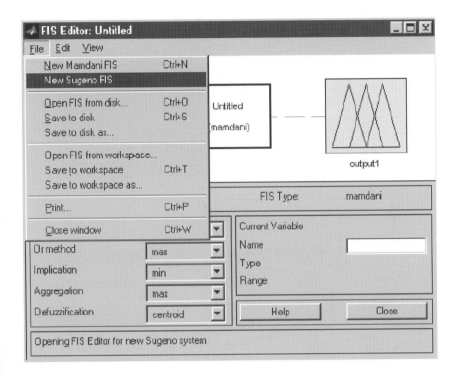

Generate a new Sugeno type fuzzy system.

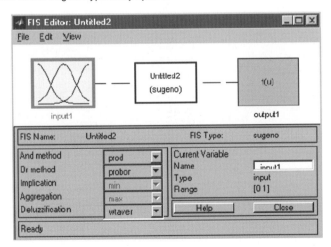

Display a new Sugeno type system.

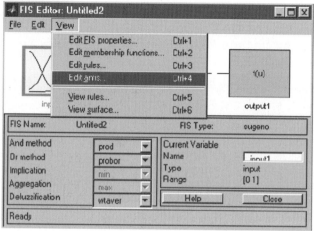

Generating anfis display.

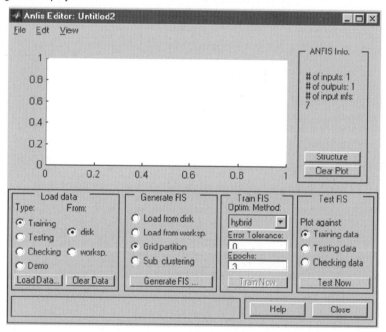

Anfis editor display.

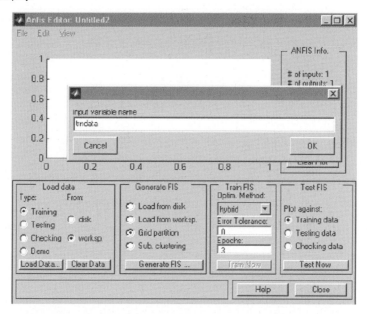

Load training data.

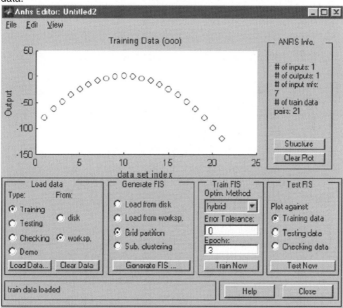

Plot of training data. The *x*-axis indicates numbering of data points rather than absolute values.

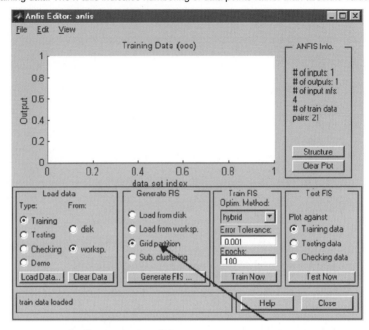

Generating initial FIS matrix using grid partition.

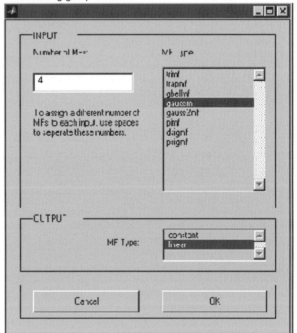

Default membership function type (gaussmf) and their number (4).

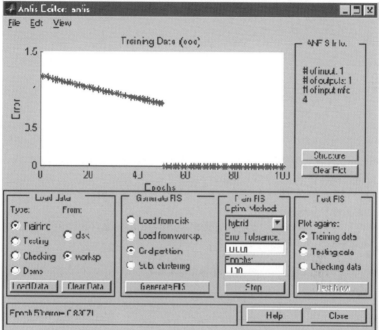

Training when error tolerance is chosen to be 0.001 and number of epochs is limited to 100.

Thus the application has been done using fuzzy toolbox GUI.

Example 6.4. Study how the nonlinearity modeled with the fuzzy system fismat1 distorts a sinusoidal signal. Assume $\sin(t)$ at the input.

Solution. Clearly higher-order harmonics are generated. Such phenomenon can be observed, e.g., in electrical transformers. This problem should be continued immediately after the example problem 6.3, because Matlab assumes that fuzzy system matrix *parab* is available. Otherwise you must repeat example problem 6.3.

Open Simulink in Matlab command window and open a new file to configure the system.

In the Simulink library open first Blocksets and Toolboxes. In the next window, open Fuzzy Logic Toolbox. Now you can choose either Fuzzy Logic Controller block or Fuzzy Logic Controller with rule viewer.

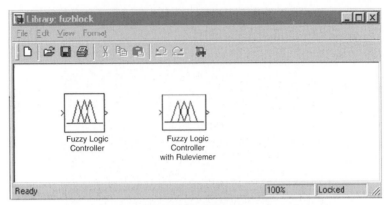

Fuzzy block library

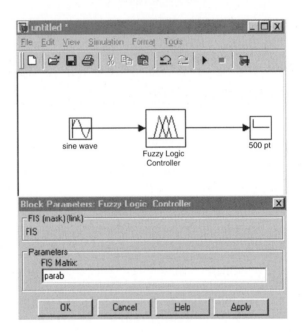

Entire system

The graphical output can be viewed through the Scope block

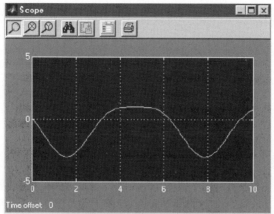

Output of the Simulink system

Comparison between the actual functions $-2x-x^2$ and the fuzzy system approximation is shown below.

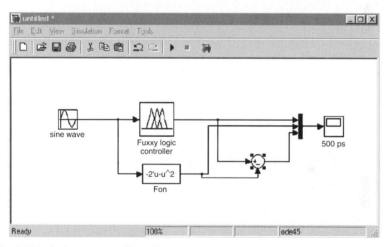

Simulink blocks for the comparison of the actual function and the fuzzy system approximation.

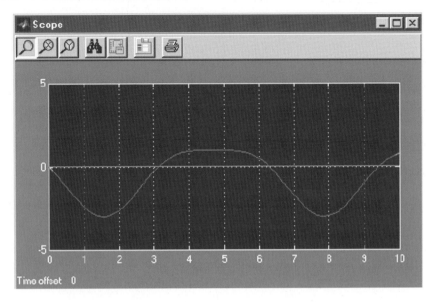

Comparison Output Responses

Combining the *Fuzzy system* with Simulink is an important feature from the user's point of view. Once the fuzzy system has been determined, it can be used in Simulink to simulate dynamical systems. This provides the user a very powerful tool to investigate behavior of complex systems. Computation by hand is tedious and in practice impossible without a computer, but it is at this point when fuzzy systems become really interesting and exciting.

Summary

In this chapter, the formation, aggregation, and decomposition of rules are explained. Fuzzy mathematical tools and the calculus of IF–THEN rules provides a most useful paradigm for the automation and implementation of an extensive body of human knowledge. The chapter is also added with the various method of inference. The comparison and the advantages of the two methods Mamdani and Sugeno model are also discussed.

Review Questions

1. In what way does the fuzzy rules play a key role in determining the output of the system?
2. What is the general format of the fuzzy rule base system?

3. Define the antecedents and consequents present in a fuzzy rule.
4. What are the most two important connectives used in fuzzy rules?
5. How are the rules formed in a fuzzy rule-based system?
6. With examples discuss about the conditional and unconditional statements used or the formation of rules.
7. How is the canonical form of rule base developed?
8. Differentiate simple and compound rules.
9. What is the purpose of decomposition of rules?
10. Discuss in detail on all the methods used for decomposition of the compound rules.
11. How are the rules aggregated to obtain the final solution?
12. Write short notes on the two methods of aggregation of rules.
13. State the properties of the rule base system.
14. Define fuzzy inference system.
15. With a suitable block diagram, explain the construction and working of fuzzy inference system
16. What are the two fuzzy inference methods?
17. Write short note on the Mamdani method of fuzzy inference system
18. Write in detail about the Sugeno method adopted in fuzzy inference system.
19. Compare Mamdani and Sugeno method of fuzzy inference system.
20. State the advantages of Mamdani and Sugeno model.

Exercise Problems

1. In a temperature controller for room, the linguistic comfort range is "slightly cold" and "not too hot" using these membership functions defined on a universe of temperature in °C.

$$\text{"Hot"} = \left\{ \frac{0}{25} + \frac{0.1}{26} + \frac{0.3}{27} + \frac{0.5}{28} + \frac{0.7}{29} + \frac{0.9}{30} \right\},$$

$$\text{"Cold"} = \left\{ \frac{1}{25} + \frac{0.8}{26} + \frac{0.7}{27} + \frac{0.4}{28} + \frac{0.3}{29} + \frac{0.2}{30} \right\}.$$

Find the membership functions for:
(a) Not very hot
(b) Slightly cold or slightly hot

2. Amplifier capacity on a normalized universe say [0,100] can be linguistically defined by fuzzy variable like here:

$$\text{"Powerful"} = \left\{ \frac{0}{1} + \frac{0.2}{10} + \frac{0.6}{50} + \frac{0.9}{100} \right\},$$

$$\text{``Weak''} = \left\{ \frac{0.9}{1} + \frac{0.8}{10} + \frac{0.2}{50} + \frac{0.1}{100} \right\}.$$

Find the membership functions for the following linguistic phases used to decrease the capacity of various amplifiers:

(a) Powerful and not weak

(b) Very powerful or very weak

(c) Very, very powerful and not weak

3. In a boiler, pressure and temperature are linguistic parameters. Nominal pressure limit ranges from 300 to 900 psi. Nominal temperature limit is 80–100°C. The fuzzy linguistic uses are as follows:

$$\text{``Low'' temperature} = \left\{ \frac{1}{80} + \frac{0.8}{82} + \frac{0.6}{84} + \frac{0.3}{86} + \frac{0.2}{88} + \frac{0}{90} \right\},$$

$$\text{``High'' temperature} = \left\{ \frac{0}{86} + \frac{0.2}{88} + \frac{0.3}{90} + \frac{0.5}{92} + \frac{0.7}{94} + \frac{0.9}{96} \right\},$$

$$\text{``High'' pressure} = \left\{ \frac{0}{300} + \frac{0.2}{500} + \frac{0.3}{600} + \frac{0.5}{800} + \frac{0.7}{900} + \frac{1}{1,000} \right\},$$

$$\text{``Low'' pressure} = \left\{ \frac{1}{300} + \frac{0.7}{600} + \frac{0.8}{700} + \frac{0.4}{800} + \frac{0.3}{900} + \frac{0}{1,000} \right\}.$$

(a) Find the following membership functions:

(1) Temperature not very low

(2) Temperature not very high

(b) Find the following membership functions:

(1) Pressure slightly

(2) Pressure fairly high $\left([\text{high}]^{2/3} \right) (\text{high}) 0$

(3) Pressure not very low or fairly low

4. In a computer system, performance depends to a large extent on relative spear of the components making up the system. The "speeds" of the CPU and memory are important factors in determining the limits of operating speed in terms of instruction executed per unit size

$$\text{``Fast''} = \left\{ \frac{0}{6} + \frac{0}{1} + \frac{0.1}{4} + \frac{0.3}{8} + \frac{0.5}{20} + \frac{0.7}{45} + \frac{1}{100} \right\},$$

$$\text{``Slow''} = \left\{ \frac{1}{0} + \frac{0.9}{1} + \frac{0.8}{4} \frac{0.5}{8} + \frac{0.2}{20} + \frac{0.1}{45} + \frac{0}{100} \right\}.$$

Calculate the membership function for the phases:

(a) Not very fast and slightly slow

(b) Very, very fast and not slow

(c) Very slow are not fast

5. The age of building is to be determined the linguistic terms are "old" and "young"

$$\text{Old} = \left\{ \frac{0}{0} + \frac{0.2}{5} + \frac{0.4}{10} + \frac{0.5}{15} + \frac{0.7}{20} + \frac{0.8}{25} + \frac{1}{30} \right\},$$

$$\text{Young} = \left\{ \frac{0.9}{0} + \frac{0.8}{5} + \frac{0.7}{10} + \frac{0.4}{15} + \frac{0.3}{20} + \frac{0.2}{25} + \frac{0.1}{30} \right\}.$$

For the building in years, find the membership functions for the following expressions:

(a) Very old

(b) Very old or very young

(c) Not very old and fairly young $\left([\text{young}]^{2/3} \right)$

(d) Young or slightly old

6. By using the canonical from rule find the volume of the cone $(1/3)\Pi r^2 h$ radius is $4\,\text{cm}$ and h is $8\,\text{cm}$.

7. The formula

$$\frac{1}{u} + \frac{1}{z} = \frac{1}{f}$$

is used in optics. The variables u, z, and f are the distance from the center of lens to the center of the object, the distance from center of lens to the center of the image and the focal length. Define canonical form of rules for this problem.

8. Given the discretized form of the fuzzy variables $\underset{\sim}{X}, \underset{\sim}{Y}, \underset{\sim_1}{Z}, \underset{\sim_2}{Z}$

$$\underset{\sim}{X} = \left\{ \frac{0.0}{0} + \frac{0.5}{1} + \frac{0.7}{2} + \frac{0.4}{3} + \frac{0.1}{4} \right\},$$

$$\underset{\sim}{Y} = \left\{ \frac{0.1}{0} + \frac{0.4}{3} + \frac{1}{4} + \frac{0.6}{5} + \frac{0.2}{6} \right\},$$

$$\underset{\sim_1}{Z} = \left\{ \frac{0.1}{5} + \frac{0.4}{6} + \frac{0.8}{7} + \frac{0.5}{8} + \frac{0.1}{9} \right\},$$

$$\underset{\sim_2}{Z} = \left\{ \frac{0.2}{10} + \frac{0.4}{11} + \frac{1}{12} + \frac{0.4}{13} + \frac{0.2}{14} \right\}.$$

(a) Form analogous continuous membership functions for $\underset{\sim}{X}, \underset{\sim}{Y}, \underset{\sim_1}{Z}, \underset{\sim_2}{Z}$

(b) A system is described by a set of three rules, using the foregoing fuzzy variables. All the rules have to be satisfied simultaneously for the system to work. The rules are these:

(1) IF $\underset{\sim}{X}$ and $\underset{\sim}{Y}$ then $\underset{\sim_1}{Z}$

(2) IF $\underset{\sim}{X}$ and $\underset{\sim}{Y}$ then $\underset{\sim_2}{Z}$

(3) IF $\underset{\sim}{X}^2$ and $\underset{\sim}{Y}^2$ then $\underset{\sim_1}{Z}$

Determine the output of the system by graphical inference, using max–min, max–product technique if $x = 3$ and $y = 4$ and use centroid method for defuzzification.

9. Let $y = f(x) = 5x+3$.
   (a) Form a fuzzy system, which approximates function f, when $x \in [-5, 5]$. Repeat the same by adding random, normally distributed noise with zero mean and unit variance.
   (b) Simulate the output when the input is $\cos(t)$. Observe what happens to the signal shape at the output.

10. Write FIS using Mamdani method for the following controlling room temperature (assume the linguistic variable yourself).

11. With FIS for controlling the speed of the motor input should armature current and torque output should be speed using Sugeno method.

12. Write FIS for the controlling the water level and temperature in the boiler using Mamdani and Sugeno models. Assume your own linguistic variables.

13. Write an FIS for controlling the temperature of an air conditioner system using any one of the inference method.

   Use a fuzzy rule base to model the ideal gas equation for a confined gas, $pV = nRT$; where p is the pressure, V is the volume, T is the temperature of the gas, n is proportional to the number of gas molecules (a constant), and R is the ideal gas constant. Assume that we allow the gas temperature to adjust to that of the surroundings. Use rules along the lines of: IF (volume is large) THEN (pressure is low) and allow the use of very for both variables. What type of membership functions would it be a good idea to use?

14. Use Matlab's Fuzzy Logic Toolbox to model the tip given after a dinner for two, where the food can be disgusting, not good, bland, satisfying, good, or delightful, and the service can be poor, average, or good. To get started, you type fuzzy in a Matlab window. Then use the fuzzy inference system and membership function editors to define and tune your rules.

15. Write down a simple fuzzy rule base by which to control the temperature of a shower, ignoring any delays, etc. (three rules are sufficient). Assume that the water is pleasant at temperatures around 35–40°C. Sketch the membership functions.

16. Consider the Takagi–Sugeno fuzzy rules:
   R_1 : IF (x is negative) THEN ($y_1 = e^{0.9z}$),
   R_2 : IF(x is zero) THEN ($y_2 = 4.2x$),
   R_3 : IF (x is positive) THEN ($y_3 = e^{-0.7x}$).

R_1: IF (x is negative) THEN ($y_1 = e^{0.9z}$),

R_2: IF (x is zero) THEN ($y_2 = 4.2x$),

R_3: IF (x is positive) THEN ($y_3 = e^{-0.7x}$).

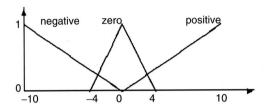

Evaluate y in case of $x = -3$, $x = 2$, and $x = 9$.

7

Fuzzy Decision Making

7.1 Introduction

Decision making is essentially an important aspect in all aspects of life. Making decisions is the fundamental activity of human beings. In any decision process we consider the information about the outcome and choose among two or more alternatives for subsequent action. If good decisions are made, then we may get a good expected output.

Decision making is defined to include any choice or selection alternatives. A decision is said to be made under certainty, where the outcome for each action can be determined precisely. A decision is made under risk. When the only available knowledge concerning the outcomes consists of their conditional probability distributions. The uncertainty existing is the prime domain for fuzzy decision (FD) making.

There are various ways in which the FD can be made. They are discussed in detail in the following sections.

7.2 Fuzzy Ordering

Fuzzy ordering involves the decision made on rank basis. Which has first rank, second rank, etc. If $x_1 = 2, x_2 = 5$, then $x_2 > x_1$, here there is no uncertainty, which is called as crisp ordering. The case where the uncertainty or ambiguity arises, then it is called fuzzy ordering or rank ordering. If the uncertainty in the rank is random, then probability density function (pdf) may be used for the random case.

Consider a random variable x_1, defined using Gaussian pdf, with a mean of μ_1 and standard deviation σ_1, also x_2, another variable which is also defined by using Gaussian pdf with a mean μ_2 and standard deviation σ_2. If $\sigma_1 > \sigma_2$ and $\mu_1 > \mu_2$, then the density functions are plotted as shown in Fig. 7.1.

The frequency of probability that one variable is greater than the other is given by

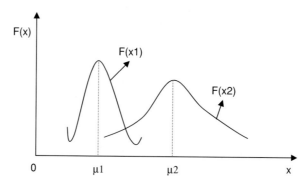

Fig. 7.1. Density function for two Gaussian random variables

$$P_{(x_1 \geq x_2)} = \int_{-\infty}^{+\infty} f_x\,(x_1\,dx_1),$$

where F_1 is cumulative distribution function.

Also, if there are two fuzzy numbers $\underset{\sim}{P}$ and $\underset{\sim}{Q}$, the ranking that $\underset{\sim}{P}$ is greater than fuzzy number $\underset{\sim}{Q}$ is given by

$$R\left(\underset{\sim}{P} \geq \underset{\sim}{Q}\right) = \underset{x \geq y}{\text{Sub min}}\left(\mu_{\underset{\sim}{P}}(x), \mu_{\underset{\sim}{Q}}(y)\right).$$

Also

$$R\left(\underset{\sim}{p} \geq \underset{\sim}{Q}\right) = 1 \ \text{ if and only if } P \geq Q.$$

Example 7.1. Consider we have three fuzzy sets, given by

$$\underset{\sim}{A} = \left\{\frac{1}{3} + \frac{0.8}{7}\right\}, \quad \underset{\sim}{B} = \left\{\frac{0.6}{4} + \frac{1.0}{6}\right\}, \quad \underset{\sim}{C} = \left\{\frac{0.8}{2} + \frac{1}{4} + \frac{0.4}{8}\right\}.$$

Make suitable decisions based on fuzzy ordering.

Solution. Using the truth value of inequality, $\underset{\sim}{A} \geq \underset{\sim}{B}$, as follows:

$$
\begin{aligned}
T(\underset{\sim}{A} \geq \underset{\sim}{B}) &= \max_{x_1 \geq x_2}\left\{\min(\mu_{\underset{\sim}{A}}(x_1), \mu_{\underset{\sim}{B}}(x_2))\right\} \\
&= \max\{\min(0.8, 0.6), \min(0.8, 1.0)\} \\
&= \max\{0.6, 0.8\} \\
&= 0.8.
\end{aligned}
$$

Similarly,

$$T(\underset{\sim}{A} \geq \underset{\sim}{C}) = 0.8, \quad T(\underset{\sim}{B} \geq \underset{\sim}{A}) = 1.0, \quad T(\underset{\sim}{B} \geq \underset{\sim}{C}) = 1.0, \quad T(\underset{\sim}{C} \geq \underset{\sim}{A}) = 1.0,$$
$$T(\underset{\sim}{C} \geq \underset{\sim}{B}) = 0.6.$$

Then,

$$T(\underset{\sim}{A} \geq \underset{\sim}{B}, \underset{\sim}{C}) = 0.8,$$

$$T(\underset{\sim}{B} \geq \underset{\sim}{A}, \underset{\sim}{C}) = 1.0,$$

$$T(\underset{\sim}{C} \geq \underset{\sim}{A}, \underset{\sim}{B}) = 0.6.$$

From this calculation, the overall ordering of the three fuzzy sets would be $\underset{\sim}{B}$ first $\underset{\sim}{A}$ second, and $\underset{\sim}{C}$ third.

Thus the fuzzy ordering is performed

7.3 Individual Decision Making

A decision situation in this model is characterized by:

- Set of possible actions
- Set of goals $p_i(i \in x_n)$, expressed in terms of fuzzy set
- Set of constraints $Q_i(j \in x_m)$, expressed in terms of fuzzy sets.

It is common that the fuzzy sets impressing goals and constraints in this formulation are not defined directly on the set of actions, but through the other sets that characterize relevant states of nature.

For the set A, then

$$P_i(a) = \text{Composition } [P_i(a)] = P_i^{\;1}(P_i(a))$$
$$\text{with } P_i^1,$$
$$Q_j(a) = \text{Composition of } Q_i(a) = Q_j^{\;1}(q_j(a))$$
$$\text{with } Q_i^1,$$

for $a \in A$.

Then the FD is given by

$$FD(a) = \min \left[\inf_{i \in N_n} P_i(a), \inf_{i \in N_m} Q_j(a) \right].$$

7.4 Multi-Person Decision Making

When decision are made by many persons, the difference of it from the individual decision maker is:

1. The goals of single decision makers differ, such that each places a different ordering arrangements.
2. The individual decision makers have access to different information upon which to base their decision.

In this case, each member of a group of n single decision makers has a preference ordering $P_k, K \in N_n$, which totally or partially orders a set x.

Then a function called "Social Choice" is to be found, given the individual preference ordering. The social choice preference function is defined by fuzzy relation as

$$S : X \times X \to [0,1].$$

which has membership of $S(x_i, x_j)$ which indicates the preference of alternative x_i over x_j.

If number of persons preferring x_i to $x_j = N(x_i, x_j)$,
Total number of decision makers $= N$.
Then,

$$S(x_i, x_j) = \frac{N(x_i, x_j)}{n}.$$

This defines the multi-person decision making also

$$S(x_i, x_j) = \begin{cases} 1 & \text{if} \quad x_i \underset{K}{>} x_j \text{ for some } k, \\ 0 & \text{other wise.} \end{cases}$$

7.5 Multi-Objective Decision Making

The process involves the selection of one alternative a_i, from many alternatives A, given a collection or set, say $\{0\}$ objectives which is important for a decision maker.

Define universe of n alternatives, i.e.,

$$A = \{a_1, a_2, \ldots, a_n\} \text{ and}$$

set of "r" objectives

$$O = \{0_1, 0_2, \ldots, 0_r\}.$$

The decision function (DF) here is given as intersection of all objectives

$$DF = 0_1 \wedge 0_2 \wedge 0_3 \wedge \cdots \wedge 0_r.$$

The membership for the alternative is given by,

$$\mu_{DF}(a^*) = \max_{a \in A}(\mu_{DF}(a)).$$

Let $\{P\} = \{b_1, b_2, \ldots, b_r\} = b_i, \quad i = 1 \text{ to } r$,
then, $DF = DM(0_1, b_1) \wedge DM(0_2, b_2) \wedge \cdots \wedge (DM(0_r, b_r))$ where $DM(0_n, b_n)$ is called decision measure (DM).

The DM for a particular alternative is

$$DM(0_i(a)b_i) = b_i \to 0_i(a) = \overline{b_i} \cup 0_i(a).$$

Thus $b_i \rightarrow 0_i$ indicates a unique relationship between preference and objective.

Thus, the DF may be given by

$$DF = \bigcap_{i=1}^{r}(\bar{b}_i \cup 0_i(a)).$$

and $a*$ is the alternative that maximizes D.

Let $p_i = \bar{b}_i \ u \ 0_i(a)$.

So,

$$\mu p_i(a) = \max\lfloor\mu_{b_i} - (a), \mu_{0_i}(a)1\rfloor.$$

The membership form of the optimal solution is:

$$\mu_{DF}(a*) = \max_{a \in A}\lfloor\min\{\mu_{p_1}(a), \mu_{p_2}(a), \ldots, \mu_{p_r}(a)\}\rfloor.$$

Thus the decision is made as discussed.

7.6 Fuzzy Bayesian Decision Method

Classical Bayesian decision methods preassumes that the future states of the nature can be characterized as probability events. The problem here in fuzzy Bayesian method is that the events are vague and ambiguous and uncertain. This is solved by the following method:

Consider the formation of the probabilistic decision method.

Assuming the set of state of nature as:

$$S = \{S_1, S_2, \ldots, S_n\}.$$

So, the probabilities that these states occur are given by

$$P = \{P(S_1), P(S_2), \ldots, P(S_n)\}$$

and

$$\sum_{i=1}^{n} p(s_i) = 1.$$

These are called as prior probabilities. If decision maker chooses m alternatives, then

$$A = \{a_1, a_2, \ldots, a_m\}$$

and for an alternative a_j, the utility value is u_{ji}, if the future state is the state S_i.

The utility values are to be found by the decision maker for each $a_j - S_i$ combination.

The expected utility with jth alternative would be

$$E(u_j) = \sum_{i=1}^{n} U_{jI} p(s_i).$$

The common decision criterion is the maximum expected utility among all alternatives

$$E(u^*) = \max_j E(u_j).$$

Which selects a_x if $u^* = E(x_k)$.

Summary

Fuzzy decision making involves various methods, which were described in this chapter. The fuzzy ordering involves the ordering formed on the rank basis. The decision situation is found to vary between the individual decision making and multi-person decision making. Fuzzy Bayesian decision making is one of the most important decision making process discussed. In the case of multi-objective decision making one alternative is found to be selected from many alternatives. Thus the various decision making process are described in this chapter.

Review Questions

1. What is meant by fuzzy decision making process?
2. What are the various methods used for fuzzy decision making?
3. Write short note on fuzzy ordering. State an example for fuzzy ordering.
4. How is the decisions made individually?
5. What are the characteristics of decision situations in individual decision making?
6. Discuss in detail on the multi-person decision making.
7. Compare the accuracy rate of individual decision making and multi-person decision making
8. What is the main aim of multi-objective decision making?
9. Derive an expression for the membership for optimal solution using multi-objective decision making.
10. What is the importance of fuzzy Bayesian decision making?
11. Define prior probabilities in fuzzy Bayesian method.

8

Applications of Fuzzy Logic

8.1 Fuzzy Logic in Power Plants

8.1.1 Fuzzy Logic Supervisory Control for Coal Power Plant

The high temperature Winkler gasification (HTW) process that was developed by Rheinbraun has been used for many years in pilot and demonstration plants to generate synthesis gas and fuel gas out of brown coal. Conventional methods were used before to control the gas throughput. While the conventional control engineering implementation was able to run the process in a stable operating point, improvements were necessary to use the HTW process in a coal power station with integrated coal gasification:

- More precise control of gas throughput under fluctuations of the coal quality
- More robust control in cases of fast load changes
- Automation of supervisory control operation

On top on the existing base level automation, a supervisory fuzzy logic control strategy was implemented on the HTW plant in Berrenrath/Germany. Fuzzy logic was used because the control problem was strongly non-linear and involves multiple measured and command variables. On the other hand, extensive operator knowledge about the process was available. The implemented fuzzy logic supervisory control strategy successfully improved throughput control quality as well as the adaptation to different coal parameters.

High Temperature Winkler Gasification

The process that is used to gasify the coal is called High temperature Winkler method (HTW). The HTW gasification method uses a high temperature fluid bed process to convert brown coal into synthesis gas, a mixture of carbon monoxide (CO) and hydrogen (H_2). This gas mix can be used to produce chemical base products like aldehydes or organic acids. Alternatively it can be used in a power plant gas turbine to generate electricity. The gas produced by the demonstration plant is used for chemical synthesizes. Later in the power plant application the gas will be used to run a gas turbine/steam turbine combination.

The HTW process has been used for the gasification of coal by the German coal company Rheinbraun since 1956. The demonstration plant started operation in 1985. It converts 720 t of coal per day into $900,000\,m^3$ (iN) synthesis gas. In 1996 Inform added a fuzzy logic supervisory control to enable the process for a power plant application. Figure 8.1 shows a photo of this plant.

The main inputs for the HTW process are coal, oxygen and steam. The coal is first ground to small pieces and pre-dried before it is fed into the bottom part of the fluid bed reactor. The steam and the oxygen are fed into the reactor on four different levels, into and above the fluid bed. In the fluid bed the coal reacts with the oxygen and the steam. This reaction takes place at a temperature of around $800°C$ and at a pressure of 10 bar. After the reaction in the fluid bed the generated gas enters the hot zone above the fluid bed. At temperatures around $1,000°C$ additional oxygen and steam is added and left over coal particles react with the gases. This way additional gas is produced and by-products like methane and other hydrocarbons are converted to carbon monoxide and hydrogen. The produced gases leave the reactor at the top through the reactor head. At this point the gas is still mixed with a lot of particles. These are filtered out and fed back into the fluid bed with a zyklon filter and a feed back tube.

Fig. 8.1. HTW plant in Berrenrath, Germany

The coal ashes accumulate at the bottom of the fluid bed reactor. They are removed from there out of the reactor by two conveyor spirals. The hot raw gas is cooled down to 270°C. Its heat is used to generate pressure steam, some of which is recycled back into the process. Ceramic filters remove remaining dust particles out of the gas. The following gas washer removes NH_3, HCL and other gas components. The following CO conversion creates the correct carbon monoxide/hydrogen mix for the methanol synthesis. After a compression to 37 bar the gas is processed in a non-selective rectisol washer (CO_2/H_2S washer). At temperatures below $-40°C$ liquid methanol is used to wash out carbon-dioxide and sulfuric components. The methanol is used again after recycling it and the purified synthesis gas is used at a nearby chemical plant. Figure 8.2 shows the process diagram of the HTW plant. The coal input, the oxygen input, the distribution of the oxygen input over the eight different nozzles and the ash removal rate have to be controlled to use the coal efficiently and to generate the correct mixture of gases. Instead of coal a mixture of coal and plastic refuse can be used in the HTW process. This way the coal consumption is reduced and the plastic refuse is recycled into synthesis gas.

Conventional Control

The HTW demonstration plant is controlled with an Eckhardt PLS-80E DCS system. This system controls over 6,000 measurements and actuators. The main control room is equipped with ten Unix-based operator consoles and four real-time servers. So far nearly all the set points of the underlying control circuits are set and adjusted manually by the operators. They constantly monitor the process condition and adjust the set points of the underlying control circuits accordingly (i.e. coal input, oxygen input). A few years ago it was tried to automatically generate some set points using a conventional PID controller, but the results were not satisfying. This supervisory control only worked fine when the coal quality was very constant. Otherwise the process quality would deteriorate significantly and the operators had to intervene and switch back to manual operation.

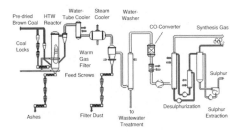

Fig. 8.2. Process diagram of HTW plant

Supervisory Fuzzy Logic Control

Two main tasks have been defined for the supervisory fuzzy logic control: regulation of the gas throughput and process stabilization.

The fuzzy logic must keep the gas throughput at the set point and it has to respond to set point changes with the correct dynamic speed. The set point can vary from 70% (partial load) to 100% (full load). The fluctuations of the gas throughput result mainly from variations of the coal quality (humidity, ash content and granularity). These effects have to be compensated by the fuzzy logic.

The process stabilization must keep several process parameters in the optimum range. The reactor load influences the optimum of these process parameters. The position of the optimum also depends on the coal quality. The following parameters were used to define the quality of the process:

– Temperature in the postgasification zone
– Height and density of the fluid bed
– Composition of the produced gas (CO, CH_4, H_2)

The process stabilization is especially difficult when the HTW process is fed with a mixture of coal and plastic refuse. This is done because plastic refuse is a very inexpensive fuel. But the addition of plastic to the coal results in drastically different process conditions. The fuzzy logic control uses the regular measurements of the process conditions to detect any addition of plastic to the coal. This will result into an adapted control strategy of the fuzzy logic control.

Fuzzy Logic Control Design

The specification of the control task resulted into a preliminary concept of the fuzzy logic controller. The operator knowledge was than used to specify the control strategy of the fuzzy logic system. Several structured audits took place to evaluate the operator knowledge systematically. The audits focused on the operators' manual control strategies and on the relationships between the inputs and the outputs of the process. This procedure is in accordance with the standardized fuzzy logic design method.

The audits resulted into the following concept for the fuzzy logic control: deviations of the gas throughput from its set point immediately result into a correction of the oxygen input. The coal input is adjusted accordingly to keep the ratio between coal and oxygen at a constant level. Changes of the reactor pressure predict changes of gas throughput. Therefore the pressure gradient is used as an early warning indicator for changes of the gas throughput.

For process stabilization and for the adaptation to different coal qualities the fuzzy logic controller uses the following parameters to keep the process in a stable operating condition:

- Coal to oxygen ratio
- Distribution of the oxygen between the fluid bed zone and the gasification zone
- Ash outtake
- Fluidification with steam and inert gas

Sometimes one and the same process input value has to be modified to keep several different critical measurements (temperature, fluid bed height) in the optimum range. This can have conflicting results. For example: a too low fluid bed is normally corrected with an increase of the coal input. A too low temperature is corrected with a reduction of the coal input and a too low dust output is also corrected with a reduction of the coal input. But a too low temperature can occur together with a too low fluid bed. The ability of the fuzzy controller to weight different conflicting indications based on their significance and to use a lot of inputs to determine the best reaction to each situation is very useful to control complex processes.

The fuzzy logic controller has a total number of 24 inputs and eight outputs. A preprocessing reduces the 24 inputs to ten characteristic descriptors. These are fed into the fuzzy logic system. The fuzzy outputs go through a post-processing step to generate the actual set points for the process inputs. Figure 8.3 shows the core structure of the fuzzy logic system.

Integration of Fuzzy Logic into the DCS

The process measurements are coming to the fuzzy logic control through the Eckhardt DCS. The fuzzy logic system generates set point values for the underlying PID controllers. The fuzzy logic controller was implemented on an OS/2 PC. Therefore the set up of the communication between the distributed process control systems (DCS) and the fuzzy logic controller was an important

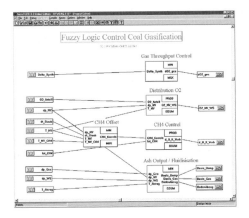

Fig. 8.3. Fuzzy controller structure

part of the whole project. The communication was implemented using Factory Link, a well-known SCADA program by US-Data. A factory link application is running on the same OS/2 PC together with the fuzzy logic controller. The fuzzy logic controller reads data out of the Factory Link Real Time DataBase and it also writes data back into it. Another Factory Link task is communicating with the DCS through the Eckhardt DCS bus. This way an image of the process measurements is created in the Factory Link RTDB and the fuzzy outputs are forwarded to the DCS.

Factory Link initiates a new fuzzy logic evaluation every 10 s. The DCS either uses the external set points generated by fuzzy logic or the internal set points entered by the operators. The operators can switch from the "manual mode" to the "fuzzy logic mode" and back. The "fuzzy logic mode" can only be activated when all the critical system variables are in a predefined safe range. The DCS automatically switches back into "manual mode" whenever a system variable exceeds the safe range. The fallback to "manual mode" also takes place if the communication between the DCS and Factory Link is interrupted.

Figure 8.4 shows the integration of the fuzzy logic control into Factory Link and the Eckhardt DCS. The OS/2 PC is also connected with a serial cable to a WIN95 PC, on which the *fuzzy*TECH development system is installed. This program was used to develop the fuzzy logic control and to generate C-Code for the implementation on the OS/2 PC. The WIN95 PC is also used for online optimization and visualization of the fuzzy logic controller. The serial link to the OS/2 PC enables the user to modify the fuzzy system on the fly from the *fuzzy*TECH development system on the WIN95 PC while the system is running and controlling the process.

Setting the Fuzzy Logic Control into Operation

The first design of the fuzzy logic control, the data preprocessing and post-processing were tested using the simulation tool VisSim. This way the concept

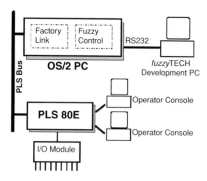

Fig. 8.4. Integration of fuzzy logic into DCS

was checked for any structural errors and an early prototype was presented to the customers. Figure 8.5 shows a test of the fuzzy logic system using simulated data as input.

After the successful completion of the simulations the fuzzy logic was tested offline with real time data from the DCS. To do this the fuzzy logic control first was implemented on the OS/2 PC. The fuzzy controller than used real time DCS measurement values to generate set points values for the DCS. But during these offline simulations the DCS was only using the internal manual set points and not the fuzzy logic set points. By comparing the external fuzzy logic set points with the internal operator set points deviations between manual and automatic operation could be detected and if necessary eliminated.

After the offline testing was finished successfully the online testing started. For the online tests the DCS activated the external set points and so the closed loop performance of the fuzzy logic controller could be tested. The fuzzy logic controller was optimized while running in the closed loop mode from the *fuzzy*TECH development tool on the WIN95 PC. To do this the OS/2 PC was connected with a serial cable to the WIN95 PC. This way the fuzzification, inference and defuzzification were visualized in *fuzzy*TECH. Modifications of the rule base or term definitions were also entered in *fuzzy*TECH and than send to the fuzzy logic controller on the OS/2 PC.

The fuzzy controller proved to be working very effectively during the first few online tests. After that a long series of evaluation tests started. During these tests the performance of the fuzzy controller was tested using a lot of different coal qualities and different loads (70–100%).

Regulation of Gas Throughput

The conventional control focussed on process stabilization not keeping the throughput at its set point. The following diagram shows the changes in the gas throughput when fuzzy controller is active. The fuzzy logic control

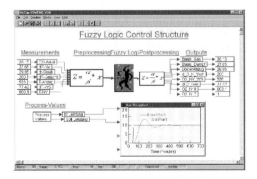

Fig. 8.5. Simulation with VisSim

clearly improves the throughput control. It compensates fluctuations in the granulation of the coal. Figure 8.6 shows the fluctuations of the synthesis gas throughput with and without fuzzy logic.

Change of Load

The fuzzy controller has to regulate the gas throughput in the range 70–100% load with a maximum load gradient of $4\%\,\mathrm{min}^{-1}$.

Figure 8.7 shows a load change from 94% to 76% and back to 94%. The resulting load gradient was $3.2\%\,\mathrm{min}^{-1}$. Currently the load change behavior is being improved by using a modified pressure evaluation.

Adaptation to Coal Add-Ons

Sometimes the HTW process is fed with a mixture of coal and plastic refuse. The resulting process parameters vary greatly from the standard operating conditions. For example the content of methane in the raw gas increases. The fuzzy controller recognizes the different operating conditions and generates a matching internal set point for methane. Figure 8.8 illustrates how this enables the fuzzy controller to stabilize the process.

Adaptation to Different Coal Qualities

The fuzzy controller also keeps the synthesis gas throughput constant when the coal quality changes. Figure 8.9 shows the results of a change of the coal's

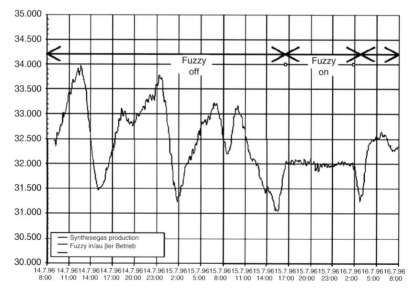

Fig. 8.6. Throughput control with and without fuzzy control

water content from 18% to 12%. The fuzzy controller reduces the coal input to compensate the coal change. Later after switching back to moist coal the coal input is increased accordingly.

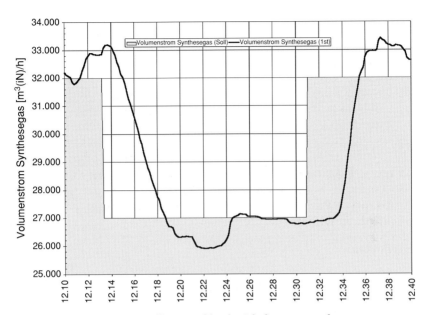

Fig. 8.7. Change of load with fuzzy control

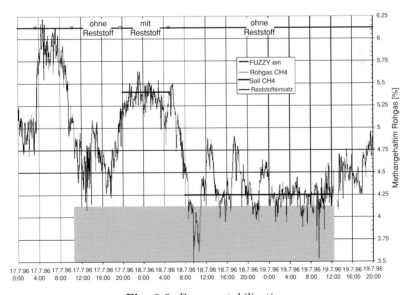

Fig. 8.8. Process stabilization

Conclusion

The fuzzy controller was implemented very quickly. The transfer of the process know-how into the fuzzy controller and its realization took 15 days. So far over 1,100 h of operating time have been evaluated. In nearly all situations the performance of the fuzzy controller was much superior to the manual control. It was able to keep the process parameters in the optimum range whenever the coal quality changed. It was also able to adjust the gas throughput with the necessary change rate. The average gas throughput was kept at the set point. The operating personal has accepted the fuzzy controller as a helpful component because its transparent integration into the PLS makes it easy for them to use it.

8.2 Fuzzy Logic Applications in Data Mining

8.2.1 Adaptive Fuzzy Partition in Data Base Mining: Application to Olfaction

Introduction

Flavor and odor remain permanent challenges in academic and industrial research. The economic impact of the olfactory field explains the large number of articles involving data analysis methods to process sensorial and experimental measurements. However, odor evaluation by man represents a special

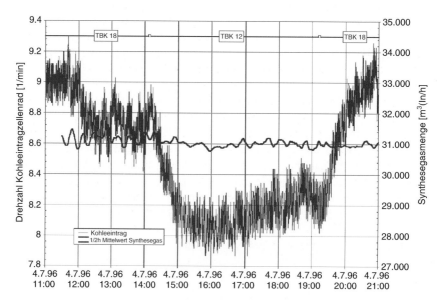

Fig. 8.9. Constant gas throughput with different coal qualities

field of research, whose specific difficulties need to be overcome to lead to robust results. The multiplicity of factors involved in the olfaction biological process prevents the derivation of efficient predictive mathematical models. Four points mainly define this complexity:

(1) A huge number of receptors is involved in olfaction
(2) Knowledge related to the 3D structure of these receptors is still missing
(3) Different types of chemical compounds can affect the same receptor
(4) One compound can exhibit simultaneously different odors

Furthermore, the importance of fuzziness linked to the expert's subjectivity has to be considered. Much progress has been made in the knowledge of physiological and psychological factors influencing the expert's olfaction evaluation, but it is not sufficient to clearly discriminate between objectivity and subjectivity in the characterization exhibited by panels of experts.

All these factors prevent the direct transposition of advances in Chemometrics and Molecular Modeling in Medicinal Chemistry into the field of olfaction. Nevertheless, the use of multivariate data analysis approaches can play an important part to improve the knowledge of the molecular descriptor role in olfaction and, then, the implementation of robust mathematical models. Traditional pattern recognition procedures, like Principal Component Analysis (PCA) (Niemi 1990), Discriminant Analysis (DA) (Hubert 1994), and Cluster Analysis (Kaufman and Rousseeuw 1990), and methods pertaining to the field of Artificial Neural Networks, like Back Propagation Neural Networks (BPNN) (Hecht-Nielsen 1989) or Kohonen Self-Organizing Maps (SOM) (Kohonen 2001), are been widely used in the development of several electronic noses and in data analysis of olfactory data sets.

These approaches offer different possibilities and objectives. PCA can be considered as being only a projective technique. It is worth using this method when clusters or classes can be visually delineated. DA is really a discriminant technique as it aims to find linear relations in the molecular descriptor hyperspace able to separate different compound categories included in the data set. Both methods, PCA and DA, work correctly if the compounds, belonging to different classes, are grouped in well separated regions, but, in more complex distributions, their classification power becomes poor.

Cluster Analysis offers a first solution to this problem. It consists of obtaining self-partitioning of the data, in which each cluster can be identified as a set of compounds clearly delineated regarding the molecular descriptor set involved. Instead of trying to inspect all the compounds in the database to understand and analyze their chemical properties, it is only required to select typical compounds representing each cluster to get a deeper knowledge of the structure of the database, i.e., of the distribution of the compounds in the derived hyperspace. The main problems related to this method are that:

(1) The number of clusters and the initial positions of the cluster centers can influence the final classification results

(2) Compound separation is based on a binary notion of belonging, for which a compound located between two clusters is included in only one cluster

SOM has been considered as an alternative method to overcome the above limitations. It integrates nonlinearity into the data set, so as to project the molecular descriptor hyperspace onto a two-dimensional map and to preserve the original topology, as the points located near each other in the original space remain neighbors in SOM. This technique has been used to process huge amounts of data in a high-dimensional space, but, like PCA, it remains an unsupervised projective method. Then, for predictive objectives, SOM has to be combined with another technique, generating a hybrid system that offers an automatic objective map interpretation.

Contrary to SOM, BPNN is a supervised predictive method. It is able to discriminate any nonlinearly separable class, relating continuous input and output spaces with an arbitrary degree of accuracy. This method, applied to several fields of chemical database analysis, has proved to be very efficient in modeling complex data set relationships. However, as in other Artificial Neural Networks techniques, the complexity of the modeling function often prevents extraction of relevant information suitable to explain the model and, therefore, to deliver a better understanding of biological mechanism.

Fuzzy concepts introduced by Zadeh (1977) provide interesting alternative solutions to the classification problems within the context of imprecise categories, in which olfaction can be included. In fact, fuzzy classification represents the boundaries between neighboring classes as a continuous, assigning to compounds a degree of membership of each class. It has been widely used in the field of process control, where the idea is to convert human expert knowledge into fuzzy rules, and it should be able to extract relevant structure–activity relationships (SAR) from a database, without a priori knowledge.

A data set of olfactory compounds, divided into animal, camphoraceous, ethereal and fatty olfaction classes, was submitted to an analysis by a fuzzy logic procedure called adaptive fuzzy partition (AFP). This method aims to establish molecular descriptor/chemical activity relationships by dynamically dividing the descriptor space into a set of fuzzily partitioned subspaces. The ability of these AFP models to classify the four olfactory notes was validated after dividing the data set compounds into training and test sets, respectively.

The aim of this work is to apply a fuzzy logic procedure, that we called AFP, to a chemical database derived from olfactory studies, in order to develop a predictive SAR model. The database included 412 compounds associated with an odor appreciation defining the presence or the absence of four different olfactory notes. A set of 61 molecular descriptors was examined and the most relevant descriptors were selected by a procedure derived from the Genetic Algorithm concepts.

Materials and Methods

Compound Selection

A database derived from the Arctander's books (Arctander 1960, 1969), including 2,620 compounds and 81 olfactory notes, was submitted to a PCA analysis, in order to determine a reduced subset of compounds representing very weakly correlated odors. The relative results allowed to determine a data set of 412 olfactory compounds homogeneously distributed in four classes: animal, camphoraceous, ethereal, and fatty odors.

Molecular Descriptors

The reduced data set was distributed in a 61 multidimensional hyperspace derived from a selected set of 61 molecular descriptors. This descriptor set includes topological, physicochemical and electronic parameters. In virtual screening, general descriptors have proved a good compromise, from an efficiency point of view, for data mining in large databases. The advantage of these descriptors is their ability to take into account not only the main structural features of each molecule, but also their global behaviors. Then, they should be able to take simultaneously into account the complexity of the olfaction mechanism and the approximation of the odor scale. Molar refractivity (MR), molar volume (MV), molecular weight (MW), and Van Der Waals volume (VdWV) were used as size descriptors.

The shape features of the molecules were characterized by topological indices which account for the ramification degree, the oblong character, etc., 20 molecular connectivity indices, a series of information content descriptors (IC0, SIC0, CIC0, IC1, SIC1,CIC1, IDW), Wiener index (W), centric index (C), Balaban index (J), Gutman index (M2), Platt number (F), counts of paths of lengths 1–4, counts of vertices with 1–4 nearest neighbors were used The number of N, O, and S atoms in a molecule was also considered. A lipophilicity descriptor represented by the octanol/water partition coefficient (log Poct/water) was calculated using the Hansch and Leo method. Another descriptor was derived from the electronegativity of molecules (EMS) by the Sanderson method.

Descriptor Selection

To select, amidst the 61 descriptors, the best parameters for classifying the data set compounds, a method based on genetic algorithm (GA) concepts was used. GA, inspired by population genetics, consists of a population of individuals competing on the basis of natural selection concepts. Each individual, or chromosome, represents a trial solution to the problem to be solved. In the context of descriptor selection, the structure of the chromosome is very simple. Each descriptor is coded by a bit (0 or 1) and represents a component of the chromosome. 0 defines the absence of the descriptor, 1 defines its presence. The algorithm proceeds in successive steps called generations. During each generation, the population of chromosomes evolves by means of a "fitness" function (Davis 1991), which selects them by standard crossover and mutation operators. The crossover phase takes two chromosomes and produces two new individuals, by swapping segments of genetic material, i.e., bits in this case. Within the population, mutation removes the bits affecting a small probability.

Genetic algorithms are very effective for exploratory search, applicable to problems where little knowledge is available, but it is not particularly suitable for local searches. In the latter case, it is combined with a stepwise approach in order to reach local convergence. Stepwise approaches are quick and are adapted to find solutions in "promising" areas that have been already identified.

To evaluate the fitness function, a specific index was derived by using a fuzzy clustering method. Furthermore, to prevent over-fitting and a poor generalization, across validation procedure was included in the algorithm during the selection procedure, by randomly dividing the database into training and test sets. The fitness score of each chromosome is derived from the combination of the scores of the training and test sets.

The following parameters were used in the data processing of the data set of 412 olfactory compounds:

(1) Fuzzy parameters – weighting coefficient $= 1.5$, tolerance convergence $= 0.001$, number of iterations $= 50$, number of clusters $= 10$.
(2) Genetic parameters – number of chromosomes $= 10$, chromosome size $= 60$ (number of descriptors used), number of crossover points $= 1$, percentage of rejections $= 0.1$, percentage of crossovers $= 0.8$, percentage of mutations $= 0.05$, time off (10,100), number of generations $= 10$, ascendant coefficient $= 0.02$, descendant coefficient $= -0.02$. Calculations were performed using proprietary software.

Adaptive Fuzzy Partition

AFP is a supervised classification method implementing a fuzzy partition algorithm. It models relations between molecular descriptors and chemical activities by dynamically dividing the descriptor space into a set of fuzzy partitioned subspaces. In a first phase, the global descriptor hyperspace is considered and cut into two subspaces where the fuzzy rules are derived. These two subspaces are divided step by step into smaller subspaces until certain conditions are satisfied, namely when:

(1) The number of molecular vectors within a subspace attains a minimum threshold number
(2) The difference between two generated subspaces is negligible in terms of chemical activities represented
(3) The number of subspaces exceeds a maximum threshold number

The aim of the algorithm is to select the descriptor and the cut position, which allows the maximal difference between the two fuzzy rule scores generated by the new subspaces to be determined. The score is defined by the weighted average of the chemical activity values in an active subspace A and in its neighboring subspaces. If the number of trial cuts per descriptor is defined by N cut, the number of trial partitions equals $(N \text{ cut} +1)N$. Only the best cut is selected to subdivide the original subspace. All the rules created during the fuzzy procedure are considered to establish the model between descriptor hyperspace and biochemical activities. The global score in the subspace Sk can then be calculated. All the subspaces k are considered and then the score of the activity O for a generic molecule is computed. The following parameters were used to process the data set of 165 pesticide compounds: maximal number of rules for each chemical activity = 35; minimal number of compounds for a given rule = 4; number of cutting for each axis = 4; $p = 1.2$ and $q = 0.8$.

Descriptor Selection

Four relevant descriptors can be selected by the GA procedure. The first three descriptors may correspond to topological indices encoding information about molecular structure. All the atoms are considered to be carbon atoms. The values for noncarbon heteroatoms are computed differently regarding the values for identically connected carbon atoms. Finally, VES, an electronic index, represents the variance of electronegativity computed by the Sanderson method (Sanderson 1976).

AFP Model

The AFP model was established on the training set compounds, defining four molecular descriptor – odor relationships, one for each olfactory note. The number of rules implemented in each relationship was dependent on the complexity of the compound distribution regarding a given odor. The animal, camphoraceous, ethereal and fatty odors were, respectively, represented by 17, 18, 14, and 24 rules. The number of rules concerning the fatty odor shows that the corresponding relationship was the most difficult to establish. A possible explanation could be found in the fact that only complex combinations of molecular descriptors can represent the distribution of the ethereal compounds, so requiring a high number of rules. Another one can be related to the cutting procedure performed by the algorithm. But this hypothesis is less probable as a different number of cuts, 3, 4, and 5 per axis, leads to similar results.

The most important ability of the AFP method is its capacity to solve such complex problems as olfaction, transcribing the molecular descriptor–activity relationships into simple rules that are directly related to the selected descriptors. The contribution of the GA procedure is obviously fundamental: it reduces the amount of information in the input step, making it easier to determine and interpret the model.

Conclusion

Data base mining (DBM) algorithms, based upon molecular diversity analysis, are becoming a must for pharmaceutical companies in the search for new leads. They allow the automated classification of chemical databases, but the huge amount of information provided by the large number of molecular descriptors tested is difficult to exploit. Then, new tools have to be developed to give a user-friendly representation of the compound distribution in the descriptor hyperspace.

Furthermore, the difficulty of data mining in olfaction databases is amplified by the fact that one compound can have different odors and its activity is usually expressed in a qualitative way. Another source of complexity derives from the fact that one receptor can recognize different chemical determinants and the same compound can be active on different receptors.

Fuzzy logic methods, developed to mimic human reasoning in its ability to produce correct judgements from ambiguous and uncertain information, can provide interesting solutions in the classification of olfactory databases. In fact, these techniques should be able to represent the "fuzziness" linked to an expert's subjectivity in the characterization of the odorous notes, computing intermediate values between absolutely true and absolutely false for each olfactory category. These values are named degrees of membership and are ranged between 0.0 and 1.0.

In this section, a new procedure, the AFP algorithm, was applied to a data set of olfactory molecules, divided into animal, camphoraceous, and ethereal and fatty compounds. This method consists of modeling molecular descriptor–activity relationships by dynamically dividing the descriptor hyperspace into a set of fuzzy subspaces. A large number of molecular descriptors may be tested and the best ones may be selected with help of an innovative procedure based on genetic algorithm concepts.

8.3 Fuzzy Logic in Image Processing

8.3.1 Fuzzy Image Processing

Introduction

Fuzzy image processing is not a unique theory. It is a collection of different fuzzy approaches to image processing. Nevertheless, the following definition can be regarded as an attempt to determine the boundaries:
Fuzzy image processing is the collection of all approaches that understand, represent and process the images, their segments and features as fuzzy sets. The representation and processing depend on the selected fuzzy technique and on the problem to be solved.

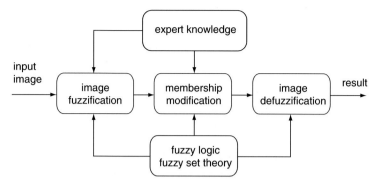

Fig. 8.10. General structure of fuzzy image processing

Fuzzy image processing has three main stages: image fuzzification, modification of membership values, and, if necessary, image defuzzification (see Fig. 8.10.).

The fuzzification and defuzzification steps are due to the fact that we do not possess fuzzy hardware. Therefore, the coding of image data (fuzzification) and decoding of the results (defuzzification) are steps that make possible to process images with fuzzy techniques. The main power of fuzzy image processing is in the middle step (modification of membership values, see Fig. 8.11). After the image data are transformed from gray-level plane to the membership plane (fuzzification), appropriate fuzzy techniques modify the membership values. This can be a fuzzy clustering; a fuzzy rule-based approach, a fuzzy integration approach, and so on.

Need for Fuzzy Image Processing

The most important of the needs of fuzzy image processing are as follows:

1. Fuzzy techniques are powerful tools for knowledge representation and processing
2. Fuzzy techniques can manage the vagueness and ambiguity efficiently
3. In many image-processing applications, we have to use expert knowledge to overcome the difficulties (e.g., object recognition, scene analysis)

Fuzzy set theory and fuzzy logic offer us powerful tools to represent and process human knowledge in form of fuzzy if–then rules. On the other side, many difficulties in image processing arise because the data/tasks/results are uncertain. This uncertainty, however, is not always due to the randomness but to the ambiguity and vagueness. Beside randomness which can be managed by probability theory we can distinguish between three other kinds of imperfection in the image processing (see Fig. 8.12):

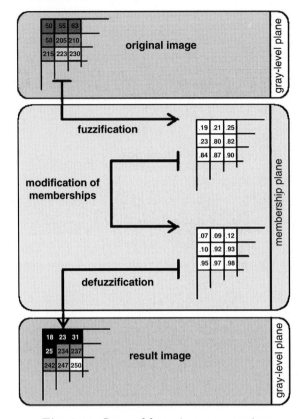

Fig. 8.11. Steps of fuzzy image processing

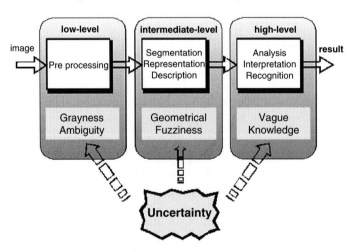

Fig. 8.12. Uncertainty/imperfect knowledge in image processing

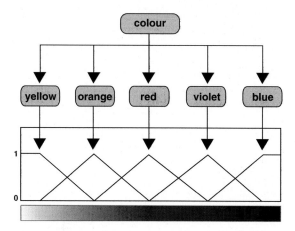

Fig. 8.13. Representation of colors as fuzzy subsets

- Grayness ambiguity
- Geometrical fuzziness
- Vague (complex/ill-defiend) knowledge

These problems are fuzzy in the nature. The question whether a pixel should become darker or brighter than it already is, the question where is the boundary between two image segments, and the question what is a tree in a scene analysis problem, all of these and other similar questions are examples for situations that a fuzzy approach can be the more suitable way to manage the imperfection.

As an example, we can regard the variable color as a fuzzy set. It can be described with the subsets yellow, orange, red, violet, and blue:
color = {yellow, orange, red, violet, blue}
The noncrisp boundaries between the colors can be represented much better. A soft computing becomes possible (see Fig. 8.13).

Fuzzy Image Enhancement

Contrast Adaptation

In recent years, many researchers have applied the fuzzy set theory to develop new techniques for contrast improvement. Following, some of these approaches are briefly described.

Contrast Improvement with INT-Operator

1. Step: define the membership function

$$\mu_{mn} = G(g_{mn}) = \left[1 + \frac{g_{\max} - g_{mn}}{F_{\mathrm{d}}}\right]^{-F_{\mathrm{e}}}$$

2. Step: modify the membership values

$$\mu'_{mn} = \begin{cases} 2 \cdot [\mu_{mn}]^2 & 0 \le \mu_{mn} \le 0.5 \\ 1 - 2 \cdot [1 - \mu_{mn}]^2 & 0.5 \le \mu_{mn} \le 1 \end{cases}$$

3. Step: generate new gray-levels

$$g'_{mn} = G^{-1}(\mu'_{mn}) = g_{\max} - F_d\left((\mu'_{mn})^{-1/F_e} - 1\right)$$

Contrast Improvement Using Fuzzy Expected Value

1. Step: calculate the image histogram
2. Step: determine the fuzzy expected value (FEV)
3. Step: calculate the distance of gray-levels from FEV

$$D_{mn} = \sqrt{|(\text{FEV})^2 - (g_{mn})^2|}$$

4. Step: generate new gray-levels

$$\begin{aligned} g'_{mn} &= \max(0, \text{FEV} - D_{mn}) && \text{if } g_{mn} < \text{FEV}, \\ g'_{mn} &= \min(L-1, \text{FEV} + D_{mn}) && \text{if } g_{mn} > \text{FEV}, \\ g'_{mn} &= \text{FEV} && \text{otherwise.} \end{aligned}$$

Contrast Improvement with Fuzzy Histogram Hyperbolization

1. Step: setting the shape of membership function (regarding to the actual image)
2. Step: setting the value of fuzzifier Beta (a linguistic hedge)
3. Step: calculation of membership values
4. Step: modification of the membership values by linguistic hedge
5. Step: generation of new gray-levels

$$g'_{mn} = \left(\frac{L-1}{e^{-1} - 1}\right) \cdot \left[e^{-\mu_{mn}(g_{mn})^\beta} - 1\right].$$

Contrast Improvement Based on Fuzzy if–then Rules

1. Step: setting the parameter of inference system (input features, membership functions)
2. Step: fuzzification of the actual pixel (memberships to the dark, gray, and bright sets of pixels)(Fig. 8.14)
3. Step: inference (e.g., if dark then darker, if gray then gray, if bright then brighter)
4. Step: defuzzification of the inference result by the use of three singletons

Locally Adaptive Contrast Enhancement

In many cases, the global fuzzy techniques fail to deliver satisfactory results. Therefore, a locally adaptive implementation is necessary to achieve better results.

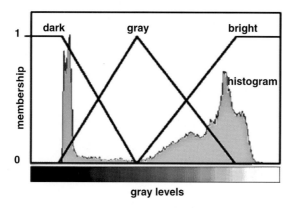

Fig. 8.14. Histogram fuzzification with three membership functions

Subjective Image Enhancement

In image processing, some objective quality criteria are usually used to ascertain the goodness of the results (e.g. the image is good if it possesses a low amount of fuzziness indicating high contrast). The human observer, however, does not perceive these results as good because his judgment is subjective. This distinction between objectivity and subjectivity is the first major problem in the human–machine interaction. Another difficulty is the fact that different people judge the image quality differently. This inter-individual difference is also primarily due to the aforesaid human subjectivity.

Following, an overall enhancement system will be described briefly. The approach is based on the combination of differently enhanced images obtained by using different algorithms each satisfying the observer's demand only partly. The fusion result should meet the subjective expectations of every individual observer.

An Overall System for Image Enhancement

The proposed enhancement system consists of two stages: an offline stage in which an aggregation matrix will be generated which contains the relevancy of different algorithms for corresponding observers, and an online stage where new image data will be enhanced and fused for a certain observer.

Offline Stage

The offline stage consists of five phases: image enhancement by means of different algorithms (or by just one algorithm with different parameters), extraction of the objective quality criteria, learning the fuzzy measure (subjective quality evaluation), aggregation (regarding to different images and different observers), and finally, a fuzzy inference (final quality measure for each image).

The result of the offline stage will be an aggregation matrix containing the relevance of all involving algorithms for each observer. The system phases can be briefly described as follows:

Phase 1 (enhancement): different algorithms Ak (or one algorithm with different parameters) enhance all test images Xi and deliver their results X'i,Ak. The selection of these algorithms is dependent on the image quality that we are interested in, e.g., contrast, smoothness, edginess, etc. At least two algorithms, or two different parameter sets for the same algorithm, should be selected.

Phase 2 (extraction): depending on the specific requirements of the application, suitable quality measures h(X'i,Ak) are extracted, e.g., contrast, sharpness or homogeneity measures. These criteria can serve as objective quality measures and will be aggregated with subjective measures in the forth phase via fuzzy integral.

Phase 3 (learning): the observer judges the quality of all enhanced images. The images are presented to the observer in random order. Moreover, the observer is not provided with any information about the algorithms used in the first phase. In order to map the subjective assessments into numerical framework, the ITU recommendation BT 500 can be used. The quality of the images generated by the kth algorithm as excellent $(= 1)$, good $(= 2)$, fair $(= 3)$, poor $(= 4)$, and bad $(= 5)$. For all M judgments pi,b of the bth observer, the mean opinion score (MOS) will be calculated.

Phase 4 (aggregation of measures/judgments): considering the objective measures and subjective judgments, one recognizes two conflicts. First, the observer judges the results of the same algorithm from image to image differently. Second, considering the divergence between objective and subjective assessments, the relevance of different algorithms is not always obvious. To solve these problems two new measures the degree of compromise m^* and the degree of compatibility g are introduced.

Phase 5 (inference): the elements of vectors G (degree of compatibility) and F (degree of compromise) are fuzzified with three membership functions. The output of the inference system is an aggregation matrix quantifying the image quality and is represented by five nonsymmetric membership functions. Then the if–then rules may be formulated.

Online Stage

In the second stage the system uses only the information stored in the aggregation matrix and an index indicating the current expert looking at the images. The image fuzzification, therefore, plays a pivotal role in all image processing systems that apply any of these components. The following are the different kinds of image fuzzification:

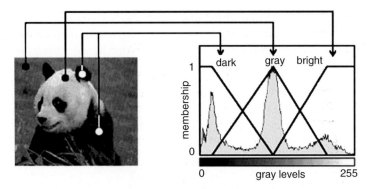

Fig. 8.15. Histogram fuzzification

Histogram-based gray-level fuzzification (or briefly histogram fuzzification)
Example: brightness in image enhancement
Local fuzzification (example: edge detection)
Feature fuzzification (scene analysis, object recognition)(Fig. 8.15)

In order to be in a form suitable for computer processing an image function
$f(x, y)$ must be digitized both spatially and in amplitude (intensity). Digitization of spatial co-ordinate (x, y) is called *image sampling*, while amplitude
digitization is referred to as *intensity* or *gray-level quantization*. The latter
term is applicable to monochrome images and reflects the fact that these
images vary from black to white in shades of gray. The terms intensity and
gray-level can be used interchangeably.

Suppose that a continuous image is sampled uniformly into an array of N
rows and M columns, where each sample is also quantized in intensity. This
array, called a *digital image*, may be represented as,

$$f(m, n) = \begin{pmatrix} x_{11} & x_{12} & x_{13} & \cdots & x_{1M} \\ x_{21} & x_{22} & x_{23} & \cdots & x_{2M} \\ \vdots & \vdots & \vdots & & \vdots \\ x_{N1} & x_{N2} & x_{N3} & \cdots & x_{NM} \end{pmatrix},$$

where m, n are discrete variables.

Each element in the array is called an *image element, picture element*, or
pixel.

There are basically two methods available for image processing. They are:

1. Frequency domain method
2. Spatial domain method

Frequency Domain technique:

It refers to an aggregate of complex pixels resulting from taking the Fourier
Transform and arises from the fact that this particular transform is composed

of complex sinusoids. Due to extensive processing requirements, frequency-domain techniques are not nearly as widely used as spatial domain techniques. However, Fourier Transform plays an important role in areas such as analysis of and object motion and object description.

Two-dimensional Fourier Transform pair of an $N \times N$ image is defined as,

$$F(u, v) = \frac{1}{N} \sum_{x=0}^{N-1} \sum_{y=0}^{N-1} f(x, y) \exp(-j2\pi(xu + vy)/N)$$

for $u = 0, 1, 2, \ldots, N - 1$.

In this method, processing is done with various kinds of frequency filters. For example, low frequencies are associated with uniformly gray areas, and high frequencies are associated with regions where there are abrupt changes in pixel brightness.

Spatial domain technique: this method refers to aggregate of pixels composing an image, and they operate directly on these pixels. Processing functions in spatial domain may be expressed as

$g(x, y) = h[f(x, y)]$
$f(x, y)$ is the input image
$g(x, y)$ is the resultant image
h is the operator on f defined over some neighborhood of (x, y)

The principal approach used in defining a neighborhood about (x, y) is to use a square/rectangular subimage area centered at (x, y). Although other neighborhood shapes such as circle are sometimes used, square arrays are by far most predominant because of their ease of implementation.

Smoothing: smoothing operations are used for reducing noise and other spurious effects that may be present in an image as a result of sampling, quantization, transmission or disturbances in the environment during image acquisition.

Mainly there are two types of smoothing techniques. They are:

1. Neighborhood averaging
2. Median filtering

Neighborhood averaging: it is a straightforward spatial domain technique for image smoothing. Given an image $f(x, y)$, the procedure is to generate a smoothed image $g(x, y)$ whose intensity at every point (x, y) is obtained by averaging the intensity values of pixels of f contained in predefined neighborhood of (x, y). The smoothed image is obtained by using the relation

$$g(x, y) = \frac{1}{P} \sum_{(n,m)\in S} f(m, n) \qquad \text{for all } x \text{ and } y \text{ in } f(x, y).$$

Median filtering: one of the difficulties of neighborhood averaging is that it blurs the edges and other sharp details. This blurring can often be reduced significantly by the use of the median filters, in which we replace the intensity of each pixel by median of the intensities in a predefined neighborhood of that pixel, instead of by the average.

Fuzzy image processing is not a unique theory. It is a collection of different fuzzy approaches to image processing. Nevertheless, the following definition can be regarded as an attempt to determine the boundaries:

Fuzzy image processing is the collection of all approaches that understand, represent and process the images, their segments and features as fuzzy sets. The representation and processing depend on selected fuzzy technique and on the problem to be solved.

Fuzzy image processing (FIP) has three main stages:

1. Image fuzzification
2. Modification of membership values
3. Image defuzzification

The general structure of an FIP is shown in the figure. The fuzzification and defuzzification steps are due to fact that we do not possess fuzzy hardware. Therefore, the coding of image data (fuzzification) and decoding of the results (defuzzification) are steps that make possible to process images with fuzzy techniques.

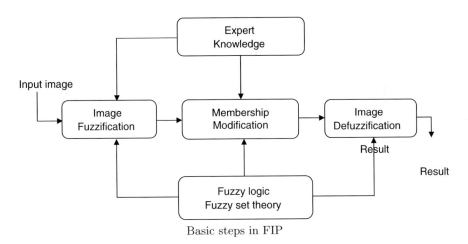

Basic steps in FIP

The main power of fuzzy image processing is in the middle step (modification of membership values). After the image data are transformed from gray-level plane to the membership plane (fuzzification), appropriate fuzzy techniques modify the membership values. This can be a fuzzy clustering; a fuzzy rule-based approach, a fuzzy integration approach and so on.

Necessity of FIP: there are many reasons for use of fuzzy techniques in image processing. The most important of them are as follows:

In many image-processing applications, we have to use expert knowledge to overcome the difficulties (e.g., object recognition, scene analysis). Fuzzy set theory and fuzzy logic offer us powerful tools to represent and process human knowledge in form of fuzzy if–then rules. On the other side, many difficulties in image processing arise because the data/tasks/results are uncertain. This uncertainty, however, is not always due to randomness but to the ambiguity and vagueness. Beside randomness, which can be managed by probability theory, we can distinguish between three kinds of imperfection in image processing.

These problems are fuzzy in nature. The question whether a pixel should become darker than already it is, the question where is the boundary between two image segments, and the question what is a tree in a scene analysis problem, all of these and other similar questions are examples for situations that a fuzzy approach can be the more suitable way to manage the imperfection.

Before one is able to conduct meaningful pattern recognition exercises with images, one may need to preprocess the image to achieve the best image possible for the recognition process. The original image might be polluted with considerable noise, which would make the recognition process difficult. Processing, reducing, or eliminating this noise will be a useful step in the process. An image can be thought of an ordered array of pixels, each characterized by gray tone. These levels might vary from a state of no brightness, or completely black, to a state of complete brightness, or totally white. Gray tone levels in between these two extremes would get increasingly lighter as we go from black to white.

Contrast enhancement: an image X of $N \times M$ dimensions can be considered as an array of fuzzy singletons, each with a value of membership denoting the degree of brightness level $p, p = 0, 1, 2 \ldots P - 1$ (e.g., range of densities from $p = 0$ to $p = 255$), or some relative pixel density. Using the notation of fuzzy sets, we can write,

$$X = \begin{pmatrix} \mu_{11}/x_{11} & \mu_{12}/x_{12} & \cdots & \mu_{1M}/x_{1M} \\ \mu_{21}/x_{21} & \mu_{22}/x_{22} & \cdots & \mu_{2M}/x_{2M} \\ \vdots & \vdots & & \vdots \\ \mu_{N1}/x_{N1} & \mu_{N2}/x_{N2} & \cdots & \mu_{NM}/x_{NM} \end{pmatrix}$$

where $0 \leq \mu_{mn} \leq 1, m = 1, 2 \ldots M, n = 1, 2 \ldots N$.

Contrast within an image is measure of difference between the gray-levels in an image. The greater the contrast, the greater is the distinction between gray-levels in the image. Images of high contrast have either all black or all white regions; there is very little similar gray-levels in the image, and very few black or white regions. High-contrast images can be thought of as crisp, and low contrast ones as completely fuzzy. Images with good gradation of grays between black and white are usually the best images for purposes of recognition by humans.

The object of contrast enhancement is to process a given image so that the result is more suitable than the original for a specific application in pattern recognition. As with all image-processing techniques we have to be especially careful that the processed image is not distinctly different from the original image, making the identification process worthless. The technique used here makes use of modifications to brightness membership value in stretching or contracting the contrast of an image.

Many contrast enhancement methods work as shown in the figure below, where the procedure involves primary enhancement of he image, denoted with an E_1 in the figure, followed by a smoothing algorithm, denoted by an S, and a subsequent final enhancement, step E_2.

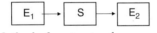

Method of contrast enhancement

The function of the smoothing operation of this method is to blur (make more fuzzy) the image, and this increased blurriness then requires the use of final enhancement step E_2. Generally smoothing algorithms distribute a portion of the intensity of one pixel in the image to adjacent pixels. This distribution is greatest for pixels nearest to the pixels being smoothed, and it decreases for pixels farther from the pixel being smoothed.

The contrast intensification operator, on a fuzzy set A generates another fuzzy set, $A' = INT(A)$ in which the fuzziness is reduced by increasing the values of $\mu_A(x)$ that are greater than 0.5 and decreasing the values that are less than 0.5. If we define this transformation T_1, we can define T_1 for the membership values of brightness for an image as,

$$T_1(\mu_{mn}) = T_1'(\mu_{mn}) = 2\mu_{mn}^2, \qquad 0 \le \mu_{mn} \le 0.5,$$
$$= T_1''(\mu_{mn}) = 1 - 2(1 - \mu_{mn})^2, \quad 0.5 \le \mu_{mn} \le 1.$$

The transformation T_r is defined as successive applications of T_1 by the recursive relation,

$$T_r(\mu_{mn}) = T_1[T_{r-1}(\mu_{mn})] \quad r = 1, 2, 3, \ldots$$

The graphical effect of this recursive transformation for a typical membership function is shown in figure below. The increase in successive applications of the transformation, the curve gets steeper. As r approaches infinity, the shape approaches a crisp function. The parameter r allows the user to use an appropriate level of enhancement for domain-specific situations.

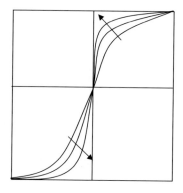

Example: Given the following 25 pixel array as shown below

110	105	140	107	110
110	132	105	115	154
140	105	105	115	154
137	135	145	150	150
140	118	115	109	148

Array of pixels with given intensities

We now scale the above values to obtain the membership functions of each of the pixel given as shown in the table below.

0.43	0.41	0.55	0.42	0.43
0.43	0.52	0.43	0.59	0.41
0.55	0.41	0.41	0.45	0.60
0.54	0.53	0.57	0.59	0.59
0.55	0.46	0.45	0.42	0.58

Scaled values indicating memberships of each pixel

Applying the formulas given before does the contrast enhancement of above array of pixels, which are stated below again.

$$T_1(\mu_{mn}) = T_1'(\mu_{mn}) = 2\mu_{mn}{}^2, \qquad\qquad 0 \le \mu_{mn} \le 0.5$$

$$= T_1''(\mu_{mn}) = 1 - 2(1 - \mu_{mn})^2 \quad 0.5 \le \mu_{mn} \le 1,$$

where μ_{mn} is the membership of the (m, n)th element in the array of pixels.

After one application of the enhancement, i.e., the INT operator on the above array of pixels we get the following results.

0.37	0.33	0.60	0.35	0.37
0.37	0.54	0.37	0.66	0.33
0.60	0.33	0.33	0.40	0.68
0.57	0.56	0.63	0.66	0.66
0.60	0.42	0.40	0.35	0.65

Membership values of pixels after application of INT operator once

Thus we see that the pixels having the membership values greater than 0.5 have been increased in intensity and those with value less than 0.5 have been decreased in intensity.

Sample calculations:

Consider the pixel of intensity 0.43, the new intensity value is, $2 \times 0.43^2 = 0.37$ as $0 < 0.43 < 0.5$.

Consider the pixel of intensity 0.55. As it is between 0.5 and 1.0 we get its new value as $[1 - 2(1 - 0.55)^2] = 0.60$.

In this way we calculate all the new intensities of the other pixels.

Smoothing: smoothing of a pixel is done by averaging the intensity values in the neighborhood of the pixel and substituting the averaged value for the intensity of the pixel. Consider the following figure.

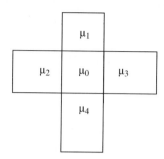

We can now write the intensity of the pixel μ_0 as

$$\mu_0 = (\mu_1 + \mu_2 + \mu_3 + \mu_4)/4.$$

Also we can find the new intensity of the pixel by substituting the median of the intensities of the pixels in place of the intensity of the pixel.

Example: Consider an array of 30 pixels denoting the letter M as shown below

220	30	40	15	250
205	230	**0**	239	230
225	20	225	20	220
217	**255**	30	10	215
220	25	15	**255**	235
210	20	10	15	220

The noisy intensities are denoted in bold. The operation of smoothing is to reduce the noise caused due to them. The scaled values giving membership of pixels is as shown below.

The scaled membership values of the pixels can be obtained by dividing the pixel intensity by 255. For example consider the pixel of intensity 220, its membership value can be determined by dividing 220 by 255. Therefore we obtain the new intensity of the pixel as,

$$220/255 = 0.86.$$

In similar manner, we calculate the membership values of other pixels also.

0.86	0.18	0.04	0.06	0.98
0.80	0.90	**0.00**	0.94	0.90
0.88	0.08	0.88	0.08	0.86
0.85	**1.00**	0.11	0.04	0.84
0.86	0.10	0.06	**1.00**	0.92
0.82	0.08	0.04	0.06	0.86

Now we apply the formula given before to each of the pixel except at the edges because at the edges we do not know the all the intensities in the neighborhood of the pixel. The application of the smoothing operation for once gives us the following results.

0.86	0.18	0.04	0.06	0.98
0.80	0.26	**0.53**	0.39	0.90
0.88	0.75	0.37	0.41	0.86
0.85	**0.45**	0.23	0.62	0.84
0.86	0.36	0.40	**0.50**	0.92
0.82	0.08	0.04	0.06	0.86

Sample calculations:

Consider the pixel in second row and second column. Its new intensity is given by

$$(0.18 + 0.80 + 0.00 + 0.08)/4.$$

The new intensity of the pixel in second row and third column is given by

$$(0.04 + 0.26 + 0.88 + 0.94)/4.$$

Here care has to be taken to incorporate the obtained new intensity of the pixel in the neighborhood, i.e., we have to substitute the new intensity of the pixel when the new intensity of the pixel in its neighborhood is to be calculated.

We see that the noise due to the pixels has been decreased very much. On further application of the smoothing operation we can decrease the noise very much.

Further examples on contrast enhancement and smoothing:

Given a 10×10 pixel array. It represents a dark square image in which there is a lighter square box that is not very apparent because the background is very nearly the same as that of the lighter box itself.

77	89	77	64	77	71	99	56	51	38
77	122	125	125	125	122	117	115	51	26
97	115	140	135	133	153	166	112	56	31
82	112	145	130	150	166	166	107	74	23
84	107	140	138	125	158	158	120	71	18
77	110	143	148	153	145	148	122	77	13
79	102	99	102	97	94	92	115	77	18
71	77	74	77	71	64	77	89	51	20
64	64	48	51	51	38	51	31	26	18
51	38	26	26	26	13	26	26	26	13

When we take the intensity values above and scale them on interval $[0,255]$, we get membership values in the density set *white* (low values are to black, high values close to white).

0.30	0.35	0.30	0.25	0.30	0.28	0.39	0.22	0.20	0.15
0.30	0.48	0.49	0.49	0.49	0.48	0.46	0.45	0.20	0.10
0.38	0.45	0.55	0.53	0.52	0.60	0.65	0.44	0.22	0.12
0.32	0.44	0.57	0.51	0.59	0.65	0.65	0.42	0.29	0.09
0.33	0.42	0.55	0.54	0.53	0.62	0.62	0.47	0.28	0.07
0.30	0.43	0.56	0.58	0.60	0.57	0.58	0.48	0.30	0.07
0.31	0.40	0.39	0.40	0.38	0.37	0.36	0.45	0.30	0.05
0.28	0.30	0.29	0.30	0.28	0.25	0.30	0.35	0.20	0.08
0.25	0.25	0.19	0.20	0.20	0.15	0.20	0.12	0.10	0.07
0.20	0.15	0.10	0.10	0.10	0.12	0.05	0.10	0.10	0.05

Using the contrast enhancement we modify the pixel values to obtain the matrix as shown below.

0.18	0.24	0.18	0.12	0.18	0.16	0.30	0.10	0.08	0.05
0.18	0.46	0.48	0.48	0.48	0.46	0.42	0.40	0.08	0.05
0.29	0.40	0.60	0.56	0.54	0.68	0.75	0.39	0.10	0.03
0.20	0.39	0.63	0.52	0.66	0.75	0.75	0.35	0.17	0.02
0.22	0.35	0.60	0.58	0.56	0.71	0.71	0.44	0.16	0.01

0.18	0.37	0.61	0.65	0.68	0.63	0.65	0.46	0.18	0.01
0.19	0.32	0.30	0.32	0.29	0.27	0.26	0.40	0.18	0.01
0.16	0.18	0.17	0.18	0.16	0.12	0.18	0.24	0.08	0.01
0.12	0.12	0.07	0.08	0.08	0.05	0.08	0.03	0.02	0.01
0.01	0.01	0.01	0.01	0.01	0.01	0.01	0.01	0.01	0.01

The point to be noted here is that the intensity values above and below 0.5 have been suitably modified to increase the contrast between the intensities.

Example on smoothing:

Consider the above example in which on repeated applications, the final enhanced image is obtained. Now some random *salt and pepper* is introduced into it. *Salt and pepper* noise is occurrence of black and white pixels scattered randomly throughout the image.

The scaled values of intensities of pixels are as shown in the matrix.

0.00	0.00	0.00	0.00	0.00	0.00	0.00	0.00	0.00	0.00
0.00	0.00	0.00	0.00	0.00	0.00	0.00	0.00	0.00	0.00
0.00	0.00	1.00	1.00	1.00	1.00	1.00	0.00	0.00	0.00
0.00	0.00	1.00	1.00	0.00	1.00	1.00	0.00	1.00	0.00
0.00	0.00	1.00	1.00	1.00	0.00	1.00	0.00	0.00	0.00
0.00	0.00	1.00	1.00	1.00	1.00	1.00	0.00	0.00	0.00
0.00	0.00	0.00	0.00	0.00	0.00	0.00	0.00	0.00	0.00
0.00	0.00	0.00	0.00	0.00	0.00	0.00	0.00	0.00	0.00
0.00	0.00	1.00	0.00	0.00	0.00	0.00	1.00	0.00	0.00
0.00	0.00	0.00	0.00	0.00	0.00	0.00	0.00	0.00	0.00

After one application of smoothing algorithm, the intensity values are as shown below

0.00	0.00	0.00	0.00	0.00	0.00	0.00	0.00	0.00	0.00
0.00	0.00	0.25	0.31	0.33	0.33	0.33	0.08	0.02	0.00
0.00	0.25	0.62	0.73	0.52	0.71	0.51	0.15	0.29	0.00
0.00	0.31	0.73	0.62	0.78	0.62	0.53	0.42	0.18	0.00
0.00	0.33	0.77	0.85	0.66	0.82	0.59	0.25	0.11	0.00
0.00	0.33	0.52	0.59	0.56	0.60	0.30	0.14	0.06	0.00
0.00	0.08	0.15	0.19	0.19	0.20	0.12	0.07	0.03	0.00
0.00	0.27	0.11	0.07	0.07	0.07	0.05	0.28	0.08	0.00
0.00	0.07	0.04	0.03	0.02	0.02	0.27	0.14	0.05	0.00
0.00	0.00	0.00	0.00	0.00	0.00	0.00	0.00	0.00	0.00

It can be seen that after application of smoothing algorithm the noise intensity has been reduced. Later we apply enhancement algorithm to obtain the figure without any noise.

We have seen two methods of fuzzy image processing namely, contrast enhancement and smoothing. There are many other techniques such as filtering, edge detection and segmentation. In contrast enhancement we improve the gradation between the black and white and are able to easily spot out the distinction between gray levels in the image. In smoothing we were able to decrease the *salt and pepper* noise in the image.

Conclusion

In this section we have seen in detail about the fuzzy image processing and the methods of image enhancement. The idea discussed can be extended even to higher dimensional problems. The process is found to operate based on the online and offline stage. Hence, this is a wide extension of fuzzy logic applications.

Adaptive Fuzzy Rules For Image Segmentation

Segmenting magnetic resonance images of the same body region taken at different times is a challenging task. Obtaining reliable data to train a classifier is difficult due the differences among subjects and even differences over time in images acquired from a single subject. Unsupervised clustering can be used to group like tissues into classes. However, clustering does not provide class labels, is time consuming, and may not always provide suitable data partitions. In this paper we show how a set of adaptive fuzzy rules can be used to identify many of the voxels from a magnetic resonance image before clustering is done. This allows clustering to be done on a subset of an image with a "good" initialization, which mitigates the time required. The identified voxels can also be used to identify clusters. The fuzzy rule based system followed by a clustering step has been applied to 105.5 mm thick, magnetic resonance images of the human brain which are taken from 15 different subjects. It is shown that the segmentations produced are approximately five times faster than those produced by fuzzy clustering alone and are comparable in the accuracy of the segmentation.

Using Fuzzy Rules for Segmentation

The fuzzy rules for partially segmenting MR images of the brain are built to operate on the T1, T2, and proton density weighted intensity feature images. The first step in developing a set of fuzzy rules to segment an image is determining the antecedent fuzzy sets. Hence, it is necessary to find thresholds that separate tissue types in each of the three feature images.

In order to build fuzzy rules that apply to a large number of images, the tissue thresholds, which determine the antecedent fuzzy sets of the rule, are found via histogram analysis applied to each image slice to which the rules will be applied. Figures 8.16–8.18 show a typical set of intensity histograms with "turning points" which can be used to approximately separate tissue types. For example, all voxels below b1 in the PD histogram are air with those between b2 and b4 generally white matter (Fig. 8.17), and voxels between a1 and a2 in the T1 histogram (Fig. 8.16) are a mixture of gray and white matter. The histogram shape remains approximately the same across normal subjects and as will be seen will have an expected set of changes for patients with brain pathology. All patients with pathology have been injected with gadolinium whose magnetic properties cause enhancement in regions where the blood, brain barrier have been breached (i.e., regions where tumor exists).
Examining the histograms for a set of training images discovered the existence of turning points. This research used six normal and four abnormal slices, which were segmented or ground truthed by expert radiologists into tissues of interest, as a training set. Projections of voxels, known to be of a given tissue type, onto one or more of the histograms shown in Figs. 8.16–8.18 allowed us to choose the turning points. The turning points in the histograms

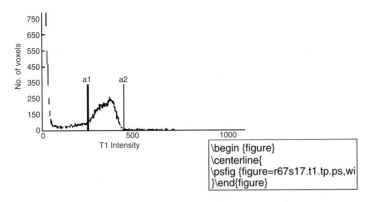

Fig. 8.16. T1 histogram with turning points

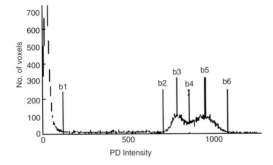

Fig. 8.17. PD histogram with turning points

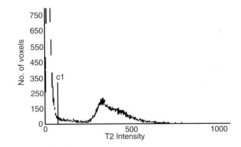

Fig. 8.18. T2 histogram with turning points

are essentially the approximate boundaries between tissue types. The turning points are automatically chosen on each test slice. From the turning points in the histogram, fuzzy rules to identify four tissue classes (white matter, gray matter, air/bone or background, and other or skull tissues such as fat, viscous fluid in the eyes, etc.) can be generated. The rules and antecedent fuzzy sets were generated by examining the intersection of tissue types in the three intensity histograms.

The rules adapt to each slice processed because they are generated from the turning points found on each slice. So, technically the rules' membership functions are automatically generated for each slice. The turning points are either peaks, valleys or the beginning of a hill in a histogram. The peaks and valleys can be found by searching for a maximum/minimum histogram value. The hill beginning is found by first creating intensity bins of width 30 (they contain 30 intensity levels). Next the approximate hill starting point is found by comparing the histogram sum in the first bin with the corresponding sum of the succeeding bin. If the ratio is greater than or equal to our ratio threshold of 1.8, then the middle intensity level of the bin was chosen as the beginning point of the hill. If the ratio is less than the threshold the next two bins are tested with the procedure continuing until a hill begin point is found. If a peak is found before a hill begin point, 0.1 reduces the ratio and the process is restarted.

Neither csf, which is a small class, nor pathology show up as a clear peak in any of the histograms. It was found that pathology and csf could be partially distinguished by viewing the voxel intensity as a percentage of the range of intensities in either T1 or T2 weighted images. This approach enabled rules to be generated for csf and pathology. The six fuzzy rules generated are shown in Fig. 8.19.

IF *voxel in T1 is Set-E*
AND *voxel in T2 is Set-F*
THEN *voxel is csf*

IF *voxel in PD is Set-C*
AND *voxel in T1 is Set-A*
THEN *voxel is White matter*

IF *voxel in PD is Set-D*
AND *voxel in T1 is Set-A*
AND *NOT (voxel in T2 is Set-F AND voxel in T1 is Set-E)*
THEN *voxel is Gray matter*

IF *voxel in T1 is Set-B*
AND *voxel in T2 is Set-F*
THEN *voxel is Pathology*

IF *voxel in T1 is Set-B*
AND *NOT (voxel in T2 is Set-F)*
THEN *voxel is Other*
IF *PD voxel intensity < b1*
AND *T2 voxel intensity < c1*
THEN *voxel is Background*
MIN was used as the fuzzy 'and' in the rules and $NOT(x) = 1 - x$.

Fig. 8.19. Fuzzy rules for MR image segmentation

The fuzzy sets used to generate the fuzzy rules are shown in Fig. 8.120 and together with the rules indicate how the turning points based on histogram shape can be used to separate voxels of different tissue types.

The rules are applied to all voxels, but will not classify all voxels. Spatial information is used to assign memberships to voxels which are unclassified. An unclassified voxel (i.e., having a zero membership in all classes) is assigned a membership that is the average membership of its eight neighboring voxels, for each of the six classes. Also in the case of isolated classifications, i.e., when a voxel has a membership of 1.0 in a class A, if all the eight surrounding voxels have zero membership in that class, then the isolated voxel's membership for class A is made zero. This step is aimed at reducing classification errors.

Finally, the voxel memberships in all classes are normalized to 1 using:

$$\mu_i(x) = \frac{\mu_i(x)}{\sum_j \mu_j(x)},$$

where (csf, GrayMatter, WhiteMatter, Pathology, Skull tissues, Background). The pathology rule applied to normal slices will incorrectly label a small number of voxels as pathology. This error will need to be corrected in later processing.

Patients with brain tumors are typically treated with radiation and chemotherapy. A side effect of treatment is that the MR characteristics of gray and white matter are changed and the PD histogram becomes something like that shown in Fig. 8.21. The "valley" shown in Fig. 8.17 is gone and "turning points" b3, b4, and b5 cannot be reliably chosen.

Our strategy is to edge detect and remove the edge voxels or sharpen the boundary between gray and white matter. The edge-value operator we used is called the DIF1 operator as described. A histogram of the voxels with low edge values will leave the peaks essentially the same and deepen the valley between the peaks. This approach can be applied to normal slices with the sole effect of deepening the already existing valley in the PD histogram.

An effective edge value threshold must be chosen to make this approach work. The initial threshold is chosen to be 5, then edge detection is done and all voxels with an edge value less than 5 are used to create a PD histogram. If two peaks are found in the histogram the turning points are created, otherwise the threshold is increased by 5 and the histogram re-created, peak detection done, etc. The process continues until two peaks are found or an edge value limit (30 here) is reached. In the case that no peaks could be found approximate peaks are chosen at 1/3 and 2/3 of the region between b1 and b2 (Fig. 8.21). Figure 8.22 shows an example abnormal slice with the PD edge value image thresholded at 15 and the histogram of all voxels with edge strengths <15.

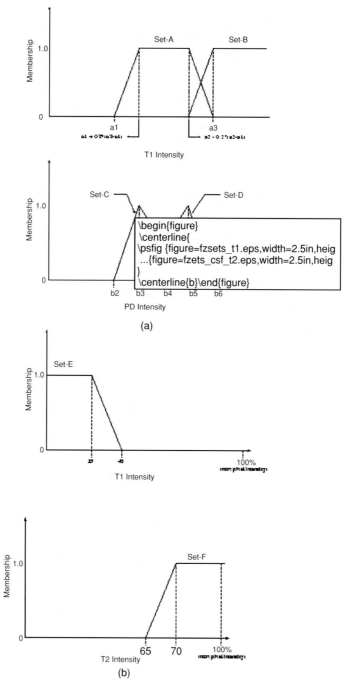

Fig. 8.20. Fuzzy sets created using turning points from histograms (**a**) and fuzzy sets created for identifying csf (**b**)

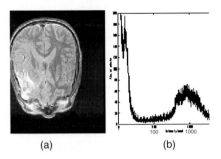

Fig. 8.21. Abnormal slice: (a) raw PD image (b) histogram

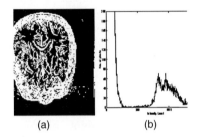

Fig. 8.22. PD edge value image: (a) Thresholded at 15 (i.e., white voxels are edge voxels with edge value >15 (b) Histogram of voxels <15

All voxels can now be classified, though imperfectly, for normal or abnormal volumes. The voxels that belong to classes with memberships greater than 0.8 are generally correctly assigned. The rest of the voxels are more problematic. Hence, we regroup them with a semisupervised clustering algorithm, ssFCM. The voxels with membership greater than 0.8 are used as training voxels for ssFCM and are weighted by a value of 100. The ssFCM algorithm works as fuzzy c-means (FCM) except that training voxels cannot change clusters and will always influence the cluster centroid to which they are assigned. When they are weighted it is the same as having w (100 here) instances of the train voxels influencing the cluster center location and hence the assignment of voxels, not in the train set, to clusters.

For a typical normal slice there will be 16,816 training voxels (memberships greater than 0.8) and 13,910 unassigned voxels. The remaining 34,810 voxels were air or skull tissue voxels and are not clustered. The clustering is done into $c = 10$ classes to allow comparisons with FCM partitions of these same images.

Experiments and Results

The fuzzy rules to identify tissues followed by an ssFCM clustering were applied to 39 normal slices from eight volunteers and 66 abnormal slices from seven patients. These slices lie in a range from near the center of the ventricles in the axial plane, characterized by a distinct X-shaped csf area and a single symmetric region of white matter to slices near the top of the brain in the axial plane, where the ventricular area is completely absent. There were six normal slices and four slices with pathology used to develop the fuzzy rule structure. These ten slices may be viewed as a set of training slices.

To approximate ground truth a set of supervised k nearest neighbor (kNN) segmentations were used. These segmentations were created by multiple observers choosing training sets for each slice. Segmentations that resulted in visually good partitions of the data are used for comparison with our unsupervised approaches. The value $k = 7$ was used.

Tables 8.1 and 8.2 summarize the comparison between the hybrid system (fuzzy rules followed by ssFCM) and regular FCM vs. pseudo ground-truth (kNN) for normal and abnormal slices, respectively. The time required is much less for the hybrid system. There are more classification differences from the kNN based "ground truth" for the hybrid system than FCM. To determine whether the differences were significant we applied a Wilcoxon's sum of ranks test. The z values obtained are shown in Table 8.3. A value $z \leq 1.64$ indicates that there is a greater than 10% chance that the observed difference is likely to occur by chance and hence cannot be proven significant. So, the z values in Table 8.3 lead us to conclude there is no significant difference in the segmentation results.

Hence a set of fuzzy rules whose antecedent fuzzy sets adapted to each image are shown to be effective in reducing the time to segment magnetic resonance images of the human brain into tissues of interest. The segmentation

Table 8.1. Mean and standard deviation of results (test slices)

$1\|c\|$	$2c\|$Regular FCM		$2c\|$Hybrid system	
	Mean	Std. Dev.	Mean	Std. Dev.
Classification differences (33 normals)	4080.3	1328.7	5076.9	1566.9
Classification differences (62 abnormals)	2376.4	1144.7	2402.5	1327.2
Execution time (33 normals)	23.1	7.2	4.8	2.0
Execution time (62 abnormals)	21.4	9.8	3.7	1.3

Table 8.2. Mean and standard deviation of results (training slices)

$1\|c\|$	$2c\|$Regular FCM		$2c\|$Hybrid system	
	Mean	Std. Dev.	Mean	Std. Dev.
Classification differences (six normals)	3986.5	846.5	3558.8	464.7
Classification differences (four abnormals)	2773.0	1039.1	2167.8	879.5
Execution time (six normals)	21.7	6.4	5.7	2.0
Execution time (four abnormals)	19.8	6.4	3.0	1.1

Table 8.3. z Values obtained from Wilcoxons' sum of ranks test

	Normal	Abnormal
Test	1.4	0.82
Train	0.48	0.29

produced by the fuzzy rules serves as an initialization to a semisupervised clustering algorithm which produces the final segmentation. The developed hybrid segmentation system is approximately five times faster than FCM clustering. It has been tested on 105.5 mm thick, magnetic resonance image slices of the human brain using T1, T2, and proton density weighted images as feature images (i.e., each voxel has three features). The images come from 15 different subjects and span a range from the ventricles (roughly the middle of the brain in the axial plane) to the top of the brain. The hybrid segmentations are insignificantly different than those obtained with FCM clustering when compared with a pseudo ground truth created from a supervised KNN segmentation.

The overall performance of the segmentation approach demands further refinement using some kind of knowledge. An example of a slice with significant extracranial tissue that is misclassified in this approach as white matter is shown in Fig. 8.23. Since the tissue is clearly outside the skull, simple knowledge about removing all tissue spatially outside the skull would prevent this tissue from being considered during processing.

The approach of using fuzzy rules whose antecedents fuzzy sets are created from intensity histograms can be applied to other domains of images taken of the same region over time as long as the shape of the histograms remains approximately constant. Such rules provide a fast initial segmentation that can be further refined via other image processing techniques or with the use of heuristics in conjunction with image processing algorithms.

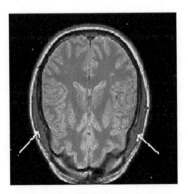

Fig. 8.23. MR image with significant extracranial tissue (run-81)

8.4 Fuzzy Logic in Biomedicine

8.4.1 Fuzzy Logic-Based Anesthetic Depth Control

Introduction

In most surgical operations, to anesthetize patients, manual techniques are used in hospitals. The manual systems work either ON or OFF situations. Because of not having interval values between ON and OFF in manual systems, anesthetic operations could not be safety and comfort. For this reason, Fuzzy logic control is applied to control anesthesia. In this paper, an objective approach of giving anesthetic to patients during surgical operation using Fuzzy logic is proposed.

Fuzzy logic theory is a general mathematical approach that allows partial memberships. Several studies have shown fuzzy logic control to be an appropriate method for the control of complex processes. The basic configuration of the logic system considered in this section is shown in Fig. 8.24.

Fuzzy logic system inputs T and N represent blood pressures (mmHg) and pulse rates (p m^{-1}), which are respectively obtained from patients during anesthesia. Anesthesia Output (AO) represents fuzzy logic system output.

The potential benefits of using fuzzy logic control during anesthesia; increasing patients safety and comfort, directing anesthetists attention to other physiological variables they have to keep under control by abating their tasks, using optimum anesthetic agent, protecting environment by using anesthetic agent and decreasing the cost of surgical operations.

Fuzzy Logic Control Application In Anesthesia

Fuzzifier

Here two fuzzy logic input sets are used. One of them is the systolic blood pressure of the patients, which are obtained in operation. The second input of fuzzy logic set is pulse rates. The minimum and the maximum values (systolic blood pressure and pulse rate) are obtained from surgical operations in 10 min intervals from 27 patients (Table 8.4).

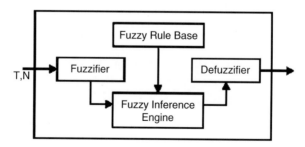

Fig. 8.24. A block diagram of basic fuzzy logic system

Table 8.4. Blood pressure, pulse rate and anesthesia ratio values accepted for fuzzy logic control

Variable	Minimum value	Maximum value
Blood Pressure (mmHg)	60	220
Pulse Rate (p m^{-1})	40	150
Anesthesia Ratio (%u)	0	4

Table 8.5. Membership function values

Linguistic variables	Very low	Low	Normal	High	Very high
Blood pressure (mmHg)	< 80	90	100–140	160–170	> 190
Puse rate (p m)$^{-1}$	< 50	60	70–90	95–110	> 120
Anesthesia ratio (%u)	0	0.5–0.8	1–2.5	3–3.6	4

The sexuality of patients that the systolic blood pressure and pulse rate values obtained are 12 women and 15 men. The age dispersion is between 3 and 77.

Aiding with the anesthetists, membership function values are formed as very low, low, normal, high and very high intervals as shown in Table 8.5.

Fuzzifier operation is applied for blood pressure and pulse rate data. This operation is realized to identify whether the input data is the member of this set or not. To fuzzify both input data, trapezoid membership set is used.

As shown in Fig. 8.25, blood pressure data membership sets between 80 and 194 mmHg are examined in groups as named T1, T2, T3, T4, T5, T6, T7, T8, T9, T10, and T11. Memberships sets for blood pressure data are computed as:
Membership function for T1:

$$\mu(x) = (80 - x)/(80 - 84) \quad 80 \le x < 84$$
$$\mu(x) = 1 \qquad\qquad\qquad\quad 84 \le x < 90$$
$$\mu(x) = (94 - x)/(94 - 90) \quad 90 \le x < 94$$

As shown in Fig. 8.26, pulse rate data membership sets between 50 and 124 are examined in groups named as N1, N2, N3, N4, N5, N6, and N7. Membership sets for pulse rate data are computed as:
Membership function for N1:

$$\mu(x) = (80 - x)/(80 - 84) \qquad 50 \le x < 54$$
$$\mu(x) = 1 \qquad\qquad\qquad\qquad 54 \le x < 60$$
$$\mu(x) = (94 - x)/(94 - 90) \qquad 60 \le x < 64$$

Fuzzy Rule Base and Data Base

Fuzzification is based on the rules that T and N inputs result in certain outputs according to the rule base. Anesthetist is consulted about input and output data in rule base.

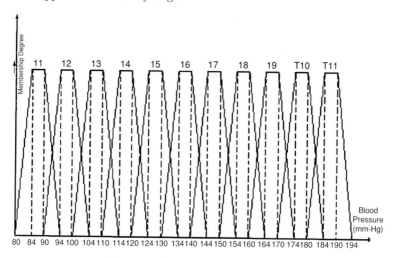

Fig. 8.25. Membership sets for blood pressure data

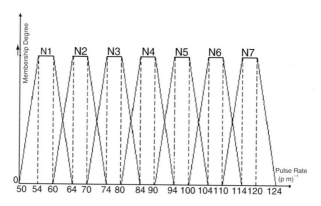

Fig. 8.26. Membership of pulse rate

The data given in Table 8.6 have impossible conditions in human beings. These values are accepted as invalid conditions (Table 8.6).

Fuzzy Inference Engine

Defining the output sets according to rule base is materialized in output unit. Contacts that are obtained according to this rule base are interpreted using minimum correlation method such as:

if T=T1 and N=N1 then A=A1.

The rule base for T and N fuzzy inputs are shown in Table 8.7.

Defuzzifier

In defuzzifier unit, fuzzified functions, obtained from fuzzy inference engine are converted into numeric values. Output membership sets A1, A2, A3, A4, and A5 are converted into numeric values using the following equation:

Table 8.6. Invalid input conditions

Blood pressure	Pulse rate	Anesthesia rate
High	Very_low	Invalid condition
Very_high	Very_low	Invalid condition
High	Low	Invalid condition
Very_high	Low	Invalid condition
Very_low	High	Invalid condition
Low	High	Invalid condition
Very_low	Very_high	Invalid condition
Low	Very_high	Invalid condition

Output membership sets according to the rule base for T and N fuzzy inputs are defined as A1, A2, A3, A4, and A5 shown in Table 8.5. Note that S is the invalid condition given in Table 8.4.

Table 8.7. Rule base for T and N fuzzy inputs

	N1	N2	N3	N4	N5	N6	N7
T1	A1	A1	A2	A2	A2	S	S
T2	A2	A2	A3	A3	A3	A4	A4
T3	A2	A3	A3	A3	A3	A4	A4
T4	A2	A3	A3	A3	A3	A4	A4
T5	A2	A3	A3	A3	A3	A4	A4
T6	A2	A3	A3	A3	A3	A4	A4
T7	A2	A3	A3	A3	A3	A4	A4
T8	S	A4	A4	A4	A4	A5	A5
T9	S	A4	A4	A4	A4	A5	A5
T10	S	A4	A4	A4	A4	A5	A5
T11	S	A5	A5	A5	A5	A5	A5

$$AO = \frac{\sum_{x=a}^{b} \mu_A(x)x}{\sum_{x=a}^{b} \mu_A(x)},$$

AO: Anesthesia output $\mu_A(x)$: Anesthetic membership function, x: Member (blood pressure, pulse rate).

Finally, anesthetic rate applied to the patient is determined from AO values. Membership function of fuzzifier is shown in Fig. 8.27.

Conclusion

Anesthetic depth control can be successfully implemented with the help of fuzzy logic. Membership function and base rules have been determined from an experimental prestudy on some patients. By this new analysis method, one can achieve better performance on how much and when to apply anesthetic agent. The main advantage of this system is that the anesthetic is given to

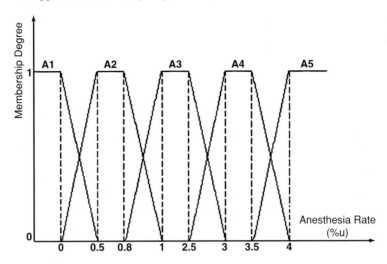

Fig. 8.27. Membership function of fuzzifier

the patient in a precise way, the anesthetist will spend less time to provide anesthetic and the patient will have a safer and less expensive operation.

8.5 Fuzzy Logic in Industrial and Control Applications

8.5.1 Fuzzy Logic Enhanced Control of an AC Induction Motor with a DSP

Introduction

Fuzzy logic is a new and innovative technology being used to enhance control-engineering solutions. It allows complex system design directly from engineering experience and experimental results, thus quickly rendering efficient solutions. In a joint application project, Texas Instruments and Inform Software have used fuzzy logic to improve AC induction motor control. The results were intriguing: control performance has been improved while design effort has been significantly reduced.

Market analysis shows that 90% of all industrial motor applications use AC induction type motors. The reasons for this are high robustness, reliability, low cost, and high efficiency. The drawback of using an AC induction type motor is its difficult controllability, which is due to a strong nonlinear behavior stemming from magnetic saturation effects and a strong temperature dependency of electrical motor parameters. For example, the rotor time constant of an induction motor can change up to 70% over the temperature range of the motor. These factors make mathematical modeling of motor control systems difficult. In real applications, only simplified models are used. The commonly used control methods are:

- Voltage/frequency control (U/f)
- Stator current flux control (I_s/f_2)
- Field oriented control

Of these approaches, the field-oriented control method has become the de facto standard for speed and position control of AC induction motors. It delivers the best dynamic behavior and a high robustness under sudden momentum changes. Alas, the optimization and parameterization of a field oriented controller is laborious and must be performed specifically for each motor. Also, due to the strong dependency of the motor's parameters, a controller optimized for one temperature may not perform well if the temperature changes. Figure 8.28 shows the demonstration of Test Motor at the Embedded Systems Conference.

To avoid the undesirable characteristics of the field oriented control approach, the companies Texas Instruments and Inform Software have developed new alternative control methods, and compared them with the field oriented control approach. The alternative methods involved two types of flux controllers enhanced by fuzzy logic and NeuroFuzzy techniques, respectively. The goal was to use fuzzy logic to improve the dynamic behavior of the flux control approach such that the robust behavior of the flux controller and the desirable dynamic properties of the field oriented controller are achieved simultaneously.

Field Oriented Control Method

Figure 8.29 shows the principle of field oriented control. It allows for control of the AC induction motor in the same way a separately exited DC motor is controlled. The flux model computes the "phase shift" between rotor flux field and stator field from the stator currents iu and iv, and the rotor angle position n. The field oriented variables of the two independent controller units are subsequently computed by the transformation of the stator currents using this "phase shift."

Fig. 8.28. Demonstration of the Test Motor at the Embedded Systems Conference, San Jose

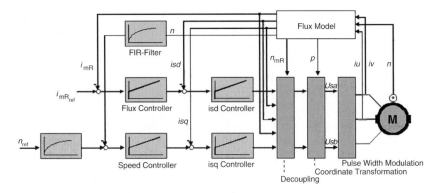

Fig. 8.29. Field-oriented control of AC induction motors

The actual control model consists of two components of cascaded standard PI controllers. The upper component comprises outer magnetizing current (i_{mR}) controller and inner i_{sd} current controller. The lower component comprises a speed controller and momentum controller. The input of the speed controller is computed as the difference between set speed n_{ref} and filtered measured speed n.

To optimize the field oriented control model, all controllers must be parameterized and optimized individually. In this application project, the method of optimized amplitude adaptation was used to tune the current controller, and the method of the symmetrical optimum was used for the velocity controller. Implementation effort for the field oriented controller was three person months, including parameterization and design of the flux model. The computation time for the inner current controllers, the flux model, and the coordinate transformation is $100\,\mu$s on a TMS320C31-40 MHz digital signal processor. When switching the set speed from $-1,000$ to $+1,000$ rpm, the new set speed is reached within only 0.25 s without any overshoot. However, this excellent performance is not always available. When the motor heats up the control performance drops significantly, and a motor with slightly different characteristics will achieve only mediocre results utilizing the same controller.

Fuzzy Flux Control Method

The conventional flux control model has been enhanced by fuzzy logic in two steps. In the first step, the nonlinear relation between slip frequency and stator current was described by a fuzzy logic system (Fuzzy Block #1). Figure 8.30 shows the principle of the resulting fuzzy flux controller. The control model consists of three inner control loops and one outer control loop. The inner control loops control the three stator phase currents using standard PI controllers. The outer control loop determines the slip frequency n_2, also using a standard PI controller. The slip frequency is the input to Fuzzy Block #1, which outputs the set value of the stator current. The primary objective for

Fuzzy Block #1 is to keep the magnetizing current constant in all operating modes. The magnetizing current is a nonlinear function of the slip frequency, the rotor time constant, the rotor leakage factor, and a nonconstant offset current.

The stator frequency n_1 is the sum of the measured rotor frequency n and the slip frequency n_2. The reference position is determined by integration of the stator frequency n_1. Modulated by sin/cos, the reference position is multiplied with the set value of the stator current, and split back into a three phase system of the stator current set values.

The rules of the fuzzy block were not manually designed, but rather generated from existing sample data by the NeuroFuzzy add-on module of the *fuzzy*TECH design software. NeuroFuzzy utilizes neural network techniques to automatically generate rule bases and membership functions from sample data. The benefit of the NeuroFuzzy approach over the neural net approach is that the result of NeuroFuzzy training is a transparent fuzzy logic system that can be explicitly optimized and verified. In contrast, the result of a neural net training is a rather nontransparent black box.

Comparison with Field Oriented Control

Figure 8.31 shows the performance of the fuzzy flux controller in comparison with the field oriented controller. The overshoot performance is almost as good as that provided by the field oriented control, however, it takes the fuzzy flux controller almost twice as long to reach the new set speed (curve *Fuzzy_1*). On the other hand, parameterization and optimization of the fuzzy flux controller only required four person days. The computation time for the entire controller is 150 µs on the TMS320C31-40 MHz digital signal processor.

To improve the performance of the fuzzy flux controller, in a second step, the standard PI controller for the outer control loop was replaced by a fuzzy PI controller (Fuzzy Block #2 in Fig. 8.31). This fuzzy PI controller does not use the proportional (P) and integral (I) component of the error signal, but rather the differential (D) and proportional (P) component then integrates the output. This type of fuzzy PI controller has been used very successfully in a number of recent applications, especially in the area of speed and temperature control. In contrast to the standard PI controller, the fuzzy PI controller implements a highly nonlinear transfer characteristic. The subwindow in the

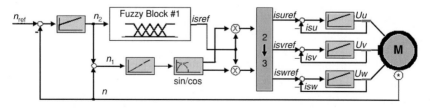

Fig. 8.30. Principle of fuzzy flux controller

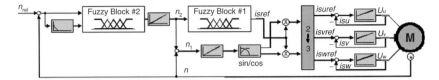

Fig. 8.31. Enhanced fuzzy flux controller

lower left part of Fig. 8.32 shows the transfer characteristics for the fuzzy PI controller implemented in this application.

The enhanced fuzzy flux controller reveals a much-improved dynamic performance. The good performance attained in this case hinges on the nonlinear behavior of the fuzzy PI controller. In contrast to the conventional linear PI controller, the nonlinearity of the fuzzy PI controller produces stronger control action for a large speed error, and a smoother control action for a small speed error. This also results a higher robustness of the enhanced fuzzy flux controller against parameter changes. The implementation of the second fuzzy block with the fuzzy flux controller only required an additional day for the fuzzy logic system itself, and two additional days for the optimization of the total system. Hence, the total development effort for the enhanced fuzzy flux controller was seven person days in comparison to three-person month for the field-oriented controller. The computation time for the entire controller is $200\,\mu s$ on the used TMS320C31-40 MHz digital signal processor.

System Simulation Using Matlab/Simulink and *fuzzy*TECH

The initial design of the system was implemented in a software simulation. The *fuzzy*TECH fuzzy-system development software was used together with the Matlab/Simulink control-system simulation software. *FuzzyTECH* allows using fuzzy blocks in Simulinks control diagrams. This tool combination allows for the design of simulations combining conventional and fuzzy logic control engineering technologies in the same software environment. Figure 8.32 shows the development of the fuzzy blocks with *fuzzy*TECH/Simulink. The differential equation used for the simulation of the AC induction motor is modeled.

Fuzzy Logic on Digital Signal Processors

Because of the increasing number of successful of applications of fuzzy logic in both control engineering and signal processing, DSP market leader Texas Instruments was looking for a software partner to implement fuzzy logic on DSP. In 1992, a formal partnership was formed with Inform Software Corp., a company specializing in fuzzy logic. One product of the partnership was the design of dedicated versions of *fuzzy*TECH that allow the implementation of fuzzy logic systems on standard TI-DSPs. The primary objective was to reach an acceptable computing performance level for fuzzy logic on DSPs, a quality

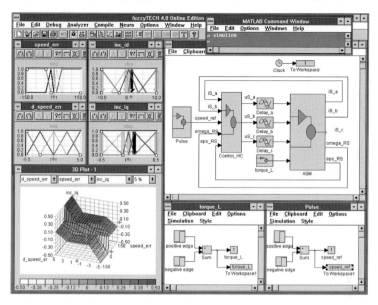

Fig. 8.32. Simulation of the enhanced fuzzy flux controller using the software products *fuzzy*TECH and Matlab/Simulink

previously unknown to software implementations of fuzzy logic. Using the *fuzzy*TECH assembly kernel for 16 bit resolution, 2.98 million fuzzy rules per second can be computed on the TMS230C52 (25 ns instruction cycle DSP), including fuzzification and defuzzification. For comparison: the most recent dedicated fuzzy processor of VLSI (VY86C500/20) only computes 0.87 million fuzzy rules per second (not including fuzzification and defuzzification) with just 12 bit resolution (VLSI data sheet). While the referenced DSP only costs a few dollars in large quantities, the fuzzy processor is quoted at $75 each. This comparison shows that in most applications, the use of dedicated fuzzy processors is not necessary.

Conclusion

The application project discussed in this section shows that even in areas where traditional control engineering already offers comprehensive solutions, fuzzy logic can deliver substantial benefits. The *fuzzy*TECH assembly kernel for DSPs developed by Texas Instruments and Inform Software Corp. allows for the integration of fuzzy logic systems together with conventional algorithms on the same chip, even when control loop times of a fraction of a millisecond are required. Texas Instruments and Inform Software Corp. now work on further enhancements of the fuzzy flux controller. The companies are currently striving for even better dynamic performance by adding a fuzzy air-gap flux observer to the system.

8.5.2 Truck Speed Limiter Control by Fuzzy Logic

Introduction

Commercial trucks having a maximum load of more than 12 tons are required to be equipped with a speed limiter that limits their maximum speed to 53.3 mph (86 km h^{-1}). This case study focuses on the electro-pneumatic design of such a speed limiter. In this design, a pneumatic cylinder mechanically limits the throttle-opening angle of the fuel pump arm. A pulse proportional electromechanical valve controls the cylinder pressure. This valve is connected to an electronic control unit (ECU) that uses a microcontroller to drive the valve according to the actual speed of the truck.

The design of an algorithm for this control problem proved to be difficult, since the same speed limiter device is used in a variety of different trucks, which exhibit different behaviors. In addition to this, the dynamic behavior of a truck differs very much depending on whether it is fully loaded or empty. Conventional control algorithms, such as PID controls, assume a linear model of the process under control and can hence not be used for a solution. A solution using a mathematical model of the truck is first laborious to build and second of prohibitive computational effort for a low-cost 8-bit microcontroller. Hence, fuzzy logic control was used to design the control algorithm.

Speed Limiter Requirements and Conventional Control

Figure 8.33 exemplifies the function of a truck speed limiter. When the truck approaches the maximum velocity, the pneumatic valve reduces the throttle opening angle of the fuel pump arm so that the maximum velocity vs. is not surpassed. If the driver pushes down the accelerator pedal even more, the speed limiter has to ensure a smooth ride at the maximum velocity.

However, due to the dead time and nonlinearities involved with this control action, an actual overshoot and hunting occurs when using a proportional or on–off controller. Adding a differential and integral part yields a PID controller model. A PID controller generates the command value as a linear combination

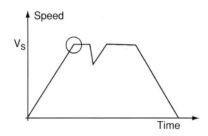

Fig. 8.33. The function of a speed limiter is to stop the truck from driving faster than the maximum allowed speed (v$_s$)

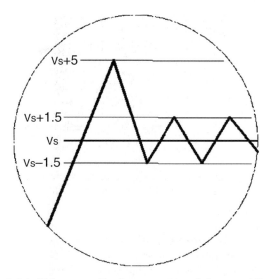

Fig. 8.34. Tolerances for the operation of the speed limiter

of the error (P), the derivative of the error with respect to time (D) and the integral of the error with respect to time (I). To tune a PID controller, the combined weights of these three (3) components must be chosen, so that they approximate the nonlinear behavior of the process under control at its operating point. While this works with most processes that are at only one operating point, it fails when the operating point moves. With a truck speed limiter, the operating point moves because of the different load situations, such as driving uphill or downhill, as well as driving empty or with a full load. Furthermore, the characteristic of the pneumatic valve and the truck fuel injection are highly nonlinear and vary from one truck to another.

Hence, if a PID control algorithm is used in a truck speed limiter, it can only be tuned well for one operation point and one type of truck. For other operation points and different truck types, overshoot and hunting occurs. The European legislation hence allows operation of the speed limiter within a certain tolerance. Figure 8.34 shows an example of this. When reaching the maximum speed, an initial overshoot of $5 \, \mathrm{km \, h^{-1}}$ is tolerated. After this, the speed must be kept constant within an interval of $\pm 1.5 \, \mathrm{km \, h^{-1}}$. Even though this overshoot and hunting is tolerated by the European legislation, it causes annoying speed fluctuations when driving.

Mechanical Design of the Speed Limiter

Figure 8.35 sketches the outline of the mechanical design of the speed limiter. An ECU compares the digital pulse signal from the speedometer with the maximum speed value preset in the device. Based on this, it computes the command value for the pulse proportional valve (PPV) that controls the air

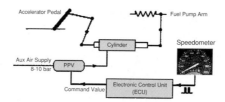

Fig. 8.35. Mechanical design of the speed limiter

Fig. 8.36. The electronic control unit of the speed limiter using a PIC 8-bit micro-controller

pressure in the cylinder. The air stems from the vehicles' pressured air system. In a nonlinear but proportional ratio, the cylinder shortens the arm linking the accelerator pedal to the fuel pump so that the fuel pump is throttled.

Design of the Speed Limiter Electronic Control Unit

The ECU itself is designed as a mixed digital and analog circuit. Speedometer signal processing, the fuzzy logic control algorithm, and diagnosis functions are computed by a PIC 8-bit microcontroller (MCU). The MCU uses an external E^{2}PROM to store parameters of the truck and speedometer, the maximum velocity, and diagnosis variables.

The MCU also generates a pulse-width modulated signal (PWM) that is amplified by a power stage to drive the PPV. The analog part is responsible for the preprocessing and filtering of the speedometer signal. Figure 8.36 shows a photo of the unit.

The Fuzzy Logic Controller

Fuzzy logic is an innovative technology to design solutions for multiparameter and nonlinear control problems. It uses human experience and experimental results rather than a mathematical model for the definition of a control

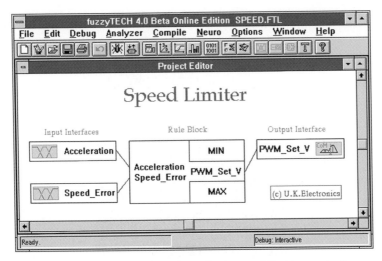

Fig. 8.37. Structure of the fuzzy logic speed limiter controller

strategy. As a result, it often delivers solutions faster than conventional control design techniques. In addition, fuzzy logic implementations on microcontrollers are very code space and execution speed efficient.

The entire fuzzy logic algorithm was developed, tested and optimized using the software tool *fuzzy*TECH. Figures 8.37–8.39 all show screen shots of the fuzzy logic speed limiter design in various editors and analyzers of *fuzzy*TECH. Figure 8.37 displays the Project Editor featuring the structure of the fuzzy logic system. On the left-side, two input interfaces fuzzify the two input variables "Acceleration" and "Speed_Error." The rule block in the middle contains all the fuzzy logic rules that represent the control strategy of the system. On the right-side, the output variable "PMV_Set_Value" is defuzzified in an output interface.

Figure 8.38 shows more details of the fuzzy logic system. Each linguistic variable is displayed in a variable editor window and the rules are displayed in the Spreadsheet Rule Editor window. Each linguistic variable contains five (5) terms and membership functions. They are connected by a total of 12 fuzzy logic rules. All membership functions are of Standard types. As a defuzzification method, the Center-of-Maximum (CoM) method is used.

All rules in the fuzzy logic system now let the designer define the best reaction (output variable value) for a given situation. The situation is described by the combination of the input variables. After such a control strategy has been defined by the designer, a number of different analyzer tools can be used to verify the system's performance. Figure 8.39 shows the 3D Plot as an example. In the 3D Plot, the two horizontal axes show the two input variables "Acceleration" and "Speed_Error." The vertical axis plots the

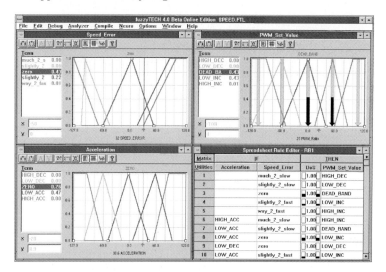

Fig. 8.38. Definition of the linguistic variables and the rule table

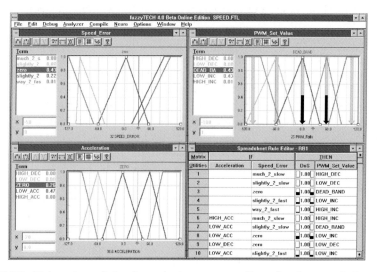

Fig. 8.39. Debugging of the fuzzy logic system – Transfer characteristics of the fuzzy logic controller under testing

output variable, "PWM_Set_Value", the set value for the PWM unit on the microcontroller. Rule 1, as shown in Figure 8.38, indicates:
IF Speed_Error=much_2_slow THEN PWM_Set_Value=HIGH_DEC
This rule represents the engineering knowledge that if the truck is way under the speed limit, no pressure should be applied to the cylinder. The membership function of the term "much_2_slow" is shown also in the respective variable editor in Figure 8.38. The 3D Plot in Fig. 8.39 plots the transfer characteristic

as a result of this. In the front part of the curve, the value of the output variable is very low (color of surface light). As you proceed to the left along the Acceleration axis, the output variable value gets higher. This is a result of Rule 6 that defines:

IF Acceleration = HIGH_ACC AND Speed_Error = much_2_slow THEN PWM_Set_Value = HIGH_INC

This rule represents the engineering knowledge that if in the same general case of the truck being way slower than the limit, the special case of a high acceleration should result in medium pressure on the cylinder. This ensures that the cylinder is already filled with some pressure if the truck reaches the limit quickly. If this rule would not have been formulated, a speed overshoot would occur.

Optimization and Implementation

A fair deal of optimization was accomplished offline on the PC since *fuzzy*TECH can simulate the fuzzy logic system without the target hardware. However, final optimization and verification of the system was conducted on real trucks.

Figure 8.40 shows the test and optimization setup. The target system, the MCU of the ECU, is mounted in the truck and connected to the development PC (notebook in the truck's cabin) by a serial cable. Such serial connection allows for modification of the running fuzzy logic controller "on-the-fly." This development technique is very efficient because it allows for the developer to analyze how a certain behavior of the fuzzy logic controller is caused by the membership function definition and the rules. Also, since modifications can be done in real time, the effects can be "felt" on the truck instantly.

One way to enable "on-the-fly" debugging is to link the *fuzzy*TECH RTRCD Module (real-time remote cross debugger) to the fuzzy logic controller that runs on the MCU and connect it to a serial driver. The RTRCD Module consumes about $1/2$ KB of ROM, a few Bytes of RAM, and some computing resources to serve the serial communication on the MCU. Since

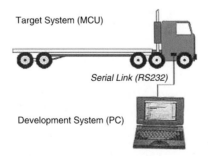

Fig. 8.40. Cross optimization of the running fuzzy logic controller by a serial link

the PIC16 MCU used in this application cannot provide such resources, the serial debug mode of *fuzzy*TECH was used rather than the RTRCD Module. In serial debug mode, the values of the input variables are sent from the MCU to *fuzzy*TECH running on the development PC via the serial cable and the results are sent back. *fuzzy*TECH shows the entire fuzzy logic computation in its editor and analyzer windows, and allows for "on-the-fly" modifications. To support serial communication, a piggyback board containing a MAX232 driver IC was mounted on the speed limiter board shown in Fig. 8.36.

The advantage of using the serial debug mode of *fuzzy*TECH over the RTRCD Module is that code size and computation effort that previously was required for the fuzzy logic computation on the MCU are now saved and can thus be used for the serial communication. The disadvantage of the serial debug mode of *fuzzy*TECH is that it computes the systems' results on the PC where real-time response cannot be guaranteed. Also, any crash of MS-Windows, the PC, or the serial communication will halt computation in serial debug mode.

Conclusion

After optimization of the fuzzy logic rule strategy on different trucks and various load conditions, the speed limiter showed a response curve as shown in Fig. 8.41. The fuzzy logic controller achieves a much smoother response, does not show overshoot behavior, and provides a higher accuracy of keeping the speed limit compared to a conventional controller.

The final fuzzy logic system was compiled to PIC16 assembly code by the *fuzzy*TECH MP-Edition and required 417 words of ROM space and 32 bytes of RAM. The RAM space can be used for other computation tasks such as preprocessing and filtering while the fuzzy logic system is not running. The entire fuzzy logic system needs less than 2 ms to compute on the PIC16 MCU.

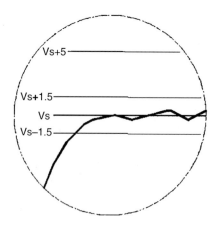

Fig. 8.41. Speed limiter performance of the fuzzy logic controller

8.5.3 Analysis of Environmental Data for Traffic Control Using Fuzzy Logic

Introduction

Traffic control is based on the analysis of traffic data and environmental conditions. Particularly bad environmental conditions may cause hazard to drivers. In this section, we discuss the use of fuzzy logic for the analysis of environmental conditions such as road surface condition, visual range and weather conditions detected by local sensor stations and road sensors. Because detection of environmental conditions involves a number of uncertainties, conventional approaches do not deliver satisfactory solutions. The fuzzy logic solution discussed in this section in contrast:

- Takes into account the different types and quality of equipment used at the different detection stations
- Uses a two-step plausibility check to determine the quality of sensor signals
- Computes substitute values for missing information using sensors from other detection stations
- Leads to more appropriate results for the evaluation of road surface conditions and visual range to indicate slippery road conditions or fog warning

Traffic Management

The first traffic management systems used in Germany were implemented on roads with frequent accidents caused by fog or icy road conditions. Later, these system were extended to detect and control traffic to increase the traffic capacity. These traffic control systems use several detection stations along the road. These stations employ magnetic sensors for traffic detection, as well as weather stations transmitting environmental data from road surface and the air layer near the ground.

A central traffic control computer collects the data transmitted from the section stations. A control strategy derives an adequate speed limit for every section. The control objectives are:

- Keep traffic flowing in case of peak traffic
- Slow down traffic at the inflow to congestion
- Warn for bad weather conditions such as fog or ice

Along the road of such an "intelligent" highway, alterable road signs posted on traffic sign gantries display speed limits for each lane and display nonregular events such as road work, warnings for traffic back-ups, breakdowns, an accident, or dangerous weather conditions (Fig. 8.42).

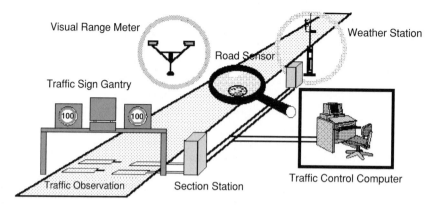

Fig. 8.42. Environmental and traffic sensors in a traffic control system

Environmental Data Analysis

The traffic situations depend highly on environmental conditions. An intelligent highway thus should warn drivers of slippery road conditions and low visibility. Sensors are used to indicate and classify icy or wet road pavement and to indicate the visual range.

An ideal weather station uses road sensors measuring road surface temperature, road surface moisture, water film depth, and salt content of water film. Near the road, the ideal weather station detects air and ground temperature, amount of precipitation, type of precipitation, wind velocity and direction, sun beam intensity and illumination.

However, most existing weather stations are not equipped with this full range of sensor equipment. In addition, some sensors do not work reliably under all conditions. For example, a standard salinometer needs a wet road surface to measure the remaining salt content on the road. Besides measurement problems, sensors frequently fail because of "biological attacks" such as spiders or butterflies that cover the surface of visual based detectors. Conventional traffic control systems misinterpret this as high rain intensity or low visual range. This can cause completely wrong traffic warnings. Because sensors often stem from different vendors, conventional systems do not use interrelationships between the signals of different detection stations to identify such implausibilities.

Fuzzy Logic System Architecture

The fuzzy logic data analysis unit was designed as part of a larger traffic control system. As shown in earlier applications, fuzzy logic is well suited to create solutions for traffic control systems. Figure 8.43 shows the architecture of the analysis unit, separating a component to verify the sensor information from components to evaluate the road surface and visual range condition.

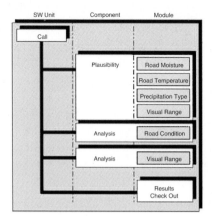

Fig. 8.43. Architecture of environmental data analysis

Sensor Plausibility Analysis

A two-step approach was used to verify the sensor signals. The first step utilizes the fact that no weather signal remains constant. In particular, if the signal jumps abruptly or remains completely constant over time, the sensor signal is considered to be faulty.

In this case, the fuzzy logic system regenerates the information from other sensors. To design this fuzzy logic system, meteorological knowledge about the maximum gradients of all sensor signals, a time frame for a required movement of the signals, and maximum jumps of the gradients to identify discontinuity were acquired from experts.

The second step uses four separate fuzzy logic modules to combine interrelated signals.

Road Moisture Module

This module, combines all data, that indicates anything about moisture or water on the road. The fuzzy logic module consists of five rule blocks (Fig. 8.44) that implement:

– A compensation rule block for the hygroscopical behavior of the road moisture sensors. For example, if the salt content is very high, the moisture sensor indicates higher values.
– A cross check rule block between detected road surface moisture, the sensors that detect a water film on the road, and the amount of precipitation ions detected during the last 30 min. For example, if strong rain was detected over the past minutes, the road must be wet.
– A cross check rule block of the humidity sensor using dew point, road temperature, and a moisture sensor. For example, if the dew point is

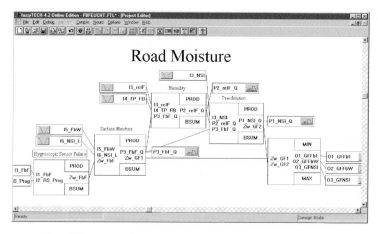

Fig. 8.44. Road moisture fuzzy logic module

lower than the road temperature and the verified moisture indicates a dry
road, the humidity signal must be wrong.
- A cross check rule block to the precipitation sensor.
- A diagnosis rule block to derive an error message from the given signal
  situation.

As result, the module produces verified signals of road moisture, amount of
precipitation, and humidity.

Road Temperature Module

This fuzzy logic module contains two rule blocks (Fig. 8.45) to compute:

- A verified value of the road surface temperature by cross check of the
  temperature signal, the gradient of this signal, and the precipitation.
  For example, the temperature signal can only decrease rapidly if a large
  amount of precipitation is detected.
- A verified value of the freezing point, taking into account salt content
  and road surface moisture. Because the salt content sensor does not work
  with dry road conditions, a salt content forecast is used when the signal
  is not available.

Precipitation Type Module

The verification of the precipitation type is the most complex verification
module. This fuzzy logic module verifies existing sensors that indicate the
precipitation type by a cross check with the verified signals of road moisture, precipitation quantity, visual range, and other environmental conditions
(Fig. 8.46). If the sensor delivers implausible results or is not available, a substitute value is computed. The module consists of a number of rule blocks
that:

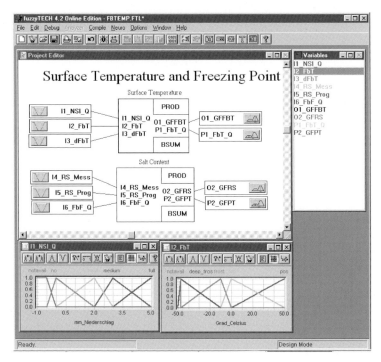

Fig. 8.45. Road temperature fuzzy logic module

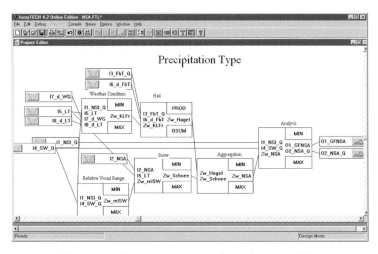

Fig. 8.46. Precipitation type fuzzy logic module

- Indicate if the weather conditions allow for hail or snow. For example, an air temperature level is defined at which snow is implausible.
- Compare the visual range with the precipitation quantity.

– Aggregate the information with the precipitation quantity and visual range.

Visual Range Module

The visual range fuzzy logic module computes a verified value of the visual range by using two rule blocks (Fig. 8.47) that:

– Cross check the visual range with the precipitation quantity. For example, there is no fog during heavy rain.
– Cross check the visual range with air humidity. For example, fog only occurs during very high humidity.

Using these verified signals, the subsequent two components conclude road and fog conditions. The Road Condition component (Fig. 8.48) uses the qualified values of precipitation type and quantity, as well as the road moisture to indicate a dangerously wet road. Precipitation type, freezing point, and road temperature indicate icy conditions. A final rule block aggregates the information to a standardized road condition classification code. An additional component aggregates the verified values of visual range and precipitation type to compute the standardized visual range classification code.

Conclusion

Conventional traffic systems are susceptible to faulty weather sensor signals. The fuzzy logic approach presented delivers more reliable results using meteorological expertise. The solution discussed was implemented using the fuzzy logic development software *fuzzy*TECH. The table in Fig. 8.49 shows the number of variables, structures, rules and memberships of each component and the complete system used.

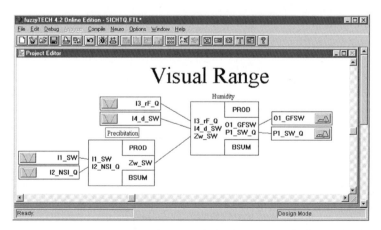

Fig. 8.47. Visual range fuzzy logic module

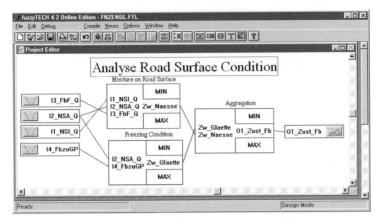

Fig. 8.48. Analysis of road surface condition

Module	Variables	Rule Blocks	Rules	MBFs
Road Moisture	16	5	489	70
Road Temperature	10	2	130	39
Precipitation Type	15	6	204	60
Visual range	7	2	142	36
Analysis road condition	7	3	60	33
Analysis visual range	3	1	35	17
Complete unit	58	19	1060	255

Fig. 8.49. Size of fuzzy logic unit and its modules

In day to day operation, the fuzzy logic solution has shown that it can prevent traffic control system malfunction in most sensor breakdown situations. In addition "biological attack" situations were detected and misleading rain or fog detection was avoided.

In a complete traffic control system, analysis of environmental conditions is only one component of its functionality. However, faulty weather detection can cause the entire traffic control system to malfunction. Thus, the enhancement of traffic control systems by fuzzy logic greatly improves the reliability of traffic control.

8.5.4 Optimization of a Water Treatment System Using Fuzzy Logic

Introduction

This case study is about a fuzzy logic solution in biochemical production at the world's largest oral penicillin production facility in Austria. After extracting the penicillin from the microorganisms that generated it, a waste water treatment system further processes the remaining biomass. Fermentation sludge

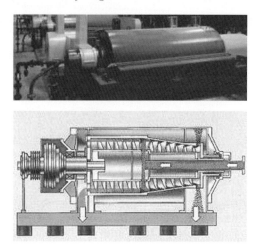

Fig. 8.50. Picture of a decanter and its schematic

obtained in the course of this treatment contains microorganisms and remnants of nutrient salts. It is the basic material for a high quality fertilizer and sold as a by-product of the penicillin production. To render the fertilizer, the sludge is concentrated in a decanter and then cleared of the remaining water in a vaporizer. In order to reduce energy costs of the vaporizing process, the separation of water and dry substance in the decanter must be optimized. Before the implementation of the fuzzy logic solution, operators controlled the process manually.

Fuzzy Logic Replaces Manual Control

Due to its complexity, operators control the process of proportioning decanter precipitants manually. Drainage control in the decanter is very crucial. To save energy cost in the vaporizing process, the decanter must extract as much water as possible. However, to achieve optimal results requires an operating point close to the point where the decanter becomes blocked. If blockage occurs the process must be stopped and the decanter manually cleaned, thus operators run the process far away from this optimal point. Alas, this results in high operational energy costs. In this case study, a fuzzy logic system replaced the suboptimal manual control of the operators (Fig. 8.50).

Fuzzy Logic Applications in Chemical Industry

A number of conventional control techniques exist to automate continuous processes in chemical industry. To keep single variables of the process constant,

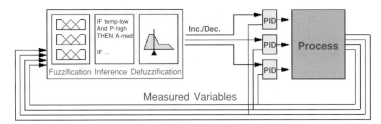

Fig. 8.51. Using a fuzzy logic controller to determine the set values for underlying PID control loops

PID controllers or the like are very common. Even though, most technical processes are nonlinear and PID controllers are a linear model, this works fine. This is because within a continuous process, the behavior of the process can be well-approximated linearly near the operation point. In processes with simple time behavior (no dead times etc.), even P (proportional) controllers and bang–bang controllers often suffice.

While keeping single process variables at their command values is rather simple, the determination of the optimal operation point as the combination of all set values of the variables is often a complex multivariable problem. In most cases a solution based on a mathematical model of the process is far beyond acceptable complexity. In some cases, the derivation of a mathematical model of the plant consumes many man-years of effort. Hence, in a large number of plants, operators prefer to control the operation point of the process manually.

In these cases, fuzzy logic provides an efficient technology for putting the operator control strategies into an automation solution with minimum effort. Figure 8.51 shows how to combine the fuzzy logic system with underlying PID controllers. In practical implementations, the PID controllers run on the same DCS as the fuzzy logic controller. Figure 8.52 shows the technical integration of fuzzy logic into DCS.

Decanter Control

The sewage of the fermenters contains about 2% dry substance. In a first step, milk of lime neutralizes the sewage. Second, a fermenter biologically degrades the sewage. Third, bentonite is added in a large reactor. This results in a precoagulation of the sludge. Next, the slurry is enriched with cationic polymer before the water is separated in the decanter. The cationic polymer discharges the surface charge of the sludge particles and hence leads to coagulation (Fig. 8.53).

Bentonite and polymer addition exert a fundamental influence on the drainage quality obtained in the decanter. In order to achieve a high drainage quality, the gradation of chemicals has to be optimized. A changed bentonite proportioning causes only a slow change in the precipitation process

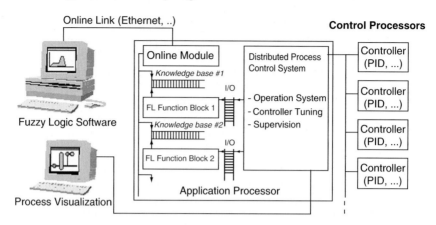

Fig. 8.52. Integration of a fuzzy logic component in a distributed control system

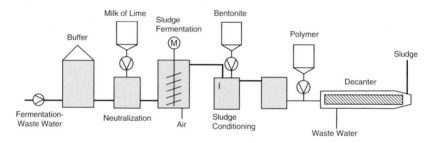

Fig. 8.53. Process diagram of sludge draining

because of the dead time of the reactor of about 20 min (reactor capacity $30\,\mathrm{m}^3$ at a volume flow of $90\,\mathrm{m}^3\ \mathrm{h}^{-1}$) and the dead time of the decanter of about 5–7 min. An increase in bentonite addition principally results in an improvement of coagulation. Implementation of just a polymer proportioning controller must only consider the dead time of the decanter. An increase in polymer addition first results in an improvement of coagulation, but with further addition coagulation impairs.

The control strategy of the operators uses two measured variables:

- A turbidity meter measures the purity of the water. High turbidity indicates a high remaining solids content of the water.
- A conductivity meter measured the draining degree of the slurry. A high conductivity of the slurry indicates a high water content.

Control Strategy Objectives

The prime objective of the control strategy is to minimize the operational cost of the total drying process. In particular, the objectives are:

- Minimize amount of precipitants. The polymer used is very expensive.
- Minimize remaining water content. The vaporization of the remaining water consumes large amounts of energy.
- The biological mass content of the sewage water of the decanter should be $0.7\,\mathrm{g}\,\mathrm{l}^{-1}$. Exceeding a value of $1.5\,\mathrm{g}\,\mathrm{l}^{-1}$ reduces the operability of the clarification plant and can even result in a breakdown of the next sewage stage because of its limited capacity to degrade biological mass.
- The third objective has an absolute priority, above other objectives, to ensure safe operation. If the biological mass content reaches its upper limits, the control strategy reduces it regardless of economic considerations. Besides the critical biological mass content of the sewage water, only the first two objectives are relevant. A rough cost estimation shows that the most effective way to reduce the expenses of the process is to minimize the energy used in the evaporators. If the dry substance content of the thick slurry is increased by 1% (this corresponds to a change of conductivity of about $0.5\,\mathrm{mS}\,\mathrm{cm}^{-1}$), the costs of energy (used in the evaporator) could be reduced by $140 per day. A reduction of bentonite addition by 10% saves about $30 per day, and a decrease of polymer addition by 10% reduces costs by $40 per day. A small decrease of chemical gradation can result in a significant reduction of dry substance contents in thick slurry. Therefore the main objective of optimization is to obtain the best possible result of draining with least use of chemicals.

Designing the Fuzzy Logic Rule Base

An increase of polymer addition results in an increase of drainage grade. This reduces the turbidity of the sewage water (degree of suspended matter) and conductivity of the thick slurry (water content). Exceeding the optimum polymer gradation, further addition of polymer leads to a decline of the separation power in the decanter. The separation power on polymer gradation $L = f(\mathrm{dmpolymer}/\mathrm{d}t)$ follows a parabolic curve (Fig. 8.54). Conductivity of the slurry is used as an indicator for the separation power obtained. Hence, the control strategy is to find the minimum of conductivity by modifying the polymer addition.

The problem with this is that the shape of the curve and hence also the level and position of the minimum may significantly vary over time. The only way to find out the direction to the minimum from the current operating point is to apply a change in the polymer gradation and evaluate its effects. If the draining of thick slurry improves, the conclusion is that this decision was correct and can be repeated. If the decision results in a decline, the fuzzy logic controller pursues the opposite strategy. The stronger the system's reaction to a change in polymer addition is, the greater is the current operation point's distance to the optimum.

Likewise, an increase in bentonite addition leads to an increase in separation power of the decanter until an optimum is reached. If more bentonite is

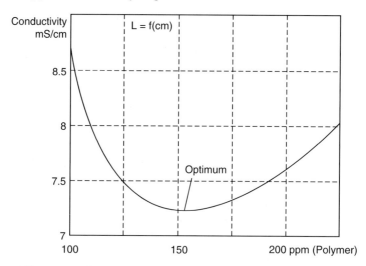

Fig. 8.54. Thick slurry conductivity over polymer proportion

added beyond this point, neither improvement nor decline occurs. Thus, an increase in bentonite addition beyond the optimal point is less critical than an increase in polymer addition. However, bentonite addition should be kept as low as possible to minimize operating cost.

Structure of the Decanter Controller

A change in bentonite addition has a greater long-term effect than a change in polymer addition, due to the larger dead time. As a correlation between bentonite and polymer addition exists, it is sensible to change the polymer addition independent of the separation quality in the short term and to let the bentonite controller run with delay.

Sewage quality is only restricted by an upper limit, whereas slurry quality is of great economical importance. For these reasons, polymer addition is controlled by a first set of rules in correlation to the slurry quality. An absolute point of reference does not exist because of the changing optimum of the polymer addition characteristics. Hence, no absolute input or output values are used. The controller determines its position in the objective function from a prior change in polymer addition and the resulting reaction of conductivity.

The first rule block of the fuzzy logic controller works on the targeting. Input values for this rule block are the increment of set point in the previous cycle (pr_decision) and the resulting change in conductivity (d_conductab). The output of the rule block is the increment (polymer1) on the set point of the polymer PID controller. Figure 8.55 shows the control strategy of this rule block as a matrix. The entries in the matrix represent the output variable polymer1 of the rule block.

Difference in Conductability
(d_conductab)

		N	Z	P
	N	N	Z	P
Previous Decision (pr_decision)	Z	Z	Z	P
	P	P	N	N

Fig. 8.55. Matrix representing the control strategy of polymer proportion. N represents "negative", Z represents "zero", and P represents "positive"

A positive change in conductivity implies a decline of the plant's operating state. If the controller in the previous cycle (pr_decision) recommended "negative" (reduce polymer addition) and conductivity increased as a reaction to this, the controller must now operate in the opposite direction and increase polymer addition. If the controller recommended "positive", polymer addition must be reduced. Conductivity may increase although the controller has not given a recommendation in the previous cycle (pr_decision = 0). In this case, the optimum moved by external influence and action must be taken.

As the fuzzy logic controller does not know in which direction the optimum moved, it has to choose one direction. For this, it increases polymer addition to test the reaction of the process. It tests with a polymer increase rather then a decrease, as the curve in the figure rises more gently into the positive direction, and an increased use of polymer is cheaper than the raised energy costs that stem from the evaporation process.

If the test result is favorable, this will be confirmed in the next cycle. If it is negative, the controller will simply change the signs of the output in the subsequent cycle (see matrix in Fig. 8.55). This only represents the basic principle of the rule block. The final implementation of the rule block contains additional rules that more finely differentiate the input value. As such, the fuzzy logic controller can differentiate between several cases and thus adapt the size of gradation change to the reaction of the process that is, to the proximity to the optimum.

The turbidity of the sewage water forms a restriction that in some cases forbids the reduction of the precipitant addition. Therefore, a second rule block evaluates the turbidity of the sewage water (turbidity) and the output of the first rule block (polymer1) to determine the final polymer addition increment (polymer2). Figure 8.56 shows the structure of the fuzzy logic system. As long as turbidity units are below 2,500 TE/F, the second rule block transfers the first block's recommendation to the output. Starting at about 3,000 TE/F, all recommendations of the first rule block are transferred, except recommendations to reduce polymer addition. If about 3,500 TE/F are exceeded, the second rule block always recommends an increase of flocculent addition. Only

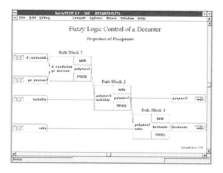

Fig. 8.56. Structure of the fuzzy logic controller

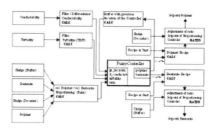

Fig. 8.57. Total controller structure with preprocessing and postprocessing

the rate of increase of flocculent addition changes, depending on the recommendation of the first rule block.

To determine the betonite addition, a third rule block uses the output of the second rule block and a fourth input variable, called ration. The ratio is calculated outside the fuzzy logic system.

Integrating the Fuzzy Logic Controller in the System

Figure 8.57 shows the total structure of the entire control system. In addition to the fuzzy logic system, function blocks (CALC, RATIO) provide preprocessing of the sensor signals, ramp generation, and ratio control. The entire control system is implemented using a standard DCS. The most important input variable of the controller is the change in the conductivity of the slurry. Alas, the conductivity measurement circuitry is subject to strong signal noise. Hence, a low pass filter is used that determines the average change in conductivity during the last 10 min at 1-min intervals. This forms the input variable d_conductab of the fuzzy logic system. Turbidity, the fuzzy system's second input, uses a similar filter.

Conclusion

Compared to the manual operation, the fuzzy logic controller saves about $70,000 in energy costs per year. Thus fuzzy logic is more efficiently used in water treatment system.

8.5.5 Fuzzy Logic Applications in Industrial Automation

Introduction

In this section, we review eight recent applications of fuzzy logic in industrial automation. All applications used the so-called "fuzzyPLC," an innovative hardware platform that merges fuzzy logic and traditional automation techniques. Following a quick overview on the fuzzyPLC, we discuss the eight applications and focus on how fuzzy logic enabled a superior solution compared to conventional techniques. Whenever possible, we quantify the benefit in cost saving or quality improvement.

In recent years, fuzzy logic has proven well its broad potential in industrial automation applications. In this application area, engineers primarily rely on proven concepts. For discrete event control, they mostly use ladder logic, a programing language resembling electrical wiring schemes and running on so-called programmable logic controllers (PLC). For continuous control, either bang-bang type or PID type controllers are mostly employed.

While PID type controllers do work fine when the process under control is in a stable condition, they do not cope well in other cases:

- The presence of strong disturbances (nonlinearity)
- Time-varying parameters of the process (nonlinearity)

Presence of Dead Times

The reason for this is that a PID controller assumes the process to behave in a strictly linear fashion. While this simplification can be made in a stable condition, strong disturbances can push the process operation point far away from the set operating point. Here, the linear assumption usually does not work any more. The same happens if a process changes its parameters over time. In these cases, the extension or replacement of PID controllers with fuzzy controllers has been shown to be more feasible more often than using conventional but sophisticated state controllers or adaptive approaches. However, this is not the only area where there is potential for fuzzy logic based solutions.

Multivariable Control

The real potential of fuzzy logic in industrial automation lies in the straightforward way fuzzy logic renders possible the design of multivariable controllers.

In many applications, keeping a single process variable constant can be done well using a PID or bang–bang type controller. However, set values for all these individual control loops are often still set manually by operators. The operators analyze the process condition, and tune the set values of the PID controllers to optimize the operation. This is called "supervisory control" and mostly involves multiple variables.

Alas, both PID and bang–bang type controllers can only cope with one variable. This usually results in several independently operating control loops. These loops are not able to "talk to each other." In cases where it is desirable or necessary to exploit interdependencies of physical variables, one is forced to set up a complete mathematical model of the process and to derive differential equations from it that are necessary for the implementation of a solution. In the world of industrial automation, this is rarely feasible:

– Creating a mathematical model for a real-word problem can involve years of work.
– Most mathematical models involve extensive simplifications and linearizations that require "fudge" factors to optimize the resulting controller later on.
– Tuning the fudge factors of a controller derived from a mathematical model is "fishing in the dark," because optimizing the system at one operating point using global factors usually degrades the performance at other operating points.

Also, many practitioners do not have the background required for rigorous mathematical modeling. Thus, the general observation in industry is that single process variables are controlled by simple control models such as PID or bang–bang, while supervisory control is done by human operators.

This is where fuzzy logic provides an elegant and highly efficient solution to the problem. Fuzzy logic lets engineers design supervisory multivariable controllers from operator experience and experimental results rather than from mathematical models. A possible structure of a fuzzy logic based control system in industrial automation applications is exemplified by Fig. 8.58. Each

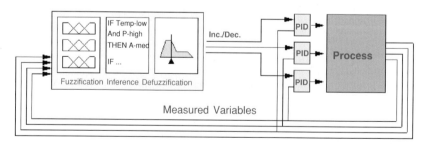

Fig. 8.58. Using a fuzzy logic controller to determine the set values for underlying PID control loops

single process variable is kept constant by a PID controller, while the set values for the PID controller stem from the fuzzy logic system. This arrangement is typical for cases like control of several temperature zones of an oven or control of oxygen concentrations in different zones of a waste water basin. In other cases, it could be reasonable to develop the complete closed loop control solution in a fuzzy system. This illustrates why it is very desirable to integrate conventional control engineering techniques, such as ladder logic or instruction list language for digital logic and PID control blocks tightly together with fuzzy logic functionality.

Merging Fuzzy Logic and PLCs

In 1990, when more and more successful applications proved the potential of fuzzy logic in industrial automation, the German company Moeller GmbH and the US/German company Inform Software created the *fuzzy*PLC based on the observation that fuzzy logic needs tight integration with conventional industrial automation techniques (Fig. 8.59).

To make it available at a low cost, the core of the *fuzzy*PLC uses a highly integrated two-chip solution. An analog ASIC handles the analog/digital interfaces at industry standard 12 bit resolution. Snap-On modules can extend the periphery for large applications of up to about 100 signals. An integrated field bus connection, based on RS485, provides further expansion by networking. The conventional and the fuzzy logic computation is handled by a 16/32 bit RISC microcontroller. The operating system and communication routines, developed by Moeller, are based on a commercial real time multitasking kernel. The fuzzy inference engine, developed by Inform Software, is implemented and integrated into the operating system in a highly efficient manner, so that scan times of less than one millisecond are possible. The internal RAM of

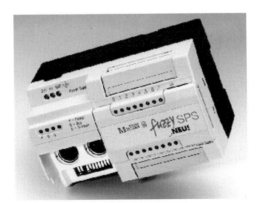

Fig. 8.59. The *fuzzy*PLC contains fuzzy and conventional logic processing capabilities, field bus connections, and interfaces

256 KB can be expanded by memory cards using flash technology. Thus, the *fuzzy*PLC is capable of solving quite complex and fast industrial automation problems in spite of its compact and low price design.

The *fuzzy*PLC Engineering Software

The *fuzzy*PLC is programed by an enhanced version of the standard fuzzy logic system development software *fuzzy*TECH from Inform Software. *fuzzy*TECH is an all-graphical, design, simulation, and optimization environment with implementation modules for most microcontrollers and industrial computers. To support the complete functionality of the *fuzzy*PLC, *fuzzy*TECH has been enhanced with editors and functions to support the conventional programing of the PLC. Thus, a user only needs one tool to program both conventional and fuzzy logic parts of the solution.

The *fuzzy*TECH software combines all necessary editors for membership functions, linguistic variables, rule tables, and system structure with analyzer functions and optimization features. The software runs on a PC and is linked to the *fuzzy*PLC by a standard serial cable (RS232) or the field bus (RS485). Through this link, the developer downloads the designed system to the *fuzzy*PLC. Because fuzzy logic systems often require optimization "on-the-fly," *fuzzy*TECH and the *fuzzy*PLC feature "online-debugging" where the system running on the *fuzzy*PLC is completely visualized by the graphical editors and analyzers of *fuzzy*TECH. Plus, in online-debugging modes, any modification of the fuzzy logic system is instantly translated to the *fuzzy*PLC without halting operation.

Application Case Studies

In this section, we review eight recent highly successful applications of fuzzy logic in industrial automation using the *fuzzy*PLC:

- Antisway control of cranes
- Fire zone control in waste incineration plants
- Dosing control in wastewater treatment plants
- Control of tunnel inspection robots
- Positioning in presses
- Temperature control in plastic molding machines
- Climate control and building automation
- Wind energy converter control

Antisway Control of Cranes

In crane control, the objective is to position a load over a target point. While the load connected to the crane head by flexible cables may well sway within certain limits during transportation, the sway must be reduced to almost zero for load release when the target position is reached. Hence, a controller must

Fig. 8.60. The 64 ton crane of Hochtief Corp. uses *fuzzy*PLC based antisway positioning control (Video)

use at least two input variables, for example position and sway angle. Thus, a simple PID controller cannot be used as it is restricted to one input. Conventional solutions of the problem require highly elaborate approaches, like model based control or state variable controllers that need intensive engineering and hardware resources. These technologies tend to push system costs into regions that make antisway systems economically unaffordable. For these reasons, most cranes are still operated manually.

In spite of the difficulties involved with automated control, human operators can control cranes quite well in most cases. Because fuzzy logic is a technology that facilitates control system design based directly on such human experiences, it has been used for crane automation for almost a decade. The types of cranes include container cranes in harbors, steel pan cranes, and cranes in a manufacturing environment. Recently, a 64 ton crane that transports concrete modules for bridges and tunnels over a distance of 500 yards has been automated with a *fuzzy*PLC in Germany (Fig. 8.60).

The benefit was a capacity gain of about 20% due to the faster transportation and an increase in safety. Accidents were frequent, because the crane operators walk parallel to the crane during operation with a remote controller. Before, when they had to watch the load to concentrate on the sway angle, they frequently stumbled over parts lying on the ground. The crane was commissioned in Spring 1995 and the fuzzy logic antisway controller has been continuously enabled by the crane operator, showing the high degree of acceptance by the operators. This fact is of special importance since not only technological feasibility but also psychological aspects are important for the success of an industrial automation solution (Fig. 8.61).

The real solution uses about ten inputs, two outputs, and four rule blocks with a total of 75 rules.

Fire Zone Control in Waste Incineration Plants

Maintaining a stable burning temperature in waste incineration plants is important to minimize the generation of toxic gases, such as dioxin and furan,

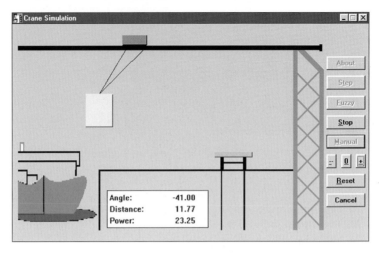

Fig. 8.61. A software simulation of the crane controller

as well as to avoid corrosion in the burning chambers. There are two primary difficulties of this temperature control process:

- The caloric value of the waste fluctuates strongly.
- The fire position and shape cannot be measured directly.

Because the heat generated from the burning process is used to produce electrical energy, a stable incineration process is also of high commercial interest.

In recent applications at waste incineration plants in the cities of Hamburg and Mannheim in Germany, fuzzy logic has been successfully applied. In Mannheim, where two *fuzzy*PLCs were used to control the burning process, the steam generation capacity of one furnace is $28 \, \text{Mg} \, \text{h}^{-1}$. Using the industry standard conventional controller, steam generation fluctuated by as much as $10 \, \text{Mg} \, \text{h}^{-1}$ in just $1 \, \text{h}$. The fuzzy logic controller was capable of reducing this fluctuation to less than $\pm 1 \, \text{Mg} \, \text{h}^{-1}$. This dramatically improved robustness and also caused the NO_x and SO_2 emission to drop slightly, and the CO emission to drop to half (Fig. 8.62).

Dosing Control in Waste Water Treatment Plants

Wastewater treatment processes are a combination of biological, chemical, and mechanical processes. This makes the creation of a complete mathematical model for their control intractable. However, there is a large amount of human experience that can be exploited for automated controller design. As such operator experience can be efficiently put to work by fuzzy logic, many plants already use this technique.

In a recent application in Bonn, dosing of liquid $FeCl_3$ for phosphate precipitation has been successfully automated using the *fuzzy*PLC. Recently

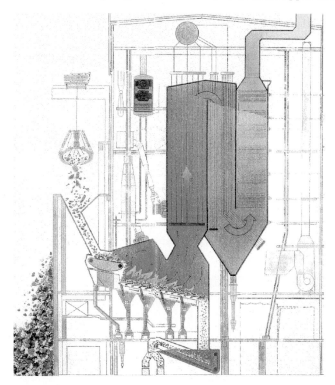

Fig. 8.62. In a waste incineration plant, a crane continuously delivers waste from the bunker to the belt running through the burning zone. The exhaust gases are cooled and cleaned

changes in legislation require water treatment plants in Germany to limit the total amount of phosphate in the released water to $1\,\mathrm{mg}\,\mathrm{l}^{-1}$. To extract the phosphate from the water, $FeCl_3$ is added, which converts the phosphate into $FePO_4$ that is sedimented with the sludge. Because a violation of the legal phosphate limit results in severe penalties, the operators tend to overdose the $FeCl_3$ (Fig. 8.63).

To optimize the $FeCl_3$ dosing, a fuzzy logic controller that uses the input variables phosphate concentration, its derivative, water flow, its derivative, and dry substance contents was designed. The output of the fuzzy logic controller is the change of the set variable for the injected $FeCl_3$. An underlying conventional PI type controller stabilizes the $FeCl_3$ flow to this set point. The PI type controller is implemented as a function block in the *fuzzy*PLC as well. This is an example of the combination of fuzzy logic and conventional control engineering techniques.

The total fuzzy logic controller uses 207 rules to express the control strategy based on the five input variables of the fuzzy logic control block. The total implementation time was three staff months and resulted in savings of about

Fig. 8.63. By injecting $FeCl_3$ into the sludge, dissolved phosphate precipitates from the waste water

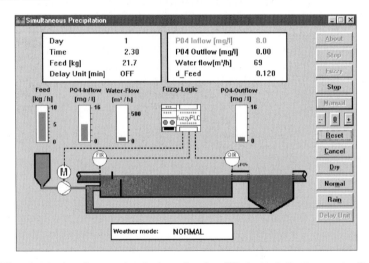

Fig. 8.64. A software simulation of a simplified precipitation controller

50% of the $FeCl_3$ compared to the manual control before. Taking implementation time and hardware/software costs into consideration with the savings on $FeCl_3$ results a return in investment time of half a year (Fig. 8.64).

Control of Tunnel Inspection Robots

The German Aerospace corporation DASA has developed a sewage pipe inspection system using two robot units and a support truck. The objective of the robots is to detect leakage in segments of the pipe by applying air pressure to the sealed space between the two robots. Because the vertical access shafts can be quite far away from each other, the robots have to operate up to 400 yards away from the truck. The robots are connected to each other and the truck by cables that provide air pressure, electrical energy, and control signals to the robots (Fig. 8.65).

When DASA developed the system, a severe control problem came up. To avoid entanglement of the cables that can result in the robots getting stuck in the pipe, cable tension must be controlled very carefully. A conventional approach using complex state variable controllers turned out to be too costly in terms of both money and design time. A control system implemented on two *fuzzy*PLCs using about 200 rules each showed very good results in a very short engineering time at less than 10% of the costs of a conventional solution.

Positioning of Presses

One area with big potential for fuzzy solutions is the control of drives. In this example, we discuss hydraulic axis control. One of the most complex fuzzy projects was done for a hydraulic press used to press laminates, printed circuit boards, and floor coverings. The task was the synchronized control of a 14-axis system. The position control of the axis, a superimposed pressure controller, the parallel running of the steel belt and the synchronization of all axes had to be solved (Fig. 8.66).

The automation system employed has a highly decentralized structure and consists of two large master PLCs, a number of smaller compact PLCs, a PC based supervisory system, and seven *fuzzy*PLCs. All units are networked using the integrated field bus interfaces. Very important for the synchronization of the entire machine is the ability of the field bus network to satisfy the real time requirements. One typical problem involved in the control of hydraulic

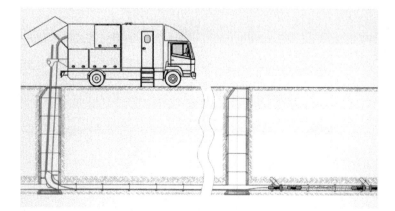

Fig. 8.65. The two robot units in the sewage pipe (right) are supplied from a specialized truck by cables

Fig. 8.66. Control of hydraulic systems is difficult as many nonlinearities, such as the "stick-slip" effect, are involved

systems is the so-called "stick slip effect." The transition of an axis from standstill to motion is highly nonlinear because of the transition from stick friction to slip friction. This makes designing a good controller for hydraulic systems difficult. In the cited application, fuzzy logic rendered a good solution technique, freeing system design from the burden of the theory of nonlinear systems synthesis. The overall design time using fuzzy logic was only a third of what a conventional approach had required in past applications of conventional control for similar presses.

Temperature Control in Plastic Molding Machines

In plastic molding machines, temperature control is crucial to achieve high and consistent product quality. This requires LABORIOUS tuning of the involved control algorithms, because the dead times involved in an extrusion machine are significant and there is significant coupling between the different temperature zones.

To cut down the commission time for these machines, KM corporation has developed a self-tuning controller using the *fuzzy*PLC. At start up time, some parameters are estimated that are used to scale the nonlinear fuzzy controller. In contrast to conventional tuning algorithms, this controller does not require a cooling down of the machine to room temperature before self-tuning can work. Even very difficult temperature zones with big dead times can be handled by this algorithm and the result is a very robust controller. This is important because the temperature properties of an empty machine and one filled with plastic material are extremely different. Compared to conventional systems, the fuzzy logic enhanced temperature controller performs with a faster response time and a significantly smaller overshoot combined with extreme robustness (Figs. 8.67 and 8.68).

Fig. 8.67. To achieve high product quality, keeping the temperature constant is critical in molding plastic

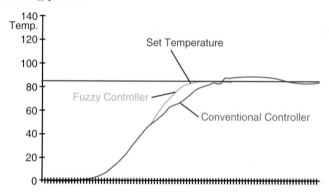

Fig. 8.68. The fuzzy logic controller in the molding machine reaches the set point faster and avoids overshoot

Climate Control Using Fuzzy Logic

Climate control systems reveal a high potential for energy savings. In a recent application at a major hospital in Europe, the integration of fuzzy logic saves about 25% on electrical energy, equivalent to the amount of $50,000 per year (Fig. 8.69).

The fuzzy logic controller outputs the set values for the coolant valve, the water heater valve, and the humidifier water valve. The fuzzy logic control strategy uses different temperature and humidity sensors to determine how to operate the air conditioning process in a way that conserves energy. Again, the capability of processing interdependent variables results in significant advantages over conventional solutions. For example, one knows that when temperature rises, relative humidity of the air decreases.

Fig. 8.69. An application of fuzzy logic in the A/C system of a large hospital in Germany saved more than 25% on energy costs

This knowledge can be exploited by implementing a fuzzy logic control strategy that allows the temperature controller "to tell" the humidity controller that it is going to activate the heater valve. The humidity controller now can respond to this before it can detect it by its sensor.

Wind Energy Converter Control

In recent years, technological advancements made the commercial use of wind energy feasible. A trend to larger plants further improved the cost/performance ratio. However, such large wind energy converters require advanced control systems both to ensure high efficiency and long life. The controller sets the angle of the rotor blades based on the wind situation (pitch control). However, wind is not a one-dimensional figure. Strength, gustiness, and fluctuation of the wind angle must be evaluated to determine the optimal rotor blade angle (Fig. 8.70).

There is a trade-off between efficiency, safety and wear of the wind energy converter. If the blade angle is set to draw the maximum amount of energy from the wind, the risk of sudden wind gusts causing excessive mechanical stress on the converter increases. For these reasons, an Aerodyn wind energy converter was enhanced with a fuzzy system based on human experience to find the best compromise to this trade-off. The first implemented system is running in a field test and shows quite promising results. The quality of the controller is not only measured in constancy of the delivered power, but also in measures of mechanical stress on the tower, the nacelle and the rotor blades. The next step will be the application of the achieved results to the first 1.2 MW systems that are to be launched in the marketplace in 1996.

Conclusions

In all eight applications, the key to success lies in the clever combination of both conventional automation techniques and fuzzy logic. Fuzzy logic by

Fig. 8.70. To maximize the efficiency of a wind energy converter, the pitch controller must consider many inputs

no means replaces conventional control engineering. Rather, it compliments conventional techniques with a highly efficient methodology to implement multivariable control strategies. Thus, the major potential for fuzzy logic lies in the implementation of supervisory control loops.

8.5.6 Fuzzy Knowledge-Based System for the Control of a Refuse Incineration Plant Refuse Incineration

Introduction

A refuse incineration plant is a complex process, whose multivariable control problems cannot be solved conventionally by deriving an exact mathematical model of the process. Because of the heterogeneous composition of household refuses, observing the combustion chamber by a human operator partly manually controlled most incineration plants. Figure 8.71 shows the diagrammatic layout of a refuse incineration plant.

The incoming refuse is initially stored in a bunker and then transported by a grapple crane into the feed hopper of the incineration plant. The refuse lands on the grate via the down shaft and feeder. The grate consists of two parallel tracks each with five under grate air zones. The optimum position of the fire is in the middle of the third grate zone, since at this point the refuse is adequately predried and thereafter there is still enough residence time in order to ensure complete burnout.

Structure of a Refuse Incineration Plant

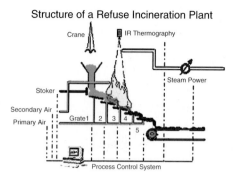

Fig. 8.71. Schematic structure of the refuse incineration plant in Hamburg stapelfeld

Although the crane operator attempts to homogenize the refuse by mixing, it is not possible to maintain a uniform feed quality. The existing automatic systems and control circuits have as their main objective the maintenance of a constant thermal output with good burnout, thereby creating the preconditions for an uniformly high energy production. In order to attain this objective the plant operators must be highly qualified. In spite of such measures, in the majority of incineration plants it is not possible to dispense with manual intervention because of the extremely inhomogeneous composition of household refuse. During the course of manual intervention the operator observes the furnace and adjusts the refuse feed and the grate-operating mode accordingly.

While the quantities of refuse to be incinerated are steadily increasing the environmental laws are becoming ever more restrictive. Recent insights on the toxicity of chlorinated organic pollutant emissions have also dictated a change in the objectives of refuse incineration; together with the ideal of constant combustion output it is becoming ever more urgent to optimize the process in ecological terms, i.e., to ensure the largest possible reduction in volume while minimizing emissions.

During combustion a control system must therefore maintain the following conditions:

1. Control of the O_2 concentration in the flue gas to keep it at a constant figure.
2. Maintanance of a uniform thermal output.
3. Maintanance of optimum flow conditions in the furnace and the first boiler pass with as little variation as possible so that the desired conditions for attaining the lowest possible
4. Degree of emissions and thus preventing corrosion are always maintained.

These conditions can only be fulfilled by optimized combustion at a stable operation point. Conventional control systems are incapable of reacting to

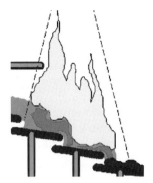

Fig. 8.72. Covered observation area of the IR-camera

the inevitable local and intermittent in homogeneities of the refuse fed to the grate, which are attributable to varying calorific values and ignition properties. It is as a result impossible to avoid pronounced variations in the combustion process and such variations are always associated with unfavorable emission figures.

The most important control variable at the plant is the steaming capacity. The primary air admitted to the various under grate air compartments mainly affects this. Disturbances here occur as a result of the inhomogeneous composition at localized points. The primary air distribution must therefore be continually matched to the requirements of the individual grate zones. Since O_2 content in the flue gas must be maintained at a constant figure the secondary and primary air are controlled in counterbalance to each other.

Between the feeder and the combustion zone there is always a quantity of uncombusted refuse whose amount varies in accordance with the feed quality. As a result of this storage effect there is no direct connection between feeder movement and the position of the fire on the grate. The position can only be registered by visual observation by the plant operators or video picture evaluation.

A possible method of automating this process is the registration of combustion by infrared thermograph. This offers the immediate advantage that the plant operator is in a position to observe combustion directly from the control room. Because of the geometry of the furnace in the plant under consideration the area, which must be observed, lies mainly in grate zone 3, but also to some extent in zones 2 and 4, which cannot be so completely observed (Fig. 8.72).

This is however more than adequate in order to determine the position of the fire. Additionally it is possible, by statistical evaluation of the infrared picture, to determine the width of the combustion zone and derive from this information on asymmetric positions of the fire or secondary combustion zones.

Even if all the information given above is available it is still necessary to development an adequate mathematical model that will allow the information to be applied to a conventional control system. However combustion processes as a rule are of a highly nonlinear nature and represent multivariant problems. For conventional control system strategies the only feasible method is therefore to influence controllers by heuristic programing.

New solutions to problems of that kind are offered by the use of advanced control techniques. In particular fuzzy logic control has been applied to similar combustion processes. Complex interactions between the various items of information evaluated require a methodology of structured information analysis.

Methods for the Development of Control Systems

Conventional technologies of process automation require a lot human resources: beside the setup of hardware and interfaces, operator's control knowledge must be acquired as well as control engineer's experience, and software specialists have to be called for the implementation. This presupposes that all these specialists communicate with each other searching for a way to translate the control strategy into the code of a programing language. For this, human ideas, concepts and causalities often must be expressed on a technical level. Fuzzy logic, uses linguistic variables with membership functions in if–then–rules, is an approved methodology to implement that kind of linguistic knowledge.

State-of-the-art in fuzzy logic control applications is the simple calculus of "fuzzy–if–then–rules" which predominantly use MIN/MAX operators. While fuzzy control algorithm using this simple calculations have successfully been applied in a variety of control problems, these approaches are only a rough approximation of the linguistic meaning they have to represent.

This can be prevented by using fuzzy operators representing linguistic conjunctions like "and" and "or" and by considering rules themselves as "fuzzy." Different advanced inference procedures were proposed by Zadeh known as the "compositional rule of inference," Kosko using so-called Fuzzy Associative Maps (FAMs) and others. These concepts allow a "degree of support" (DoS) (also called "degree of plausibility") to be associated with any rule. Zadehs concept enables a very fine tuning of the rules but involves a large computational effort that forbids its usage in most real-time systems. In order to that, Koskos FAMs only require a small computational effort but only introduce a "weight" factor to each rule.

For both computational efficiency and appropriateness, a combination of these methods was found: the degree to which every rule fires is determined by aggregating the degree to which the premise is fulfilled with its DoS. Of course, this operation can be computed using a fuzzy operator. Applying the product operator for this item, the DoS can be interpreted as a "weight" for every rule. This method is rather simple to use: first, define degrees of support of only either zero or one. Second, for fine-tuning, use values between 0 and 1.

For larger systems handling 50 fuzzy rules and more, rule based systems are no longer lucid and easy to comprehend, if rules are represented in a simple list. Additional structural features enabling for example the classification of rules in rule blocks coming with appropriate rule representations must be used.

The choice of appropriate methods for the knowledge representation is due to the fact that appropriate software tools for the development of control systems are available. Although fuzzy control systems are sometimes programed in a conventional programing language, for complex application, it is necessary to use a tool, which prevents repeated programing fuzzy methods like inference strategies or defuzzification methods.

For the definition of a fuzzy control strategy, a variety of software tools based on the concept of compiling/precompiling, which already exist in the market, can be used. Although this concept works well for the development of conventional software, it has several inherent drawbacks for the construction of fuzzy control systems.

This work investigates which advanced methods render the application of fuzzy control in complex problems possible and how fuzzy controllers may be built more efficiently. The results of this theoretical work have already been implemented in a professional fuzzy logic development system.

First one has to define an initial control strategy as a prototype. Containing the complete structure of the desired system, the prototype is representing all items of the control strategy. For the given process, the prototype was set up with 18 linguistic variables used in 70 fuzzy rules, with nine rule blocks. The control strategy is compiled and linked to the process or its simulation for testing.

Improvement of the Control System

If at first the controller does not work perfectly – which happens often – the control strategy must be revised. For the revision, existing debuggers are clearly inappropriate since they either work on the compiled code or do not work in real-time. Additionally, when a fuzzy control strategy is being designed for a continuous process or its simulation, trying small definition changes and then analyzing the subsequent reaction of the control loop do optimization. To recompile the controller, and thereby interrupting the continuity of the process, a small change is always made, development time is increased considerably.

For these reasons, online-technology has to be used for further optimization. With this tool, a fuzzy logic control strategy can be graphically visualized while the fuzzy system is actually controlling the process in real-time. This enables the engineer to understand the dynamic behavior of both the controller and the process.

In the optimization step, one often wants to try out little modifications to the rule strategy or the membership functions to subsequently increase system performance. If a code generating approach such as a fuzzy-precompiler is used, whenever a change is done, the controller has to be put offline and the controller to be recompiled. In addition to being very inefficient, this approach has got another drawback: a continuous process is always put back to manual control out of its operating point by the recompilation. Hence, the engineer can no longer visualize the effect of the control strategy modification. This makes efficient optimization next to impossible. The incineration plant is a perfect example for such an application.

Once the system is readily optimized, a precompiler or compiler can be used for a code-optimized implementation of the final system on the controller hardware. The development approach for complex fuzzy logic control systems must cover the following steps:

Design: Definition: the system design contains the definitions of linguistic variables, fuzzy operators, the fuzzy rule base and the defuzzification method. Graphical design tools should support this. The step results in a first prototype of the controller.

Offline optimization: to check the controller's static performance, one can either test the controller interactively by applying input values and analyzing the information flow in the system or one can simulate the controller's performance on pre-recorded process data or a mathematical model of the plant, if available.

For all debugging, simulation and analyzing steps, a software implementation of the controller is necessary. Graphical analyzing tools and debug features connected to graphical design tool ease the error detection and optimization. This step ends with a refined prototype.

Online optimization: The refined prototype is now optimized on the running process. To allow for online development, the workstation/PC running the development tool must be connected to the process controller hardware by just a serial cable. This step establishes the final optimized system ready for implementation.

Implementation: The optimized system is code-optimized for the final system. Highly optimizing precompiler and compiler specialized for the applied hardware can be used. Result is a controller software, code size and runtime optimized.

A New Structure for the Control System

The application of advanced fuzzy logic development techniques has led to a new structure for the desired control system. The system is divided into three stages in each of which a short-term and a long-term strategy is operated. The first stage corresponds to a steaming capacity control circuit, subdivided into:

- A short-term control cycle for the steaming capacity and
- A long-term control cycle influencing the O_2 concentration in the flue gas

- The set value for the long-term control system is calculated from a combination of the O_2 content together with the information obtained from the short-term control system on system behavior

The second stage controls the throughput of refuse and here:

- The grate movement is responsible for the short-term control of the position of the fire on the grate and
- The feeder is utilized for the long-term adjustment of the amount of refuse fed to the grate
- Characteristic parameters from the IR thermography are utilized to determine the position of the fire on the grate at any given time.

The third stage, which serves to optimize combustion, additional information from IR-thermography is employed. The optimization consists of the following steps:

- Control of the primary air for the various undergrate air zones.
- Control of the length of the fire by governing the feed velocity in the individual grate zones.

The utilization of fuzzy logic permits the integration of many disparate items of information. Besides directly measurable parameters such as, for example, the steaming capacity, various identifying parameters for the position of the fire and its lenghtare utilized which can be obtained from the IR thermography data.

The result is a hybrid system in which conventional calculation mthods are combined with the methods of fuzzy logic. This permits the uncertainties which ineevitably occur to be taken into account in an appropriate way. In order to model the control strategy from existing experts knowledge the individual components of the structure indicated were implemented using linguistic variables and fuzzy logic rules.

The graphic development tool *fuzzy*TECH was employed to implement this control concept. Figure 8.73 shows a section of the main worksheet in which the firing capacity control system is depicted. The symbols here represent rule blocks and interfaces. Each rule block stands for a set of fuzzy rules and the interfaces represent the data transfer to the upstream or down-stream mathematical operations.

Conclusion

This optimized fire capacity control system fundamentally ensures a more even combustion pattern. As a result it is possible to reduce the CO content in the flue gas and improve the burnout parameters while simultaneously reducing emissions and in particular the chlorinated organic pollutants therein.

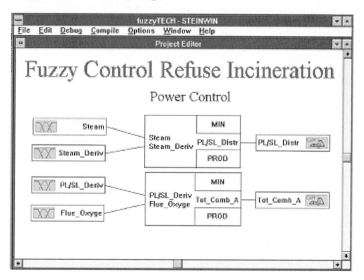

Fig. 8.73. Part of the structure of the fuzzy project: "power control"

8.5.7 Application of Fuzzy Control for Optimal Operation of Complex Chilling Systems

Introduction

The optimization potentials for the operation of chilling systems within the building supervisory control systems are limited to abilities of PLC functions with their binary logic. Little information about thermal behavior of the building and the chilling system is considered by operation of chilling systems with PLC-solutions. The aim of this project introduced in this paper is to replace inefficient PLC-solutions for the operation of chilling system by a fuzzy control system. The focus of the optimization strategy realized by fuzzy control is to ensure an optimal operation of a chilling system. Optimal operation means:

– Reducing operation time and operation costs of the system,
– Reducing cooling energy generation and consumption costs.

Further requirement on the optimization strategy is providing a user net chill water supply temperature with a set point error as little as possible. This feature of the chilling system is important, in order to ensure research and working conditions in the building. Analysis of the online thermal behavior of the building and the chilling system is necessary, in order to find the current efficient cooling potentials and methods during the operation. The thermal analysis also focuses the measurement of important physical values of the system as input variables for different fuzzy controllers, since no expert knowledge exists for optimally operation of the system. This realized fuzzy control system would open new application fields for fuzzy technology within the building automation engineering.

The designed fuzzy controllers are software solutions, in order to use the existing building supervisory control system with its interface units, connected to the chilling system. Three different fuzzy controllers have been developed with a total rule number of just 70. Comparison of the system behavior before and after the implementation of fuzzy control system proved the benefits of the fuzzy logic based operation system realized here.

Description of the Chilling System

The chilling system described here supplies chill water to the air conditioning systems (AC-systems) installed in the Max Plank Institute for Radio astronomy in Bonn. The AC-systems ensure the research conditions by supplying conditioned air to the building. The amount of cooling power for the building is the sum of internal cooling load (produced by occupants, equipment and computers) and the external cooling load, which depends on out door air temperature (T_{out}) and sun radiation through the windows. The cooling machines installed here, uses the compression cooling method. The principle of a compression cooling machine can be described in two thermodynamically processes.

In the first step of the cooling process, the heat energy will be transferred from the system to the heat exchanger (evaporator) of the cooling machine, and therefore the liquid gas will evaporate by absorbing the heating energy. After the compression of the heated gas, in the second part of the process, the gas condenses again by cooling the gas through the air-cooling system. In that step of the process, the heat transfer is from the condensation system to the out door air space. The process is continuous, and based on the second law of the thermodynamics.

Figure 8.74 presents the chilling system as a schematic diagram. The whole system consists of the following components:

- three compression-cooling machines
- three air cooling systems and
- two cooling load storage systems

During the operation of the cooling machines, the air cooling systems will be used, in order to transfer the condensation energy of the cooling machine to the out door air space. If the out door air temperature is much lower than user net return temperature on heat exchanger one, the air cooling system should serve as a free cooling system and replace the cooling machine. The additional cooling load storage systems are installed in order to fulfill the following requirements: firstly to load cooling energy during the night time, and therefore reduce the cost of electrical power consumption (by using cheap night tariff for electrical power), and secondly supplying cooling energy during

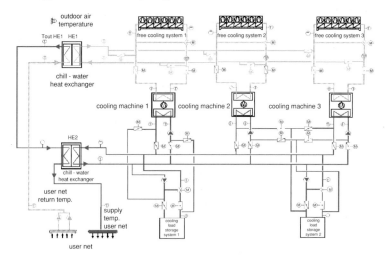

Fig. 8.74. Schematic diagram as a part of the chilling system with simplified instrumentation

the operation time, if a maximum cooling energy is needed and cannot be provided by existing cooling machines. In both cases the cooling storage system does not reduce energy consumption, but the cost of energy production and consumption.

State of the Control Engineering for Operation of Chilling Systems

The heart of a building energy management system is the Building Supervisory Control System, which consists of a hierarchically organised, function orientated control system having separate intelligent automation units. A clearly defined division of functions by hierarchical levels with extensive communication horizontally and vertically across all levels is an essential aspect of perfect operational efficiency. The building supervisory control system with its "distributed intelligence" is configured into four hierarchical information-processing levels, as shown in Fig. 8.75.

Supervisory Level for Implementation of the Fuzzy Control System

The initial function of this Level is to analyze the operating status of the systems. The main function of this level is to Control, monitor and log the processes within the Building as a whole but serves also for configuring of the automation units at the automation level. The supervisory control level has access to all physical data points of the chilling system. The fuzzy control system for optimization strategy realized here is a software solution and is implemented into the supervisory control system. The designed software Fuzzy control has to be translated into a system orientated mathematical, and logical

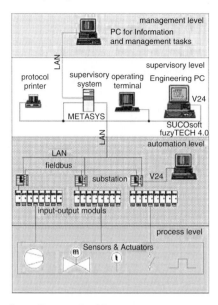

Fig. 8.75. Distributed intelligence building supervisory systems with implemented fuzzy controller

programming language (GPL). All the operation instruction formulated in the supervisory level will be transferred to the chilling system through the automation level as shown in Fig. 8.75.

Automation Level for the Operation of the Chilling System

The automation level houses the distributed intelligence for mathematically and physically based operation functions as multicontrollers. The purpose of the D^3-C (Distributed Direct Digital Control) systems is to monitor and control the most important status and processes within the building. The D^3-C system, which also provides PLC functions, allows a logical link to be set up in the form of time or status elements, in order to guarantee optimum performance. The control strategy for the inner control loops of the chilling system has been realized on this level. Therefore the supervisory level sends the set points and the start/stop instructions as result of fuzzy controllers for each unit of the system to this level.

Fuzzy Control System

The aim of the fuzzy control system which has to be developed, and implemented into the existing building supervisory system as shown in Fig. 8.75 is to run the chilling system in such a way that the following requirements for the operation of the system will be fulfilled: regarding the cooling potential of the out door air, the air cooling system should serve as a cooling power generator as long as possible. The free cooling system (FC system) should run before the cooling load storage system (CLS-system), and cooling machines. This has to be considered by the fuzzy controller for the operation of cooling machines. The CLS-system should run during the daytime before any cooling machine, if the cooling load of the building is expected to be low.

Optimization strategy for the discharge of CLS-system will ensure that there will not be a peak in the electrical power consumption, and reduce the cost of electrical power consumption, by keeping of low price tariffs for electrical power. The cooling machines should run at their lowest possible level. The fuzzy control system must ensure supply of the needed cooling power during the operation time of the building by lowest cost and shortest system operation time with a low range of set point error for the supply temperature.

A concept of knowledge engineering by measuring and analyzing of system behavior is necessary, since no expert knowledge exists for the formulate of the fuzzy rules. Measuring of two physical values of the system is necessary, in order to consider system behavior for an online optimization strategy. These process values are: the out door air temperature T_{out}, which partially presents the thermal behavior of the building, and the user net return temperature (T_{run}), which contains the total cooling load alternation of the building. These requirements focus on three different fuzzy controllers for the different components of the chilling system as shown in Fig. 8.76.

Conclusions

An optimization strategy for the operation of a complex chilling system is realized by Fuzzy control system, and implemented into an existing building automation system. The focus of the optimization strategy by fuzzy control is to ensure an optimal operation of a chilling system. Optimal operation means: reducing operation time and operation costs of the system, reducing cooling energy generation, and consumption costs. Few rules for each controller were necessary, in order to have the fine-tuning of the fuzzy control system. Three fuzzy controllers were necessary in order to reach maximum efficiency by operation of different components of the chilling system. This realized fuzzy control system is able to forecast the maximum cooling power of the building, but also to determine the cooling potential of the out door air. The operation of systems by fuzzy control enormously reduced the cost of cooling power.

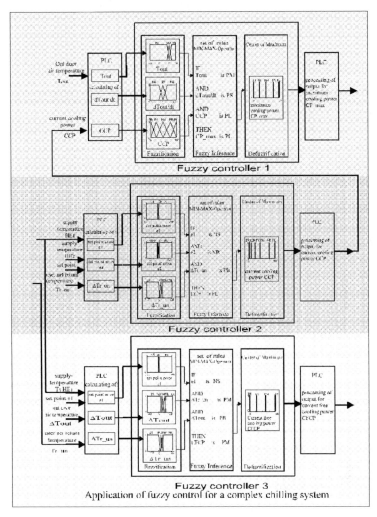

Fig. 8.76. Combination of three fuzzy controllers for the operation of the chilling system

8.5.8 Fuzzy Logic Control of an Industrial Indexing Motion Application

Introduction

The use of closed loop control systems in factory automation continues to increase as manufacturing tolerances tightens and consumers are demanding ever-higher quality products. Closed loop systems, by their very nature, provide tighter control and increased robustness compared to open loop systems. However, most industrial closed loop control algorithms used in today's factories are based on the traditional proportional-integral-derivative, or PID, controller. The PID controller has provided adequate control for years dating back to the 1920s, but does have some weaknesses that make it vulnerable in the face of ever more demanding factory automation applications.

Background

Most PID controllers used in industry are tuned by a trial and error method. Even if a more systematic design based on classical control theory is used, it is still limited by the assumption that the plant is linear, or at least a plant that operates in a linear region, and its dynamics do not change over time. When nonlinear effects arise in the system to be controlled, the PID controller may have difficulty controlling it. This nonlinearity includes friction, saturation, backlash, hysteresis, etc. Ultimately, all physical systems will display nonlinear, time-varying behavior as inputs grow unbounded and as the systems are operated over extended periods of time. The PID controller may have difficulty controlling these types of plants whereas a more robust type of controller will provide better performance. One such type of control is fuzzy logic control.

Fuzzy logic control is an intelligent control technique that uses human expert knowledge of the system to be controlled and incorporates it into a series of control rules. It is, however, more than just a series of if–then rules that operate on variables belonging to sets described by binary logic. These rules operate on variables that will have varying degrees of membership in fuzzy, or multivalued, sets depending upon the operating region of the system. This method of control mimics that of human reasoning. We investigate the increased servo motor controller robustness potential through the use of fuzzy logic control for a particular industrial indexing operation.

Problem Description

The hardware setup of an electric servomotor that feeds paper to a specific length in a manufacturing process is illustrated in Fig. 8.77. Its exact duplicate was built in the laboratory for this investigation. The paper is cut and used as a component in a consumer product. Repeatable and accurate feed length is critical as it affects the quality of the final product. The motor is directly coupled to a knurled drive roller via a flexible coupling. The flexible coupling allows for any slight misalignment between the motor shaft and the drive roller. It provides radial flexibility while minimizing torsional flexibility. A second roller, namely the tension roller, provides the friction that pulls the paper between the two rollers. Tension adjustment is provided via a tensioning screw above the tension roller.

The present method of controlling this feed-to-length servo application is by commercially available controller, motor drive and servomotor. This configuration is illustrated in Fig. 8.78. The controller uses PD position control

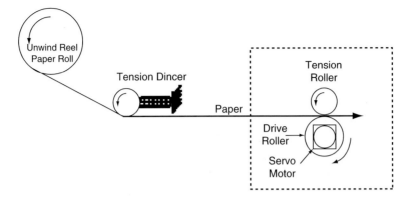

Fig. 8.77. Paper feed system

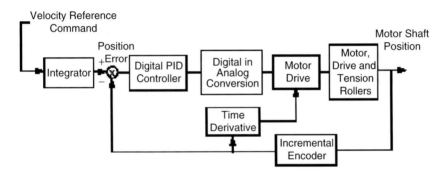

Fig. 8.78. Control system configuration

based upon a quadrature incremental encoder feedback signal. The motor drive also has its own inner velocity control loop that provides additional stability.

The present PD control provides adequate performance and can meet the present manufacturing specifications. The motivation behind this research is to find an improved method of control. Why attempt to improve upon an adequate control solution? There are several reasons.

There is a need to retune the present PD controller whenever a plant parameter changes. This can be as small as a change in paper thickness or it could be a parameter that slowly changes with time such as bearing friction. If a more robust means of control is used, it may require less tuning of the controller in the field. This then provides improved standardization of identical machines in the field because all machines will use similar or identical controller parameters.

This type of mechanical configuration (motor coupled to load via a flexible coupling) is very common. There are plants with this configuration that are very likely poorly controlled by PID control and that could benefit from this even more than the present plant.

Fuzzy Logic Position Loop Design

In designing an FLC for servo motor position control, several key items must be clearly defined. First of all, as in the design of any control system, the performance specifications must be specified. These may include:

– Time response
– Stability and robustness
– Disturbance rejection

Once these are established, the FLC design proceeds with the assignment of:

– A universe of discourse for each input and output variable
– Linguistic labels for the variables and their values
– Membership functions for each variable
– Rule base
– The method for combining membership functions from different universes of discourse
– The implication method
– The aggregation method
– The defuzzification method

Simulation and tuning then follow to complete the design.

Performance Specifications

The performance requirements of the paper feed system are:

1. Feed length: $0.680'' + 0.010''$
2. Feed time: 98 ms
3. Overshoot: $<0.007''$

These machine performance requirements translate into closed loop servo control requirements:

1. Feed length: 866 servomotor encoder counts ± 12 counts
2. Settling time: 98 ms
3. Resolution: 0.000785 in per count
4. Percent overshoot: <8 encoder counts
5. Disturbance rejection: in this system, frictional torque can be considered a low frequency disturbance

Obviously, it is desirable to attenuate any low frequency disturbance but some friction may actually help dampen this system.

High Frequency attenuation: The actual plant is known to have a resonant dip at $700 \, \text{rad s}^{-1}$ and a resonant peak at $1,000 \, \text{rad s}^{-1}$. These frequencies and higher must be attenuated to ensure the magnitude does not approach and climb above the 0 dB line.

Structure of the Fuzzy Logic Position Loop

Many servo motion control applications do not fully utilize all three parameters of the PID controller. Instead, only the proportional and derivative gains are used. Proportional gain adjustments vary the bandwidth to meet the settling time specification. Derivative gain adjustments vary the system response to meet overshoot requirements. Increasing derivative gain will reduce the control signal magnitude when the error rate is high, thereby reducing or eliminating overshoot. Integral gain is normally only used to reduce steady state error caused by friction, gravity, etc. However, the misapplication of integral gain can cause stability problems.

Likewise, when designing an FLC, we begin by looking at the error signal and the rate of change of the error. With these two pieces of information, we can determine the state of the system and hopefully control it. Therefore, the FLC designed here will be a two input one output controller; the inputs being the error and error rate and the output being the control signal to the plant.

The FLC in this section is positioned as shown in Fig. 8.79, directly where a PD controller would be positioned in the control loop. This configuration is the most straightforward and allows a direct comparison between the FLC and PD controllers.

Note in Fig. 8.79 that the second input, y, is the discrete time approximation of the rate of change of the inverted output signal. y is used instead of the time derivative of the error because the error can change instantaneously with a step input. This instantaneous change produces an impulse in the error rate, which adversely affects the FLC performance. The output, however, cannot change instantaneously and is therefore used as one of the FLC inputs. However, with a continuously differentiable error signal, the error rate should be used instead.

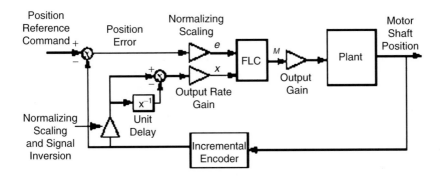

Fig. 8.79. FLC control configuration

Table 8.8. Membership parameters

$e, \Delta y, u$ – Triangular membership functions

$$\mu_A(e) = \begin{cases} \frac{e-a_1}{a_2-a_1} & \text{if } a_1 \le e \le a_2 \text{ and } a_1 \ne a_2 \\ \frac{a_1-e}{a_3-a_2} & \text{if } a_2 \le e \le a_3 \text{ and } a_3 \ne a_2 \\ 1 & \text{if } a_1 = a_2 = a_3 \\ 0 & \text{Otherwise} \end{cases}$$

Linguistic variable	e			Δy			u		
	a_1	a_2	a_3	a_1	a_2	a_3	a_1	a_2	a_3
NX	–	–	–	$-\infty$	-1.10	-1.00	–	–	–
NL	$-\infty$	-1.00	-0.40	-1.10	-1.00	-0.70	-1.30	-1.00	-0.70
NM	-0.85	-0.33	-0.15	-0.80	-0.60	-0.25	-0.90	-0.60	-0.30
NS	-0.40	-0.12	-0.15	-0.30	-0.17	0.00	-375	-175	0.00
ZE	-0.07	0.00	0.07	-0.10	0.00	0.10	-0.225	0.00	0.225
PS	0.015	0.12	0.40	0.00	0.17	0.30	0.00	0.175	0.375
PM	0.15	0.33	0.85	0.25	0.60	0.80	0.30	0.60	0.90
PL	0.40	1.00	∞	0.70	1.00	1.10	0.70	1.00	1.30
PX	–	–	–	1.00	1.10	∞	–	–	–

Setting Up the Fuzzy Logic Controller

There are three essential ingredients in a FLC: fuzzification, rule base, and defuzzification. Fuzzification is accomplished through the definition of the fuzzy membership functions, shown in Table 8.8. Here, the numerical values of e and y are translated to symbolic ones, such as positive medium (PM) or negative large (NL) etc. with an associated grade. This grade is defined by using the triangular membership functions. These triangular membership functions for $e, y,$ and u are defined as:

$e = (a_1\ a_2\ a_3) \in \mathbb{R}_3$, with $a_1 \le a_2 \le a_3$. The value of a_2 is the center of e where the linguistic label has its maximum membership grade of one, that is $\mu_e(a_2) = 1, a_3$, and a_1 are the other two defining membership function points that lie to the left and right of a_2. These membership functions are fully described by the equation at the top of Table 8.8. The membership functions for y and u are defined in the same manner. The final parameters for each of the membership functions are based on the fine-tuning of the controller.

Once the input signals are converted to symbolic variable through fuzzification, they are ready to be processed by using an English-like rule base, consisting of if–then statements. Generating the rules for a fuzzy inference system is often the most difficult step in the design process. It usually requires some expert knowledge of the plant dynamics. This knowledge could be in the form of an intuitive understanding gained from an operator who has experience manually controlling it. Or, it could come from a plant model, which is then used in a computer simulation. This latter method was used in this study. The rule base for the FLC proposed in this study is listed in Table 8.9. These rules follow from an initial rule base that was modified in the tuning process

Table 8.9. Rule base

		Output rate								
		NX	NL	NM	NS	ZE	PS	PM	PL	PX
Error	NL					NL				
	NM		NL		NM	NL	NM	NS	ZE	
	NS		NM			NL		NS	PS	
	ZE	NL	NS			ZE*		PS		PL
	PS		NS	PS	PL		PM			
	PM		ZE	PS	PM	PL	PM	PL		
	PL					PL				

*Weighted 0.1 of other rules

in order to achieve the desired control and robustness. Multiple simulations were run to arrive at these final rules. The total number of rules in this FLC is 39 out of a possible 63.

The final step in FLC is to compute the numerical value of the controller output based the outcomes rule inference. Using the common centroid method, also known as the center of gravity (CoG) method, does this.

FLC Tuning

Tuning an FLC may seem at first to be a daunting task. There are many parameters that can be adjusted. These include the rules, membership functions and any other gains within the control system. The approach used here is to reduce the fine-tuning effort to the adjustment of two gains; the output gain and the *output rate* gain.

As shown in Fig. 8.79, the output gain adjusts the control signal, $u(t)$, and the output rate gain adjusts the magnitude of the output rate signal, $y(k)$. One method of tuning is, the error signal, e, will always be automatically normalized based on the present reference signal magnitude and no further adjustment of this gain is allowed nor needed. The output gain is in some ways analogous to the proportional gain in a PD controller and the output rate gain is analogous to the derivative gain. This analogy breaks down when the FLC is grossly mistuned. Adjustment of the output gain changes the system bandwidth and adjustment of the output rate gain will change the overshoot characteristics.

The overall FLC tuning process is as follows:

- Gross adjustment of the system is done by iteratively adjusting rules, membership functions, the output gain and the output rate gain. Rule optimization is accomplished through computer simulation on a trial and error basis. If there is inadequate knowledge represented in the rule base to properly control the plant, additional membership functions and rules must be added to the FLC.

– Once gross tuning is accomplished, the FLC is finely tuned. This involves slight adjustments of individual membership functions and the output and output rate gains.

To better understand the tuning process requires an understanding of how membership function parameters affect the system response. The reason this is important is because the adjustment of the output gain and the output rate gain are actually an adjustment of the control signal's membership functions and the output rate's membership functions.

Let us begin with the output gain. Increasing the output gain effectively stretches the control signal membership functions in proportion to the output gain. As the membership functions widen and move away from the origin, it makes the meaning of their associated linguistics quantify larger numbers. Imagine the control signal membership functions moving away from the origin and widening. The effect will be that for a constant value generated in the antecedent, the consequent becomes larger. The opposite is true when the output gain is decreased. Keep in mind that increasing or decreasing the output gain is not actually stretching or contracting the membership functions. The change in gain merely has the same effect as if the gain was left unchanged and the membership functions were changed.

Moving or changing the width of individual control signal membership functions will have a similar effect. The membership function whose position or width has changed relative to its adjacent membership functions will have a greater or lesser effect on the control signal. A simple example is the ZE control signal membership function used in this study. As this membership function is widened relative to its adjacent membership functions, it will move the COG closer to zero during defuzzification. This then has the effect of reducing the control signal gain when both the error and output rate are small. Narrowing the ZE membership function will move the COG further away from zero and therefore increase the gain when both the error and error rate are small.

Changing the output rate gain has the effect of stretching or contracting the output rate membership functions. This is similar to the effect the output gain has on the control signal membership functions. However, there is one major distinction. The output rate membership functions are scaled inversely proportional to the output rate gain. As the gain is increased, the membership functions effectively narrow and move closer to the origin hence making the meaning of their associated linguistics quantify greater numbers during input fuzzification. In this study, increasing the output rate gain will slow and dampen the system response. The opposite is true when the output rate gain is decreased.

Remarks

There are some essential points to keep in mind when designing an FLC:

- Thoroughly study the physical problem at hand before beginning FLC design.
- Initially, keep it simple. Start with the minimum number of inputs, outputs and membership functions (perhaps five or less). This will hopefully keep the maximum number of rules manageable.
- Initially, use triangular (or similarly simple) membership functions and space them uniformly over the universe of discourse. Always normalize the universe of discourse between minus one and one or zero and one.
- Initially, add rules to the rule base such that they monotonically increase or decrease across rows and down columns (i.e., avoid rows and columns such as NL, NM, NS, NM, ZE).
- Initially, scale the input and output signals to determine their effect on the overall response. Perhaps this will suffice in designing and tuning the FLC. If the response is poor and does not meet specifications, then modify the rules. Modify membership functions only for fine-tuning when output gain and output rate gain adjustment is inadequate.
- Keep in mind that it will be relatively straightforward to design an FLC for a plant for which the designer understands the nominal conditions. The challenge and difficulty is designing a robust FLC that is relatively insensitive to changing plant parameters encountered in the factory environment. In this case, membership function positions may need to be dramatically changed. Also, the rules may need modification such that they do not necessarily monotonically increase or decrease across rows and down columns.
- Designing an FLC, as with most controller designs, is an iterative process that may require many cycles before specifications are met. Fewer cycles should be needed on subsequent designs as the designer gains experience.

Conclusions

An FLC has been developed to control the position of a servomotor in an industrial indexing application. The FLC's performance can be compared to a PD position controller. This can be done by software and hardware prototype demonstrations that disturbances and parameter variations have less effect on the performance of the FLC than the PD controller. These parameters include step size, load inertia, friction and input disturbances. There is, however, a tradeoff between the robust performance of FLC and the complexity of the controller and control design. The decision on whether to adopt an FLC method will have to be made based on the improvement/cost ratio for each application.

8.6 Fuzzy Logic in Automotive Applications

8.6.1 Fuzzy Antilock Brake System

Introduction

In recent years fuzzy logic control techniques have been applied to a wide range of systems. Many electronic control systems in the automotive industry such as automatic transmissions, engine control and Antilock Brake Systems (ABS) are currently being used. These electronically controlled automotive systems realize superior characteristics through the use of fuzzy logic based control rather than traditional control algorithms.

ABS is implemented in automobiles to ensure optimal vehicle control and minimal stopping distances during hard or emergency braking. The number of cars equipped with ABS has been increasing continuously in the last few years. ABS is now accepted as an essential contribution to vehicle safety. The methods of control utilized by ABS are responsible for system performance.

Intel Corporation is the leading supplier of microcontrollers for ABS and enjoys a technology agreement with Inform Software Corporation the leading supplier of fuzzy logic tools and systems. The combination of Intel ABS architecture and fuzzy logic is a result of long-term investment and exploration of new technologies and ideas. The increasing automotive customer awareness of ABS has greatly increased the demand for this technology. Improving ABS capability is a mutual goal of automotive manufacturers and Intel Corporation. The growing interest in the automotive community to implement fuzzy logic control in automotive systems has produced several major automotive product introductions.

Fuzzy Logic Overview

Formal control logic is based in the teachings of Aristotle, where an element either is or is not a member of a particular set. Since many of the objects encountered in the real world do not fall into precisely defined membership criteria, some experimentation was inevitable. Zadeh was one of those who investigated alternative forms of data classification. The result of this investigation was the introduction of fuzzy sets and fuzzy theory at the University of California Berkeley in 1965. Fuzzy logic, a more generalized data set, allows for a "class" with continuous membership gradations. This form of classification with degrees of membership offers a much wider scope of applicability, especially in control applications.

Although fuzzy logic is rigorously structured in mathematics, one advantage is the ability to describe systems linguistically through rule statements. One such control rule statement for an air conditioning unit might be:

"If temperature is Hot and Time of Day is Noon then air conditioning equals very high."

Several rules, similar to the example, could be used to describe a system and controlled response. The parameters of Hot, Time and Very High are defined by membership functions. As linguistic descriptions of a system are much easier to produce than complex mathematical models, fuzzy logic has great appeal for controlling complex systems as changes in the system have little if any effect upon the algorithm.

Fuzzy ABS would require more complex control constructs than simple "if–then" rules. In this type of control system, input variables map directly to output variables. This simple mapping does not provide enough flexibility to encode a complex system such as an ABS system. However, more complex techniques are available which can be applied to fuzzy logic systems. For example, it is possible to build a control with intermediate fuzzy variables, or systems, which have memory. With these constructs, it is possible to build rules such as...

"If the rear wheels are turning slowly and a short time ago the vehicle speed was high, then reduce rear brake pressure."

Such rules lend themselves to development of an ABS braking system based on fuzzy logic. The output of a fuzzy logic system is determined in one of several ways. The COG technique will be discussed in this paragraph. Once all rules are evaluated, their outputs are combined in order to provide a single value that will be defuzzified. This output calculation is performed as follows. The control rule output value is multiplied by its position along the X-axis, yielding position times weight for the rule. This calculation is repeated for all control rules. These position/weight products are combined to form the sum of products. This sum of the products is divided by the sum of output values to determine the COG output along the X-axis. COG is the final system output in a control algorithm.

Fuzzy ABS

ABS systems were introduced to the commercial vehicle market in the early 1970s to improve vehicle braking irrespective of road and weather conditions. However, due to the technical difficulties and high cost of early systems, ABS was not recognized by automakers as an advantage until the mid-1980s. The ABS market has rapidly grown and is forecast to be $5 billion yearly by 1995 and $10 billion or more by the year 2000. Experts predict that 35–50% of all cars built worldwide in 5 years will have ABS as standard equipment.

Electronic control units (ECUs), wheel speed sensors, and brake modulators are major components of an ABS module. Wheel speed sensors transmit pulses to the ECU with a frequency proportional to wheel speed. The ECU then processes this information and regulates the brake accordingly. The ECU and control algorithm are partially responsible for how well the ABS system performs.

Since ABS systems are nonlinear and dynamic in nature they are a prime candidate for fuzzy logic control. For most driving surfaces, as vehicle braking force is applied to the wheel system, the longitudinal relationship of friction between vehicle and driving surface rapidly increases. Wheel slip under these conditions is largely considered to be the difference between vehicle velocity and a reduction of wheel velocity during the application of braking force. Brakes work because friction acts against slip. The more slip given enough friction, the more braking force is brought to bear on the vehicles momentum. Unfortunately, slip can and will work against itself during cornering or on wet or icy surfaces where the coefficient of surface friction varies. If braking force continues to be applied beyond the driving surface' useful coefficient of friction, the brake effectively begins to operate in a nonfriction environment. Increasing brake force in a decreasing frictional environment often results in full wheel lockup. It has been both mathematically and empirically proven a sliding wheel produces less friction a moving wheel.

Inputs to the Intel Fuzzy ABS are derived from wheel speed. Acceleration and slip for each wheel may be calculated by combining the signals from each wheel. These signals are then processed in the Intel Fuzzy ABS system to achieve the desired control. Unlike earlier 8-bit microcontroller architectures with limited math capability, the Intel Fuzzy ABS example utilizes a high performance, low cost, 16-bit 8XC196Kx architecture to take advantage of improved math execution timing.

Model **BUILDER**

Unlike a conventional ABS system, performance of the Intel Fuzzy ABS system can be optimized with less detailed knowledge of the internal system dynamics. This is due to the process used to refine the rule base and in the initial development of the system using Inform Software Corporation *fuzzy*TECH(R) 3.0 MCU-96 software tuned for the Intel Architecture with optimized code output and the associated Real Time Cross Debugger. The software tool set combined with a linguistic approach to control implemented in the Intel Fuzzy ABS solution allows for rapid development. A cornerstone of this rapid development is the Intel fuzzy logic modeling software kit called *fuzzy*BUILDER.

The development system, called *fuzzy*TECH(R) MCU-96, is specifically optimized for the MCS(R) 96 architecture. It contains:

- A fully graphical CASE tool that supports all design steps for fuzzy system engineering.
- A simulation and optimization tool for fuzzy systems. This tool displays system performance and can be interfaced to conventional simulators to obtain performance data.
- A code generator which generates complete C-Code for the fuzzy system. The C-Code calls optimized assembly routines on the target controller for fast performance.

Table 8.10. Performance of a 20 MHz 8XC196Kx device

7 Rules	20 Rules	20 FAM rules	80 FAM rules
2 in/1 out	2 in/1 out	2 in/1 out	3 in/1 out
0.22 ms	0.33 ms	0.34 ms	0.50 ms

The following Table 8.10 shows the performance of several test systems on a 20 MHz 8XC196Kx device. All times shown are worst-case execution results. Note FAM rules are individually weighted as opposed to a system in which all rules have identical weight:

Conventional ABS control algorithms must account for nonlinearity in brake torque due to temperature variation and dynamics of brake fluid viscosity. Also, external disturbances such as changes in frictional coefficient and road surface must be accounted for, not to mention the influences of tire wear and system components aging. These influential factors increase system complexity, in turn effecting mathematical models used to describe systems. As the model becomes increasingly complex equations required to control ABS also become increasingly complicated. Due to the highly dynamic nature of ABS many assumptions and initial conditions are used to make control achievable. Once control is achieved the system is implemented in-vehicle and tested. The system is then modified to attain the desired control status. However, due to the nature of fuzzy logic, influential dynamic factors are accounted for in a rule based description of ABS. This type of "intelligent" control allows for faster development of system code.

ABS Block diagram:

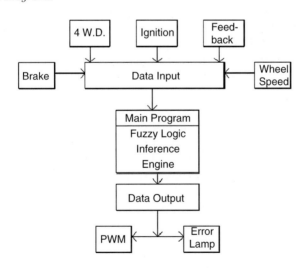

Inputs:

The Inputs to the Intel Fuzzy ABS are represented in the diagram above and consist of:

1. The brake: this block represents the brake pedal deflection/assertion. This information is acquired in a digital or analog format.
2. The 4 W.D: this indicates if the vehicle is in the 4-wheel-drive mode.
3. The ignition: this input registers if the ignition key is in place, and if the engine is running or not.
4. Feed-back: this block represents the set of inputs concerning the state of the ABS system.
5. Wheel speed: in a typical application this will represent a set of four input signals that convey the information concerning the speed of each wheel. This information is used to derive all necessary information for the control algorithm

*Fuzzy*TECH(R) 3.0

The proposed system shown above has two types of outputs. The PWM signals to control ABS braking, and an Error lamp signal to indicate a malfunction if one exists.

Intel Fuzzy ABS Features

In the Intel Fuzzy ABS an embedded 87C196JT microcontroller (a member of the 8XC196Kx family) is used in conjunction with Inform Software Corporation *fuzzy*TECH(R) software. Rules constitute the base of the algorithm and are evaluated in sequence, one after the other.

In contrast, if a custom dedicated fuzzy parallel processor were to be used, rules could be evaluated in parallel. The parallel processing method suggests a fast processing cycle. However, in this case data acquisition and data output continues using conventional peripherals. The time gained in parallel rule processing can be lost in acquiring and manipulating data via external peripherals. The best solution continues to use a software fuzzy algorithm on a microcontroller with fast internal peripherals. In this case, sequential rule processing is transparent to the system and the process appears to have been done in parallel. The MCS(R) 96 family of microcontrollers is equipped with high performance internal peripherals that make data acquisition and data conditioning of outputs fast and easy to handle. This, and the wide range of addressing modes, broad availability of interrupts and a powerful set of instructions make Intel microcontrollers immanently suitable for fuzzy logic applications.

For an ABS implementation, the MCS(R) 96 family is also a perfect match. The High Speed Input Output unit can be used to effectively handle I/O without impacting precious on-chip timer resources. Most microcontrollers in the Intel 16-bit family have also incorporated on-chip Analog-to-Digital converters with 1,024 discrete codes (10-bit resolution). The use of on-chip A/D reduces chip count. The A/D can be used to sense *braking action* taken by the driver. In addition, there is a large set of both direct and indirect interrupts to deal with real-time events and exceptions. The priority scheme of the interrupts can be modified dynamically in software.

For outputs the on-chip pulse width modulator (PWM) unit is available for use in providing variable output signals to the individual wheels. Changing the frequency and/or the duty cycle of the PWM can be done simply with a very fast register write operation.

In addition to the peripherals, microcontrollers in the Intel 16-bit MCS(R) 96 family have internal RAM and ROM. Program instructions and data can be stored on-chip for optimized execution. No long external bus cycles are required to read data due to the large register based architecture. This feature is extremely beneficial to fuzzy logic. The knowledge base, i.e., the rules and the membership functions can be stored on-chip. Thus, rules can be evaluated in a very short amount of time.

Conclusion

The use of fuzzy-logic in conjunction with microcontrollers is a fairly new development in automotive applications. Fuzzy logic and or neural networks are used to control automotive applications like ABS, automatic braking for collision avoidance, adaptive cruise control and chassis control.

8.6.2 Antilock-Braking System and Vehicle Speed Estimation Using Fuzzy Logic

Introduction

Vehicle dynamics and braking systems are complex and behave strongly nonlinear which causes difficulties in developing a classical controller for ABS. Fuzzy logic, however facilitates such system designs and improves tuning abilities. The underlying control philosophy takes into consideration wheel acceleration as well as wheel slip in order to recognize blocking tendencies. The knowledge of the actual vehicle velocity is necessary to calculate wheel slips. This is done by means of a fuzzy estimator, which weighs the inputs of a longitudinal acceleration sensor and four-wheel speed sensors. If lockup tendency is detected, magnetic valves are switched to reduce brake pressure. Performance evaluation is based both on computer simulations and an experimental car.

Fuzzy control, a relatively new, intelligent, knowledge based control technique performs exceptionally well in nonlinear, complex and even in not mathematically describable systems. Thus the use of fuzzy logic for an ABS seems to be promising.

Antilock-Braking Systems

The aim of an ABS is to minimize brake distance while steerability is retained even under hard braking. To understand the underlying physical effect, which

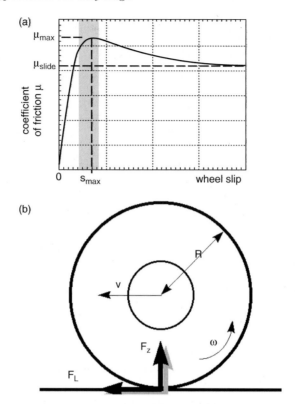

Fig. 8.80. (a) Friction characteristics; (b) Wheel model
F_z: Wheel load; R: Wheel radius; ω: Angular wheel frequency; v: Velocity of wheel center; F_L: Longitudinal force

leads to wheel blocking during braking, consider Fig. 8.80a: coefficient of friction is shown as a function of wheel slip, relating to the terms given in Fig. 8.80b.

Calculating the wheel slip by

$$s = \frac{v - \omega R}{v} 100\%,$$

the longitudinal wheel force results in $F_\perp(s) = \mu(s)F_z$.

At the beginning of an uncontrolled full braking, the operating point starts at $s = 0$, then rises steeply and reaches a peak at $s = s_{max}$. After that, the wheel locks within a few milliseconds because of the declining friction coefficient characteristic which acts as a positive feedback. At this moment the wheel force remains constant at the low level of sliding friction. Steering is not possible any more.

Therefore a fast and accurate control system is required to keep wheel slips within the shaded area shown in Fig. 8.80a.

Vehicle Speed

A crucial point in the development of wheel slip control systems is the determination of the vehicle speed. There are several methods possible: until now the velocity is measured with inductive sensors for the wheel rotational speed. Especially in the case of brake slips the measured speed does not correspond with reality. To obtain very accurate results, optical or microwave sensors take advantage of a correlation method. However, these sensors are very expensive and will not be used for ABS.

Sensors and Actuators

The experimental car was fitted with sensors and actuators shown in Fig. 8.81. Each wheel is connected to a metallic gearwheel, which induces a current within an attached sensor. The frequency of the rectangular shaped current is proportional to the angular frequency $w_{i,j}$ and can be evaluated by a microcontroller. In addition to common ABS fitted cars, a capacitive acceleration sensor for measuring the longitudinal acceleration a_x is implemented.

Furthermore Fig. 8.81 depicts the hydraulic unit including main brake cylinder, hydraulic lines and wheel brake cylinders. By means of two magnetic two-way valves each wheel, braking pressure $p_{i,j}$ is modulated. Three discrete conditions are possible: decrease pressure, hold pressure firm, and increase pressure (up to main brake pressure level only). Each valve is hydraulically connected to the main brake cylinder, to the wheel brake cylinders and to the recirculation.

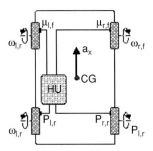

Fig. 8.81. Sensors and actuators of the experimental car
CG: Center of gravitiy; a_x: Longitudinal acceleration; $w_{i,j}$: Angular wheel frequency; HU: Hydraulic Unit; $p_{i,j}$: Wheel brake pressure; i: l=left, r=right; j: f=front, r=rear

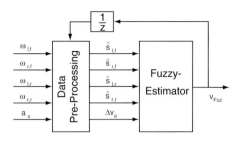

Fig. 8.82. Estimation of car velocity

Estimation of Vehicle Speed Using Fuzzy Logic

There exists an estimation system based on Kalman-Filter, which performs well, but is not suitable because of very high performance requirements. In this approach the speed estimation uses multisensor data fusion that means several sensors measure vehicle speed independently and the estimator decides which sensor is most reliable. Figure 8.82 represents the schematic structure of the fuzzy estimator. The signals of the four wheel speed sensors $\omega_{i,j}$ are used as well as the signal of the acceleration sensor a_x.

In a data preprocessing block the measured signals are filtered by a lowpass and the inputs for the fuzzy estimator are calculated: four wheels slip $\omega_{i,j}$, and an acceleration value Δv_a. The applied formulas are:

$$\exists \omega_{i,j}(k) = \frac{v_{\text{Fuz}}(k-1) - \omega_{i,j}(k-1)R}{v_{\text{Fuz}}(k-1)}100\%$$

and

$$\Delta v_a(k) = \frac{(a_x(k) - a_{\text{Offset}}(k))T}{v_{\text{Fuz}}(k-1)}100\%$$

whereby a_{Offset} is a correction value consisting of an offset and a road slope part. Comparing the measured acceleration with the derivative of the vehicle speed v_{Fuz}, which is calculated with the fuzzy logic system, derives it. After this subtraction, the signal is low pass filtered to obtain the constant component a_{Offset} $v_{\text{Fuz}}(k-1)$ is the estimated velocity of the previous cycle. A time-delay of T is expressed by the term $1/z$.

The fuzzy estimator itself is divided into two parts. The first (Logic 1) determines which wheel sensor is most reliable, and the second (Logic 2) decides about the reliability of the integral of the acceleration sensor, shown in Fig. 8.83.

This cascade structure is chosen to reduce the number of rules.

Starting at block "Logic 1" and "Logic 2" the crisp inputs are fuzzificated. Figure 8.84 shows the input-membership-functions (IMF) with four linguistic values (*Negative, Zero, Positive* and *Very_Positive*).

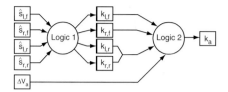

Fig. 8.83. Stucture of the fuzzy estimator

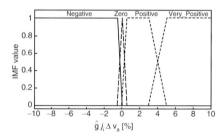

Fig. 8.84. Input membership functions

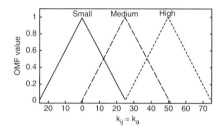

Fig. 8.85. Output membership functions

The rule base consists of 35 rules altogether. To classify the present driving condition vehicle acceleration is taken into consideration. This should be explained for three situations:

1. Δv_a Positive: braking situation, all wheels are weighted low because of wheel slips appearing.
2. Δv_a Zero: if wheel speeds tend to constant driving the acceleration signal is low weighted in order to adjust the sensor.
3. Δv_a Negative: the experimental car was rearwheel driven therefore rear wheels are less weighted than front wheels.

Figure 8.85 depicts the output-membership-functions (OMF). Here, three linguistic values are sufficient. The output of the estimation is derived as a weighted sum of the wheel measurement plus the integrated and corrected acceleration:

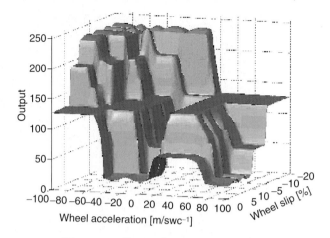

Fig. 8.86. Fuzzy characteristic surface

$$v_{\text{Fuz}}(k) = \frac{\sum_{i=1}^{4} k_i\, \omega_i(k)R + k_{\text{a}}\left[v_{\text{Fuz}}(k-1) - Ta_{x,\text{Corrected}}(k)\right]}{\sum_{i=1}^{4} k_i + k_{\text{a}}}$$

The Fuzzy-ABS Algorithm

The Fuzzy-Controller uses two input values: the wheel slip S_{B}:

$$S_{\text{B}} = \frac{v_{\text{Fuz}} - \omega R}{v_{\text{Fuz}}} = \frac{v_{\text{Fuz}} - v_{\text{Wheel}}}{v_{\text{Fuz}}} \quad \text{and the wheel acceleration:}$$

$$a_{\text{Wheel}} = \frac{\partial v_{\text{Wheel}}}{\partial t} \approx \frac{\Delta v_{\text{Wheel}}}{\Delta t},$$

with wheel speed v_{Wheel} and vehicle speed v_{Fuz}, which is given by the Fuzzy-Estimator.

The input variables are transformed into fuzzy variables *slip* and $\mathrm{d}v_{\text{wheel}}/\mathrm{d}t$ by the fuzzification process. Both variables use seven linguistic values, the slip variable is described by the terms
slip = {*zero, very small, too small, smaller than optimum, optimum, too large, very large*},
and the acceleration $\mathrm{d}v_{\text{wheel}}/\mathrm{d}t$ by
$\mathrm{d}v_{\text{wheel}}/\mathrm{d}t$ = {*negative large, negative medium, negative small, negative few, zero, positive small, positive large*}.

As a result of two fuzzy variables, each of them having seven labels, 49 different conditions are possible. The rule base is complete that means, all 49 rules are formulated and all 49 conditions are allowed. These rules create a nonlinear characteristic surface as shown in Fig. 8.86.

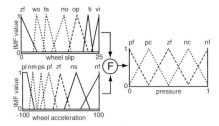

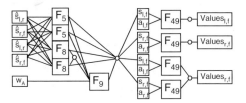

Fig. 8.87. Structure of the fuzzy ABS controller

Fig. 8.88. Fuzzy calculations

Using this characteristic surface, the two fuzzy input values *slip* and dv_{wheel}/dt can be mapped to the fuzzy output value *pressure*. The labels for this value are:

pressure = {positive fast, positive slow, zero, negative slow, negative fast}

The structure of the fuzzy ABS controller is shown in Fig. 8.87.

The optimal breaking pressure results from the defuzzification of the linguistic variable *pressure*. Finally a three-step controller determines the position of the magnetic valves, whether the pressure should be increased, hold firm or decreased.

Figure 8.88 summarizes the total amount of fuzzy calculations. Numbers within a rectangle indicate the quantity of fuzzy rules.

Simulation of a Full Braking

After implementation of the whole system in SIMULINK, a full braking on high-m-road was carried out, with and without the fuzzy ABS. Without fuzzy ABS the braking pressure reaches a very high level and the wheels block within short. This results in an unstable behavior, the vehicle cannot be steered any more and the stopping distance increases.

With fuzzy ABS controller activated, steerability is not only retained during the whole braking maneuver, but the slowing down length was considerably shortened as well. The following graphs show the steady decline of the vehicle speed, the fluctuating decline of the wheel speed of the left front wheel as an example and the fluctuating level of the wheel slip. The applied braking pressure is depicted in the last diagram. The other wheels behave approximately similar (Fig. 8.89).

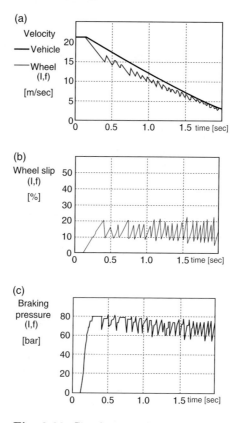

Fig. 8.89. Simulations of a full braking

Implemenation of the Fuzzy ABS Controller

The fuzzy ABS controller uses the microprocessor SAB 80C166 together with the fuzzy coprocessor SAE 81C99A. Due to the implementation of Fuzzy algorithms into the hardware of the coprocessor, the calculation speed of the host processor increased significantly. While the control cycle time was set to a standard value of 7 ms, the computation time was only 0.5 ms! This offers facilities for implementation of extended vehicle dynamics control. The flexibility of the coprocessor is considerable, up to 64 rule bases are possible, each of them having up to 256 inputs and rules. Furthermore an interface to most commonly used microprocessors is available. Arbitrary shapes of membership functions, different defuzzification modes including "Center of Gravity," an enormous rule engine with up to 10 million rule calculations per second makes this device a very interesting product in the field of real time fuzzy control.

Conclusion

The basis of the controlling algorithm consists of a nonlinear characteristic surface, which was created by fuzzy logic. The convincing advantage of fuzzy logic is the ability to modify and tune certain parts of this characteristic surface easily and carefully. Just the linguistic rules or variables need to be varied. This simplifies the development and shortens the development time considerable. Implementation of the fuzzy ABS leads to excellent results of braking behavior of the test vehicle. The deceleration level and steerability is comparable to commercially available systems.

8.7 Application of Fuzzy Expert System

8.7.1 Applications of Hybrid Fuzzy Expert Systems in Computer Networks Design

Introduction

The task of designing and configuring large computer networks most suited to a certain application and environment is difficult, as it requires highly specialized technical skills and knowledge, as well as a deep understanding of a dynamic commercial market. Current expert systems have made solid achievements in supporting decision makers; they use prior experience to solve problems in different domains. Hybrid fuzzy expert systems have appeared all over the world proving that integrated fuzzy expert systems/neural networks methods replaces classical hard decision methods and providing better performance.

The current most significant trend in the computing world is the growth of distributed processing, a technique that puts computing power closer to users rather than in large, central mainframes. Computer communication networks are the key for such distributed systems as they ease the share of information between cities, building complexes, buildings, departments, and networked nodes. Computer communication networks are generally classified into three broad categories each differentiated primarily by the distances they span:

- Local area networks (LANs) are short-distance networks (usually with a range of less than 1 mile) typically used within a building or building complex for high-speed data transfer between computers, terminals, and shared peripheral devices,
- Metropolitan-area networks (MANs) are medium-distance, high-speed networks with range of from 1 mile to 50 miles. MANs often transmits voice and video in addition to data, and
- Wide area networks (WANs) are primarily long-distance networks used for the efficient transfer of voice, data, or video between local, metropolitan, campus, and premise networks. WANs typically use lower-transfer

rates, and common-carrier services or private networking via satellite and microwave facilities.

In this age of internet working the ability to effectively communicate is the key to the business success. Efficient and optimal network design is necessary to make communication networks usable and affordable. By network design, we mean the selection of various network devices and connections to accomplish an organization's operational objectives. A network's configuration can greatly affect its performance and cost. It is, therefore, vital that the best combination of equipment, connections, and placement of network connections for end-user nodes be made to satisfy an organization's objectives.

These objectives may include a multitude of factors other than the prices of the computers and the networks, such as the reliability, the response, the availability, and the serviceability.

Professional designers with intensive knowledge and experience are needed for large computer communication network design, modeling and simulation. Such designers must be well informed about the most recent updates in this rapidly advancing field to be able to handle the available state of-the-art technologies. Since designers of such caliber are difficult to find and usually very expensive, we proposed the use of hybrid fuzzy expert systems to play their role and/or assist them in their task.

In this section, we focus on presenting the design, the knowledge representation, and the operation of a network design hybrid fuzzy expert system (FES).

Expert Systems in Network Design

ELAND, an Expert Design of Local Area Networks, has been the first activity in applying expert systems in network design. The ELAND's problem decomposition approach to the computer network design problem was suitable for solving the problem 5–8 years ago.

In COMNED, we proposed a modern approach for using expert systems for network design, keeping in mind (1) the system openness and modularity in-order to allow the system future updatability and functionality and (2) the usage of the available powerful network simulation tools.

COMNED has been fully implemented in the Telecommunication and Networking Laboratories, of the University of Miami, the expert system recommends the network feasible solutions most suited to the user's application and environment. A network simulation package receives the configuration of the network solutions from the expert system to be modeled, simulated and evaluated, after which the values of the performance indices are reported back to the expert system. One of the most significant properties of the computer communication market is its rapid advancement and change in a very short period of time.

Using classical expert systems and machine learning approaches to learn new emerging technologies, we have to learn the values of all the design variables for this technology and create all the facts/rules required to make the technology available as a design option the next time the system is used.

Consider learning the value of a design variable like the "**noise_resistance**" for a newly emerging cabling system, then the knowledge engineer should enter the degree of truth for the low, medium, high, and very_high values of this design variable for this cabling system. The problem with this approach is that when using the experience of different design experts or knowledge engineers their estimates for the degrees of certainty could highly differ, then with this approach there will be multiple basis for the estimation process.

A second problem emerged with this approach. Consider, for example, that a 1.0 degree of truth was given to the ATM topology support for 500 nodes, and now the system was required to learn a new topology which has better capabilities in supporting the 500 nodes, then the degree of truth of the ATM topology support for this number of nodes should change to a lower value for the 1.0 to be given to the newly learned topology. On what basis this change should happen? If we depend on the knowledge engineer estimate then we are magnifying the problem of multiple estimation basis. If we consider a fixed change like 0.1 decrease (the degree of truth of the ATM topology support for 500 nodes will change to 0.9), we then find that many cases are far from realistic.

Also during the development of COMNED, a significant potential was found in integrating fuzzy sets, fuzzy logic, and neural networks into the knowledge representation and the reasoning process of the expert system. In particular we realized that:

1. The nature of knowledge representation and its suitableness to undergo the well-founded FES theory. Sixty to 70% of the used expert system facts/rules were found to be of fuzzy nature.
2. Hybrid fuzzy reasoning, was found to give better reasoning performance.

In addition, neural networks could be integrated with fuzzy expert systems to tune the shapes of fuzzy membership functions of the different design variables, this will improve the reasoning and confidence performance of the expert system and make the FES approach more justifiable.

For all the above reasons, we found ourselves motivated to present a hybrid FES approach for network design, keeping all the objectives of COMNED, and using fuzzy logic/neural networks to improve the system reasoning, manage the confidence calculation and estimation, and solve the above mentioned machine learning problem.

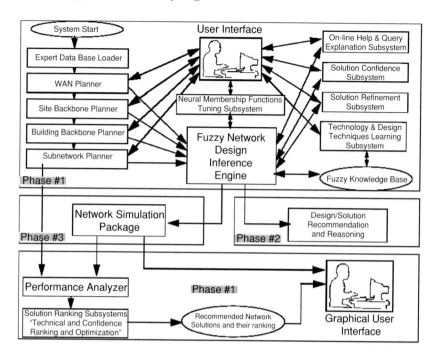

Fig. 8.90. END architecture and operation

Network Design Hybrid Fuzzy Expert System: Architecture and Operation

The hybrid FES proposed in this paper is a part of a global network design system called "END," An Expert Network Designer. The internal structure of the hybrid FES is shown in Phase #1, Phase #2, and Phase #4 areas of Fig. 8.90. END is divided into four distinct Phases – the configuration entry phase, solution recommendation phase, model simulation phase, and the performance analysis phase.

Configuration Entry Phase

In this phase, the hybrid FES interacts with the user through a user interface to obtain a general description of the networking project. The description is obtained through a group of planners which issue a number of hierarchical questions going from the highest possible network level, which is the number of network sites, the WAN interconnectivity between the different sites, passing by the number of buildings in each site, the number of floors in each building. etc., and ending with the number of workstations and servers in the departmental LANs. The system questions are designed to be as simple as possible for any person, not necessarily a network specialist, who is aware of the network general functionality and layout. The system uses a fuzzy knowledge base consisting of a database of information necessary for providing expert advice, a set of assertions relating these pieces of information, domain-specific information about the network configuration, traffic characteristics, and measures for determining the suitability of various models in its model library, as well as information about modeling, analysis, and simulation in general.

The answers of the configuration questions and the other design guidelines are used to find the network feasible solutions through a fuzzy network design inference engine. This inference engine uses fuzzy rule-based heuristics on an if–then formalism. In the case of finding multiple solutions for a certain user configuration, the system interacts with the user, in a solution refinement session, with a new set of questions, depending on the different solutions obtained, to be able to filter the solutions to the most suitable solutions for the user environment.

The user answers are used to revise the design guidelines initially assumed by the system. The Technology and Design Techniques Learning is a neural network/knowledge acquisition learning subsystem used to improve the time-efficiency of END's network design problem solver and allow the hybrid FES to learn the new emerging network technologies, modern network design techniques, and the updated specifications of the existing technologies. Finally, the Neural Membership Functions Tuning subsystem is a neural network to tune the membership functions of the fuzzy network design variables.

Solutions Recommendation Phase

In this phase, the hybrid FES reports the best feasible topologies and cabling systems most suited to the user's application and environment (entered in the last phase) for each subnet, backbone and WAN, in addition to the confidence rank in each solution. At the end of the user interaction session, a graphical layout of each global feasible network solution is given to the user on a GUI. Optional full solution reasoning is available if the operator is interested in knowing why these solutions were chosen.

Model Simulation Phase

If the operator chooses to run the optional network simulation ranking, END will generate a separate model for each feasible solution under a network simulation package with the aid of a communications – oriented simulation language, run the simulations, and report the simulation results to the expert system part.

Solutions Analysis Phase

In this phase a Performance Analyzer receives the simulation results from previous phase, in conjunction with the global network solutions from the Solution Recommendation Phase, to start classifying the different solutions with respect to their significance in each measured performance parameter.

The hybrid FES is designed to have two solution ranking subsystems:

1. A *Technical Ranker* which ranks the solutions according to the measured performance parameters.
2. A *Confidence Ranker* which simply ranks the solutions according to the user confidences in their satisfaction to the solution properties.

All solutions with confidence less than a preassigned value are initially eliminated as feasible solutions.

All the other solutions are ranked first according to the Technical Ranker, and second according to the Confidence Ranker.

Network Design Problem

The network design problem solving structure is a tree traversal. Figure 8.91 presents the decomposition of the network design problem, which introduces in each step a sequence of subproblems that must be solved. The global physical network problem "root node" is decomposed into subproblems for each network site and the WAN connecting these sites. In the design tree lower levels each network site problem is decomposed into subproblems for each building in the site and the site's backbone. Then each building problem is further decomposed into subproblems for each subnetwork in the building and the building's backbone. Finally, all the backbone and the subnetwork problems are decomposed into topology and cabling system subproblems.

The global network design problem is finally decomposed at the tree leaves into a number of topology and cabling system subproblems. For solving these two problems, COMNED uses a single general purpose subnetwork design inference engine which is invoked every time the system reaches the design

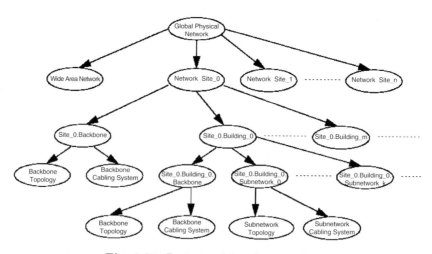

Fig. 8.91. Design problem decomposition

tree leaves to obtain the suitable topology and cabling system for the currently active problem. This problem decomposition represents the formalization of the practical approach used by design experts for computer networks design.

Operation and Knowledge Representation

The expert system uses fuzzy fact-like and constraint-like structures to represent the specification of the different network topologies and cabling systems. In addition, it uses conceptual application-oriented design rules to represent the logic of network design.

There are three main fuzzy inferencing rules, the first for finding feasible cabling systems, the second for finding feasible topologies, and the third for finding WAN feasible solutions. A prolog-like format of the main inference rules and design decision trees are shown in Fig. 8.92. The subnetwork general specifications (obtained from the user and represented by the expert planners as design guidelines) are passed as arguments to these three inference rules which therefore will obtain, using the fuzzy predefined knowledge, the cabling systems "Cable_media," the network "Topology," and WAN types "WAN_Type" satisfying the passed subnetwork specifications. It is clear from the cabling system design decision tree and the inferencing rule shown in Fig. 8.92a, that the chosen cabling system depends on three design variables, the level of noise resistance "noise_resistance," the budget required to connect one node using this cabling system "cable_with_budget_per_station," and the distance supported by the cabling system "cable_with_distance." The values of such variables for the chosen cabling system (Cable_noise_resistance, Cable_budget, and Cable_distance) should be, compared to the general specifications obtained from the user (Noise, Budget, and Distance), of equal or higher noise resistance "Noise=<Cable_noise_resistance," of less budget requirements "Budget>=Cable_budget" and able to support wider network span "Distance=<Cable_distance."

Such design variables are chosen in our system as fuzzy variables with different membership functions. For example, the "noise_resistance" is a fuzzy variable with four membership functions low, medium, high, and very_high.

$$\text{low(noise\_resistance)} = F_1(\text{cabling\_sytem})\text{medium(noise\_resistance)}$$
$$= F_1(\text{cabling\_system})$$
$$\text{high(noise\_reistance)} = F_3(\text{cabling\_system})\text{very\_high(noise\_reistance)}$$
$$= F_4(\text{cabling\_system})$$

The four membership functions F_1, F_2, F_3, and F_4 were chosen to be gaussian functions, as shown in Fig. 8.93. The values of such functions represent the degree of truth of the satisfaction of these membership functions of the "noise_resistance" fuzzy design variable. The membership functions are chosen such that the functions will have a contour similar to that estimated by a single knowledge engineer or a design expert.

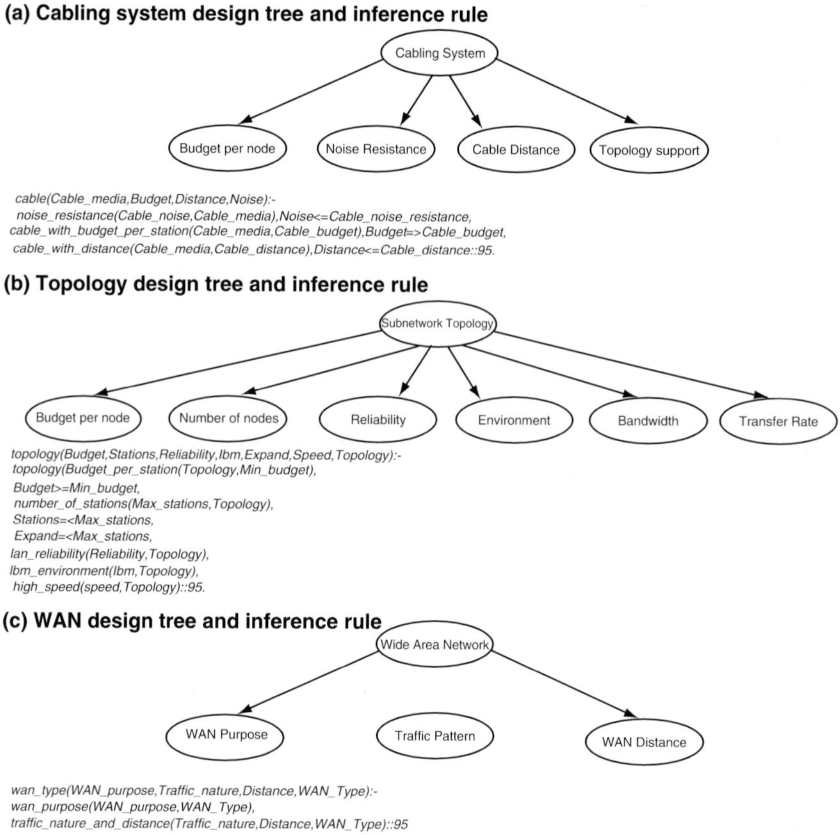

(a) Cabling system design tree and inference rule

```
cable(Cable_media,Budget,Distance,Noise):-
   noise_resistance(Cable_noise,Cable_media),Noise<=Cable_noise_resistance,
   cable_with_budget_per_station(Cable_media,Cable_budget),Budget=>Cable_budget,
   cable_with_distance(Cable_media,Cable_distance),Distance<=Cable_distance::95.
```

(b) Topology design tree and inference rule

```
topology(Budget,Stations,Reliability,Ibm,Expand,Speed,Topology):-
topology(Budget_per_station(Topology,Min_budget),
Budget>=Min_budget,
number_of_stations(Max_stations,Topology),
Stations=<Max_stations,
Expand=<Max_stations,
lan_reliability(Reliability,Topology),
Ibm_environment(Ibm,Topology),
high_speed(speed,Topology)::95.
```

(c) WAN design tree and inference rule

```
wan_type(WAN_purpose,Traffic_nature,Distance,WAN_Type):-
wan_purpose(WAN_purpose,WAN_Type),
traffic_nature_and_distance(Traffic_nature,Distance,WAN_Type)::95
```

Fig. 8.92. Inference engine design trees and rules

Consider that the user entry for a subnetwork media noisiness (part of any subnetwork general specification obtained from the system user during the *Configuration Entry Phase*) was level high. From Fig. 8.93, a cabling system like the unshielded_twisted_pair will have a degree of truth equal to 0.09 while another cabling system like the thick_coaxial will have a degree of truth equal to 0.82. Similar membership functions are used for the other two fuzzy design variables "cable_with_budget_per_station" and "cable_with_distance." Obtaining the degree of truth for each of the three fuzzy design variables (Cable_noise_resistance, Cable_budget, and Cable_distance) say d1,d2, and d3, for a specific cabling system, we can realize that the degree of truth of choosing this particular cabling system "Cable_media" (which is the same like the degree of linguistic certainty of the rule used in its choice) is:

Conf = min(min(d1, d2, d3), 95),

where 95 is the linguistic certainty of the premise of the rule "cable (Cable_media, Budget, Distance, Rule)."

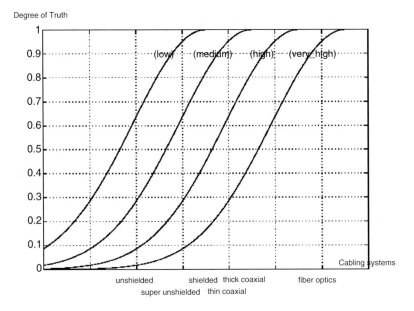

Fig. 8.93. Membership functions of the noise_resistance fuzzy variable

The above method is used in the MILORD computation method for calculating the degree of linguistic certainty of fuzzy rule.

All the above applies for all the design variables in the "topology" and "wan_type" design inference rules shown in Fig. 8.92b,c. For example, the membership functions of the fuzzy design variable "ibm_environment," which shows the extent of suitableness of a specific networking topology for an IBM mainframe environment (used as a fuzzy design variable in the last raw of the topology design inference rule), has two membership functions yes and no.
yes(ibm_environment)=G_1(topology)
no(ibm_environment)=G_2(topology)

The two membership functions G_1 and G_2 were chosen as gaussian and constant functions, respectively. The two functions are shown in Fig. 8.93. The values of such functions represents the degree of truth of the satisfaction of these membership functions of the "ibm_environment" fuzzy design variable.

Expert System Shell and Solution Reasoning

The hybrid FES uses an optimized FES shell based on MILORD system shell which provides an inference engine to supervise the execution of the system rules. It uses standard backward chaining with uncertain reasoning capabilities based on fuzzy logic to satisfy the top-level goal. The engine allows the application of degrees of certainty by means of expert-defined linguistic statements.

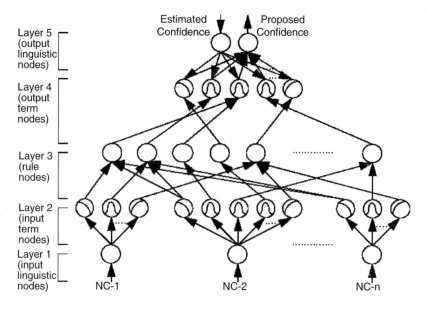

Fig. 8.94. Connectionist neural model for fuzzy membership function optimization

It also provides the features of accumulating proofs for query explanation and solution reasoning.

The system stores proofs and explanations for the system predefined knowledge, the user collected knowledge, and the system queries. The explanations for the system predefined knowledge and the user collected knowledge are used by the design/solution reasoning subsystem to explain the system decisions. The system query explanations are used by the Online Help and Query Explanation subsystem to explain the purpose and the role of these queries to the system operators whenever requested.

System Output

The hybrid fuzzy rule-based expert system reports the best feasible topologies and cabling systems for each subnet, backbone, and WAN during its operation, in addition to the confidence rank in each solution. At the end of the user interaction session, a graphical layout of each global feasible network solution is given to the user on the GUI. Finally, if the operator chooses to run the optional network simulation ranking, the system will generate a separate model for each feasible solution under the network simulation package, run the simulations, and report the best solution to the operator. In addition an optional full solution reasoning is available if the operator is interested in knowing why these solutions were chosen.

Sensitivity to Changes in the Membership Functions

Using different shapes, centers and values for the membership function, it was found that there is a general increase in the number of feasible solution with the increase in the values of the membership functions of the different fuzzy design variables and vice versa. The change in the number of solutions is not directly proportional to the change in the membership functions but it depends on many other factors, like the user's network configuration, the nature of the feasible solutions and the operating point in the formalized network design space. Minimum inferencing was found to be less sensitive to the changes of the membership functions. Product and Average inferencing were found to be more sensitive to such changes.

Membership Functions Optimization in the Network Design Fuzzy Expert System

In the previous sections a FES was successfully applied in computer communication network design. The only problem with this approach was the choice of the membership functions of the different network fuzzy design variables and their shapes.

A neural network connectionist model could be used to solve the problem of choosing the optimal shape of the fuzzy membership functions in the network design FES. The neural network connectionist model has just one fuzzy output which is the network feasible solutions for the input network configurations. Figure 8.94 shows the proposed neural-network connectionist model:

Layer 1: the nodes in this layer transmit network configuration input values (NC-1 to NC-n) to the next layer directly.

Layer 2: the output function of this node is the membership function of the different fuzzy design variables.

Layer 3: the links in this layer are used to perform the precondition matching of the fuzzy logic rules of the network design FES.

Layer 4: the nodes in this layer have two operation modes: down–up transmission and up–down transmission modes. In the down–up transmission mode, the links at layer four should perform the OR operation to integrate the fired rules which have the same consequence.

Layer 5: there are two kinds of nodes in this layer. The first kind of node performs the up–down transmission for the estimated confidence in the obtained feasible solutions (training data from a design expert based on the input network configuration and the output feasible solutions) to be fed into the network. The second kind of node performs the down–up transmission for the proposed confidence in the same obtained feasible solutions.

Again a two-stage hybrid learning algorithm is used. This learning algorithm will determine optimal centers and widths of the membership functions of the fuzzy design variables used in Layers 2 and 4.

In phase one of the hybrid learning algorithm, a self-organized learning scheme is used to locate initial membership functions. In phase two, a supervised learning scheme is used to optimally adjust the membership functions for desired outputs. To initiate the learning scheme, training data (the estimated confidence in each network feasible solution) and the desired or guessed coarse of fuzzy partition (i.e., the general membership function shapes of the different fuzzy network design variables) must be provided from the outside world. Before this network is trained, an initial form of the network is first constructed. Then, during the learning process, some nodes and links of this initial network are deleted or combined to form the final structure of the network.

With the connectionist neural network model for fuzzy membership function optimization, the category of the network design FES will move from Loosely Coupled hybrid FES model to Fully Integrated model.

Conclusion

Fuzzy expert systems have proved to be very successful in formalizing the practical rules used by the design experts for computer networks design, formalizing the logic of solving computer network design problems, and initially choosing the most suitable solutions for a certain networking requirement. By using fuzzy expert systems to generate the network models/simulations the user is not required to have any kind of background of the simulation package operation. Neither is the user required to bean expert in networks design. The user is only required to answer a group of general questions about the network requirement: he/she will get a network design, a description of the design, why it was chosen, a graphical diagram of the design, a design simulation, vector simulation results (curves), and even results analysis by the expert system. Automating the process of network modeling and simulation generation is very important as it can save the user time and expense.

Fuzzy logic addresses several problems with current expert systems like, providing better knowledge representation, better reasoning performance, and better management of confidence factors. Fuzzy logic was found to aid the easiness of the learning process, in this paper, it was clear how fuzzy logic solved the problem of multiple estimation basis by the usage of predefined membership functions and how the problem of saturating the membership functions was eliminated by the usage of the membership functions axis-shifting. Proposing the full integration of neural networks to tune the shapes of fuzzy membership functions will improve the performance of the FES and make the FES approach more justifiable.

8.7.2 Fuzzy Expert System for Drying Process Control

Introduction

The problems during the synthesis of the automatic control systems for diversified processes in agriculture and industry cannot be sufficiently solved using classical control methods. The difficulties occur when a number of mutually dependent variables, having different values under diversified conditions, are to be controlled. In many of the controlled processes, the system operator knows the way to change the control variables in the case of altered conditions on the basis of the experience. In order to realize the automation of a certain process, it is necessary to implement the information based on experience into the control system. This is very difficult to achieve in the classical control systems.

Mentioned problems can be solved through integration of the expert system and the fuzzy control. The expert systems are based on the computer control of the processes. Every expert system includes existence of a knowledge basis, which is in fact implementation of the experience, obtained by the system investigation, into the process control. The form, the structure, and the components of the knowledge basis, depend on the controlled process itself. Using the set of measured data, the demand for doing an action in the system as well as the action character, can be established (qualitatively and quantitatively) according to the knowledge basis.

The knowledge basis handling is very often based on the fuzzy logic. In most of the cases the expert system is in fact a fuzzy system used for inclusion of the working regimes into the control system. The fuzzy logic is very successfully applied in the process control itself, i.e., in the control of the measured values. The advantages of the fuzzy logic are stated when the control of a number of the mutually dependent values, influenced by the same controlling variables, is concerned. The fuzzy logic can achieve that error signals of all the controlled values are taken into account at the same moment. Based on the fuzzy system control rules, which are obtained by the investigation of the controlled process in different working conditions, the outputs that can bring the system very fast into the desired condition are generated.

Expert System Development

An optimum regime in the knowledge basis should be defined as one enabling the highest proportion of the obtained final product and used fuel, for the given values of input parameters, including the adequate output material moisture content. During the operation, the control system selects, in this way, the nominal values obtained by the fuzzy controller of the fuel flow-rate and the belt-conveyer speed. It is important to notice that the belt-conveyer speed and the fuel flow-rate are now changing simultaneously, i.e., it is not necessary to wait for the variation of the output temperature in order to change the fuel flow rate. The structure of the described conception of the control system is shown in Fig. 8.95.

The synthesis of this fuzzy control system can be divided into the five steps:

1. Creation of the knowledge basis

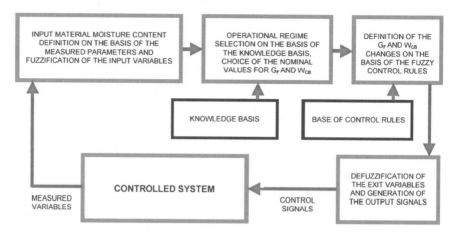

Fig. 8.95. Structure of the fuzzy control system for direct rotary dryer

2. Definition of the input material moisture content and presentation of the input variables using fuzzy variables
3. Selection of the operational regimes
4. Selection of the control rules and synthesis of the fuzzy controller
5. Choice of the fuzzy controller parameters

The knowledge basis should ensure the selection of the nominal values for the fuel flow-rate and the belt-conveyer speed on the basis of the measured input material moisture content, height of the material layer, and nominal temperature at the dryer output (between 80°C and 90°C).

In order to implement into the expert system and to obtain the data from the knowledge basis, it is necessary to present the measured input material moisture content in the form of the triangular fuzzy numbers:

$$A = (a_1, a_2, a_3) \qquad (8.1)$$

where a_1, lower range of the fuzzy number; a_2, nominal value of the fuzzy number corresponding to the higher degree of membership; a_3, upper range of the fuzzy number.

Graphical interpretation of the input moisture content using fuzzy numbers is shown in Fig. 8.96.

Every measured value of the input material moisture content will always have exactly two corresponding fuzzy numbers having the membership functions different from zero. Two nominal values of the fuel flow-rate and the belt-conveyer speed, along with the given nominal output temperature and height of the material layer, are chosen for these two fuzzy numbers out of the knowledge basis. Based on the membership function for these two fuzzy numbers, and on these two nominal values for the fuel flow rate, the fuel flow-rate nominal value is calculated. Situation is a little bit different if the

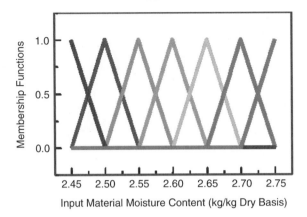

Fig. 8.96. Interpretation of the input material moisture content using fuzzy numbers

belt-conveyer speed is considered. The belt-conveyer speed is changing in precisely determined shift quanta. If the two different values for the belt-conveyer speed are chosen from the knowledge basis as the nominal values, one of these values must be adopted as the nominal value for the belt-conveyer speed. The difference between these two values is one shift quantum.

Fuzzy Controller Synthesis

For the process control itself it is necessary to present the output temperature and the fuel flow-rate deviations from theirs nominal values in the form of fuzzy variables. In the case of the belt-conveyer speed, situation is again different (the belt-conveyer speed is changed in the shift quanta), so the deviation of the belt-conveyer speed from the nominal value can always be presented by a certain number of the shift quanta. Concerning the process control, the belt conveyer-speed can be changed for the exactly determined number of quanta. The belt-conveyer speed preservation (influenced by the input material moisture content) at the nominal value is a part of the process control with applied expert system. On the other hand, maintenance of the fuel flow-rate, and the output temperature, at the nominal values is carried out using fuzzy controller.

Deviations of the output temperature and the fuel flow-rate from the nominal values can be expressed as:

$$\Delta T_o = T_o - T_{on} \tag{8.2}$$

$$\Delta G_f = G_f - G_{fn} \tag{8.3}$$

It is necessary to express the deviations T_o and G_f in the form of fuzzy numbers. T_o is expressed by the linguistic fuzzy variables having the following

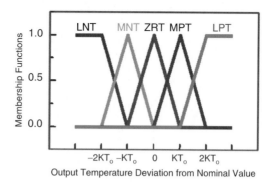

Fig. 8.97. Interpretation of the output temperature deviation by the linguistic fuzzy variables

meaning: LNT – large negative deviation of the temperature, MNT – mean negative deviation of the temperature, ZRT – zero deviation of the temperature, MPT – mean positive deviation of the temperature, LPT – large positive deviation of the temperature.

The membership functions of these fuzzy variables are shown in Fig. 8.97.

It can be noticed from Fig. 8.97 that membership functions of the fuzzy variables, describing the deviation of the output temperature, depend on a single parameter, marked with KTo. This parameter remains changeable in the program for the system control. The variations of this parameter make possible that the user is able to adjust the performances of the fuzzy controller according to his demands. The parameter KTo has immense influence at the fuzzy controller behavior. The proper work of whole the control system depends on the proper selection of this parameter. Deviation of the fuel flow-rate from the nominal value is represented by the fuzzy variables in a similar way. The membership functions of the linguistic variables describing deviation of the fuel flow-rate are shown in Fig. 8.98 and have the following meanings: LNG – large negative deviation of the fuel flow-rate, MNG – mean negative deviation of the fuel flow-rate, ZRG – zero deviation of the fuel flow-rate, MPG – mean positive deviation of the fuel flow rate, LPG – large positive deviation of the fuel flow rate.

In the case of presenting the fuel flow-rate deviation from the nominal value by the fuzzy variables, one parameter remains also unfixed (KGf). In order to avoid the possibility of the steady-state error in the case of the measurement error, the uneven distribution of the points having maximum degree of membership for the linguistic variables is adopted.

Defuzzification

Generation of the signal controlling the fuel-flow valve is obtained using the defuzzification method. The output (controlling) signal, which is brought to

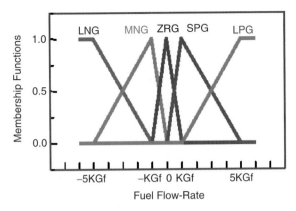

Fig. 8.98. Interpretation of the deviation of the fuel flow rate by the fuzzy variables

Table 8.11. Fuzzy control rules

$\Delta G_f \backslash \Delta T_o$	LNT	MNT	ZRT	MPT	LPT
LNG	1: LPO	2: LPO	3:ZRO	4: ZRO	5: MNO
MNG	6: LPO	7: MPO	8: ZRO	9: MNO	10: MNO
ZRG	11: LPO	12: MPO	13: ZRO	14: MNO	15: LNO
MPG	16: MPO	17: MPO	18: ZRO	19: MNO	20: LNO
LPG	21: MPO	22: ZRO	23: ZRO	24: LNO	25: LNO

the valve, is proportional to the necessary change of the fuel flow-rate. The corresponding fuzzy variables are as follows: LNO – large negative output, MNO – mean negative output, ZRO – zero output, MPO – mean positive output, VPO – large positive output.

The nominal value for the zero output is 0, while the nominal values for LNO, MNO, MPO, and VPO are chosen by the user, so that the performances of the fuzzy controller can be adjusted by changing these parameters. It is adopted that MPO $=$ $-$MNO $=$ Kos and LPO $=$ $-$LNO $=$ Kob. Kos and Kob are two more parameters that influence the performance of the fuzzy controller. The degree of membership of the output fuzzy variables is generated on the basis of the fuzzy control rules. Fuzzy control of this one system is based on 25 control rules, which can be represented as in Table 8.11. The rules are shown in such a way that rows present the deviation of the fuel flow-rate, and columns present the deviation of the output temperature. Each table field corresponds to the single control rule, i.e., to the fuzzy variable, which describes the system output for the corresponding input fuzzy variables. The rules can be numbered in order to enable easier analysis. This is also done in Table 8.11.

Every decision rule represents the one fuzzy relation between the temperature deviation, the fuel flow rate deviation, and the system output. The decision rules are given in the form of the logical implications:

If $\Delta T_{\mathrm{o}} = $ (Linguistic variable LVT)

And $\Delta G_{\mathrm{f}} = $ (Linguistic variable LVG)

then

Output = (Linguistic variable LVO)

All the control rules can be considered as the fuzzy phrases in the form of the fuzzy implications. If one of the control rules is marked with N as fuzzy phrase, according to the Min–Max-Gravity Method its membership function would be:

$$\mu N(\Delta T_{\mathrm{o}}, \Delta G_{\mathrm{f}}, \text{output}) = \text{Min}\left\{\mu LVT(\Delta T_{\mathrm{o}}), \mu LVG(\Delta G_{\mathrm{f}}), \mu LVO(\text{output})\right\}$$
(8.4)

LVT, LVG, and LVO are the linguistic variables describing input and output. The total value of the fuzzy output membership function is given by the expression:

$$\mu \text{OUTPUT}(\Delta T_{\mathrm{o}}, \Delta G_{\mathrm{f}}, \text{output}) = \text{Max}\left\{\mu N1(\Delta T_{\mathrm{o}}, \Delta G_{\mathrm{f}}, \text{output}), \ldots, \mu Nn \right.$$
$$\left. \times (\Delta T_{\mathrm{o}}, \Delta G_{\mathrm{f}}, \text{output})\right\}$$
(8.5)

Therefore, the fuzzy output membership function is determined by the maximum degree of membership of a fuzzy phrase, from the set of the control rules. This means in practice, that the degree of membership of the certain output linguistic variable LVO is equal to the maximum degree of membership among all the fuzzy implications, which implicate the control rule LVO. In this manner, the degrees of membership of all the output linguistic variables taking the values LNO, MNO, ZRO, MPO, and LPO, are defined. The output itself is calculated as the center of gravity from the following expression:

$$\text{output} = \frac{\sum_i \text{LVO}_i \mu \text{OUTPUT}(\Delta T_{\mathrm{o}}, \Delta G_{\mathrm{f}}, \text{LVO}_i)}{\sum_i \mu \text{OUTPUT}(\Delta T_{\mathrm{o}}, \Delta G_{\mathrm{f}}, \text{LVO}_i)}$$
(8.6)

Conclusions

The explained process control is especially convenient to use in the systems where more dependent variables have to be controlled. It is also applicable in the systems where it is desirable to change the operational regimes and where the nominal values of the controlled variables are changeable during the system operation on the basis of the input parameters. The concept of the knowledge basis is in operation with the fuzzy variables. In other words, the knowledge basis is developed as the fuzzy system and used for the operational regime definition. The model of fuzzification and defuzzification was worked out.

The linguistic variables of the model have different distance between the points having maximum degree of membership. In this manner, by using relatively low number of the linguistic variables, rough and fast control is obtained for the large disturbances, while slow and precise control is obtained in the case of small disturbances. Compared with classical control, the result is in the highly improved response to the disturbances of the system.

The fuzzy logic can find an adequate application in the drying process control because of the almost unpredictable way of the input material modification. In order to achieve the high quality of the final product, rational energy consumption and the increment of productivity, it is necessary to vary working regimes, i.e., to control a large number of mutually dependent variables. It is also necessary to implement the working regime selection experience into the control system. Considering the above facts, it is reasonable to expect that expert system with the fuzzy control can give very good results in the drying process automation.

8.7.3 A Fuzzy Expert System for Product Life Cycle Management

Introduction

The real-world decision-making is too much complex, uncertain and imprecise to lend itself to precise, prescriptive analysis. It is this realization that underlies the rapidly growing shift from conventional techniques of decision analysis to technologies based on fuzzy logic. It was originally proposed as a means for representing uncertainty and formalizing qualitative concepts that have no precise boundaries. So far, engineering applications of fuzzy logic have gained much more attention than business and finance applications, but an even larger potential exists in the latter fields.

Fuzzy logic is an excellent means to combine Artificial Intelligence methods. The advantage of fuzziness dealing with imprecision fit ideally into decision systems; the vagueness and uncertainty of human expressions is well modeled in the fuzzy sets, and a pseudo-verbal representation, similar to an expert's formulation, can be achieved. Fuzzy logic avoids the abrupt change from one discrete output state to another when the input is changed only marginally. This is achieved by a quantization of variables into membership functions.

Expert systems were designed to reason through knowledge to solve problems using methods that humans use. A FES is an expert system that utilizes fuzzy sets and fuzzy logic to overcome some of the problems, which occur when the data provided by the user are vague or incomplete.

In this section, we illustrate that the fuzzy approach may be useful in industrial economics. In particular a FES is adapted for product life cycle management. All products have certain life cycles. The well-known *product life cycle* approach describes the changing features of markets during their evolution. It may therefore serve as the theoretical framework within which the market changes can be explained. The life cycle refers to the period from product's first launch into market until its final withdrawal and it is split up in phases.

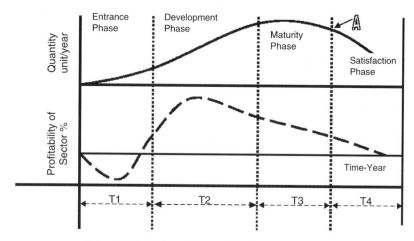

Fig. 8.99. Life cycle period of a new product

Although life cycle varies in accordance with the product and sector base, usually there are four phases in life cycle period as shown in Fig. 8.99. First period is the Entrance Phase, second period is the Development Phase, third one is the Maturity Phase and fourth period is the Satisfaction Phase. Entrance Phase is the period of a product presentation to market and effort spent for acceptance and in general it is the period of catching up at par point. Development Phase is the best step in which the product has reached maximum profit and has been through the brightest period. In the Maturity Phase problems come up gradually and decrease in sales starts. Decrease in sales start but companies try to keep sales high by some other marketing activities, which are called as other sales efforts. In that period increase in sales like jumping sales (comb tooth) occur. It is generally agreed that innovation, performance, and competition depend significantly on the maturity of markets. Satisfaction Phase is the period that the companies would not prefer to be in and will start to lose in a while.

During maturity period significant changes are made in the way that the product is behaving into the market. Since an increase in profit is the major goal of a company that introduces a product into a market, the product's life cycle management is very important. Presentation of a new product to the market at the best time shall provide advantage to companies in competition and increase in share in the market.

In the conventional product life cycle, introduction of new product to market corresponds to the point shown as point "A" in Fig. 8.99. When the company comes to this point in the end of maturity period, it has to choose one of the alternatives of new product, new market, or withdrawal of goods from market, so as not to enter into the fourth period, the regression. Depending on the structure in which the company is, new product alternative can be the new product in physical/functional context, new product in the consumer's view or alternative usage.

Point chosen as point "A" in Fig. 8.99 in the existing systems is considered to be late for the new product to enter to market. Because, this point is the period in which the company withstands a number of costs called other sales efforts (promotion, excess goods, discount, etc.) to keep the sales active. It is plain to see from review of the conventional life cycle that profit has started to fall in spite of the increase in sales.

It is suggested in the proposed system to determine the point specified as point "A" in Fig. 8.99 by means of the expert system. In this proposed structure, "A" point can be taken to an earlier time than in the existing policies. In operation of the system, product life cycle maturity period characteristics will be reviewed and efforts will be made to determine the most suitable time for presentation of the product to market by evaluation of the factors named as macro and micromarket indicators.

To support the decision process, a FES is designed to determine whether the entrance of a new product into market or not. Finally, when operating the FES, three different deductions can be made, as preservation of the present status, introduction of new product to market and withdrawal of product from market.

Fuzzy Expert Systems

Expert systems were designed to reason through knowledge to solve problems using methods that humans use. Expert systems use heuristic knowledge – rather than numbers – to control the process of solving the problem. Expert systems have their knowledge encoded and maintained separately from the computer program, which uses that knowledge to solve the problem. Expert systems are capable of explaining how a particular conclusion was reached, and why requested information is needed.

A FES is an expert system that utilizes fuzzy sets and fuzzy logic to overcome some of the problems, which occur when the data provided by the user are vague or incomplete. The power of fuzzy set theory comes from the ability to describe linguistically a particular phenomenon or process, and then to represent that description with a small number of very flexible rules. In a fuzzy system, the knowledge is contained both in its rules and in fuzzy sets, which hold general description of the properties of the phenomenon under consideration.

One of the major differences between a FES and another expert system is that the first can infer multiple conclusions. In fact it provides all possible solutions whose truth is above a certain threshold, and the user or the application program can then choose the appropriate solution depending on the particular situation. This fact adds flexibility to the system and makes it more powerful. Fuzzy expert systems use fuzzy data, fuzzy rules, and fuzzy inference, in addition to the standard ones implemented in the ordinary expert systems.

The Fuzzy Expert System Design

The fuzzy expert system design steps are shown as following:

1. Identification of the problem and choice of the type of fuzzy system, which best suits the problem requirement. A modular system can be designed consisting of several fuzzy modules linked together. A modular approach, if applicable, may greatly simplify the design of the whole system, dramatically reducing its complexity and making it more comprehensible.
2. Definition of input and output variables, their linguistic attributes (fuzzy values) and their membership function (fuzzification of input and output).
3. Definition of the set of heuristic fuzzy rules. (if–then rules).
4. Choice of the fuzzy inference method (selection of aggregation operators for precondition and conclusion).
5. Translation of the fuzzy output in a crisp value (defuzzification methods).
6. Test of the fuzzy system prototype, drawing of the goal function between input and output fuzzy variables, change of membership functions and fuzzy rules if necessary, tuning of the fuzzy system, validation of results.

In building FES, the crucial steps are the fuzzification and the construction of blocks of fuzzy rules. These steps can be handled in two different ways. The first is by using information obtained through interviews to the experts of the problem. The second is by using methods of machine-learning, neural networks and genetic algorithms to learn membership functions and fuzzy rules. The two approaches are quite different. The first does not use the past history of the problem, but it relies on the experience of experts who have worked in the field for years. The second is based only on past data and project into the future the same structure of the past. The first approach seems preferable for our purpose, because no systematic past data on industrial districts are available and because the empirical identification of the industrial districts requires a careful assessment of their characteristics that only experts in this field can make.

We can formalize the steps in the following manner.
For each linguistic variable, input $x_i (i = 1 \ldots m)$ and output y, we have to fix the one's range of variability U_i and V.
$\forall i, (i = 1 \ldots m)$, if n_i is the number of the linguistic attribute of the variable x_i and $\hat{n}^{MAX}_{i=[1,m]} n_i$; We define the set where, $\forall j_i \in n_i, \forall n_i \in [1, \hat{n}] A^i_j$ are the fuzzy numbers describing the linguistic attributes of the input variable x_i, in the same way, we define the set $B = \{B_1, B_2, \ldots, B_k, \ldots, B_\gamma\}$, where $\forall k \in [1, r] B_k$ are the fuzzy numbers describing the linguistic attributes of the output variable y.

At every elements of A^i and B is associated a membership function; $\mu A^i_1(x) : U_i \to [0, 1]$ and $\mu B_1 : V_i \to [0, 1]$.

The elements of A^i and B overlap in some "grey" zone, which cannot be characterized precisely. Many phenomena in the world do not fall clearly into one crisp category or another.

Experts that use abstraction as a way of simplifying the problem can contribute to identify these "grey" zones. The choice of the slopes of the elements of and B is a mathematical translation of what the experts think about the single terms. The second step is the block-rules construction. We define the set of L fuzzy rules where,

$$L \leq \prod_1^m n_i, \quad \forall j_i \in [1, n_i], \quad \forall n_i \in [1, \check{n}], \quad \forall k \in [1, r], \tag{8.7}$$

$$IF((x_1 \text{ is } A_{j1}^1) \otimes (x_2 \text{ is } A_{j2}^2) \otimes \otimes(x_m \text{ is } A_{jm}^m)) \quad THEN(y \text{ is } B_k). \tag{8.8}$$

The relations above are called "precondition" and the symbol represents one of the possible aggregation operators. In practical applications, the MIN and MAX operators, or a convex combination of them, are widely used and so a "negative" or "positive" compensation will occur for different values of

$$\gamma MIN + (1 - \gamma)MAX \quad \text{with } \gamma \in [0, 1]. \tag{8.9}$$

Instead of MIN and MAX, it is also possible to use other t-norms or conorms, which represent different ways of linking the "and" with the "or". More generally, indicating with $\mu_{A \cap B}$ a general membership of the intersection and with $\mu_{A \cup B}$ a general membership of the union, we can define as membership of the aggregated set *with* $A \ominus B$

$$\mu_{A \ominus B} = \mu_{A \cap B}^{1-\gamma} * \mu \gamma_{A \cup B} \quad \text{with } \gamma \in [0, 1]. \tag{8.10}$$

This is not, in general, a t-norm or a t-conorm. In particular, if we use the algebraic product and sum as intersection and union, we obtain the Gamma operator [8]

$$\mu = \left(\prod_1^n \mu_1\right)^{(1-\gamma)} * \left(1 - \prod_1^n (1 - \mu_1)\right)^{\gamma} \tag{8.11}$$

The parameter denotes the degree of compensation. As it is shown in some recent work, this aggregator concept can represent the human decision process more accurately than others.

The aggregation of precondition and conclusion can be made in several ways. The more used are the MAX and the BSUM methods. The choice depends of the type of application. The MAX has the meaning of keeping as "winner" the strongest rule, in the sense that if a rule is "firing" (activated) more then one time, the result is the maximum level of firing. In the BSUM case, all the firing degree is considered and the fial result is the sum of the different level of activation (not over one). In any case, the two methods produce a fuzzy set, which has membership function $\mu_{\text{agg}}(y)$.

Now we have a result of the fuzzy inference system, which is a fuzzy re-play. We need to return to a "crisp" value, and this step is called "defuzzifica-tion." This operation produces a "crisp" action y that adequately represents the membership function $\mu_{\mathrm{agg}}(y)$. There is no unique way to perform this operation. To select the proper method, it is necessary to understand the linguistic meaning that underlies the defuzzification process. Two of these different linguistic meanings are of practical importance: the "best compro-mise" and the "most plausible result." A method of the first type is the Center of Area (CoA) that produces the abscissa of the center of gravity of the fuzzy output set

$$\bar{y} = \frac{\int y \mu_{\mathrm{agg}}(y)\mathrm{d}y}{\int \mu_{\mathrm{agg}}(y)\mathrm{d}y}.$$

A method of the second type is the "Mean of Maximum" (MoM). Rather then balancing out the different inference results, this method selects the typical value of the terms that is most valid.

Marketing Decision Model

The system structure identifies the fuzzy logic inference flow from the input variables to the output variables. The fuzzification in the input interfaces translates analog inputs into fuzzy values. The fuzzy inference takes place in rule blocks, which contain the linguistic control rules. The outputs of these rule blocks are linguistic variables. The defuzzification in the output interfaces translates them into analog variables.

$$Performance\_of\_product \begin{cases} Global\_Market & \begin{cases} Economic\_conditions \\ Political\_Circums\tan ces \end{cases} \\ Manufacture & \begin{cases} Competition \\ Other\_Selling\_Efforts \\ \Pr oportional\_Increase\_in\_Sells \end{cases} \\ Target\_Market & \begin{cases} Manufacture\_Po\,\mathrm{int} \\ \mathrm{Re}\,newal \end{cases} \end{cases}$$

The following Fig. 8.100 shows the whole structure of this fuzzy system including input interfaces, rule blocks and output interfaces. The connecting lines symbolize the data flow. The fuzzification method, "Compute MBF," is the standard fuzzification method used in almost all applications. This method only stores the definition points of the membership functions in the generated code and computes the fuzzification at runtime.

For output variables, different defuzzification methods exist as well. The most often used method is CoM, which delivers the best compromise of the firing rules.

In Fig. 8.100, rule block of the structure of the fuzzy logic system is shown. This block contains the rules of the system describing the control strategy. The rule blocks contain the control strategy of a fuzzy logic system.

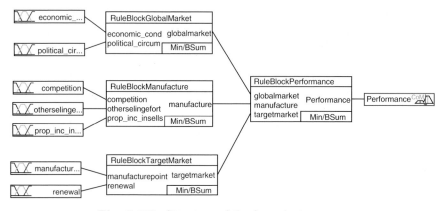

Fig. 8.100. Structure of the fuzzy logic system

Table 8.12. Rules of the rule block "Rule Block Global Market"

IF			THEN
Economic cond.	**AND**	**Political circums.**	**Global market**
Negative		Negative	Pessimistic
Negative		Ineffective	Pessimistic
Negative		Positive	Pessimistic
Ineffective		Negative	Pessimistic
Ineffective		Ineffective	Pessimistic
Ineffective		Positive	Optimistic
Positive		Negative	Pessimistic
Positive		Ineffective	Optimistic
Positive		Positive	Optimistic

Each rule block confines all rules for the same context. A context is defined by the same input and output variables of the rules. The rules' "IF" part describes the situation, for which the rules are designed. The "THEN" part describes the response of the fuzzy system in this situation. The DoS is used to weigh each rule according to its importance. Factors named as global market indicators; overall economic situation and legal and political circumstances prevailing in the market are reviewed. The rules in DT of the Global Market, can be summarized in production rules as following (see Table 8.12).

Factors named as target market indicators; combination of product is reviewed through comparison of performances of the product and its rival. Result of review reveals the probability that performance of the product can be lower or higher than or equal to that of the closest rival product. The condition of "manufacture point" has three condition domain factors: $m.p.<c.m.p.$, $m.p.=c.m.p.$ and $m.p.>c.m.p.$

Above, "m.p." as manufacture point of our product, "c.m.p." as manufacture point of competitor product. The Fuzzy expert rules in the target market can be summarized as following (see Table 8.13).

Table 8.13. Rules of the ruleblock "Rule Block Manufacture"

IF					THEN
Competition	AND	Other selling efforts	AND	Prop. increase in sells	Manufacture
Decreased		Decreased		Decreased	Poor
Decreased		Decreased		Increased	Good
Decreased		Increased		Decreased	Good
Decreased		Increased		Increased	Very good
Increased		Decreased		Decreased	Good
Increased		Decreased		Increased	Very good
Increased		Increased		Decreased	Very good
Increased		Increased		Increased	Very good

Table 8.14. Rules of the ruleblock "Rule Block Target Market"

IF			THEN
Manufacture point	AND	Renewal	Target market
Mp<cmp		not_ok	Wait
Mp<cmp		ok	Medium
Mp=cmp		not_ok	Wait
Mp=cmp		ok	Impulsive
Mp>cmp		not_ok	Wait
Mp>cmp		ok	Impulsive

Factors of innovation in micromarket indicators; technological novelty significant in competitiveness is reviewed in the form of physical and functional production novelty, alternative usage and new markets. The fuzzy expert rules in the manufacture can be summarized in production rules as following (see Table 8.14).

As the result of sales rates' decrease, the company will initiate other sales efforts to increase sales. These efforts will escalate cost of other sales efforts. Thus, profit rate will drop as a big portion of the profit is used to finance other sales efforts. The fuzzy expert rules in the performance can be summarized in production rules in Table 8.15.

As the result of operating the expert system, three different deductions can be made, as preservation of the present status, introduction of new product to market and withdrawal of product from market, i.e.:

If PERFORMANCE = Active *Then* "Preserve the present status"
If PERFORMANCE = Passive *Then* "Introduce new product to market"
If PERFORMANCE = Bad *Then* "Withdraw the product from market"

MBF of Performance is represented in Fig. 8.101.

Conclusion

This section proposed an effort on the issue of the life cycle management. The idea, surely new, is to reproduce, in a structural way, what the experts do when

Table 8.15. Rules of the ruleblock "Ruleblock performance"

	IF		THEN
Global market AND	**Manufacture AND**	**Target market**	**Performance**
Pessimistic	Poor	Wait	Bad
Pessimistic	Poor	Medium	Bad
Pessimistic	Poor	Impulsive	Passive
Pessimistic	Good	Wait	Bad
Pessimistic	Good	Medium	Passive
Pessimistic	Good	Impulsive	Passive
Pessimistic	Very good	Wait	Passive
Pessimistic	Very good	Medium	Passive
Pessimistic	Very good	Impulsive	Active
Optimistic	Poor	Wait	Passive
Optimistic	Poor	Medium	Passive
Optimistic	Poor	Impulsive	Active
Optimistic	Good	Wait	Passive
Optimistic	Good	Medium	Active
Optimistic	Good	Impulsive	Active
Optimistic	Very good	Wait	Active
Optimistic	Very good	Medium	Active
Optimistic	Very good	Impulsive	Active

Table 8.16. Project statistics

	Input variables	Output variables	Intermediate variables	Rule blocks	Rules	Membership functions
Global Mar.	2	1	1	1	9	6
Manufacture	3	1	1	1	8	7
Target Mar.	2	1	1	1	6	5
Performance	7	1	3	1	18	3
Result	7	1	3	4	41	21

have to decide a new product's market entering time. This study is applied on liquid detergent production. Experts' accumulated experience is translated in the decision tree and in the ruleblocks. As the result of the performed study, the most suitable time for introduction of the product to the market can be determined, instead of withstanding the costs of other sales efforts and losing profit as well as risking the loss of market share in the product's maturity period. When operating the expert system, three different deductions can be made, as preservation of the present status, introduction of new product to market and withdrawal of product from market. This structure, which has been designed solely for a liquid detergent producing company, can be conveniently used for different sectors too with new rule bases to be obtained from experts of the other sectors. The information about the variables, rules, membership functions for different parameters in the project is as shown in Table 8.16.

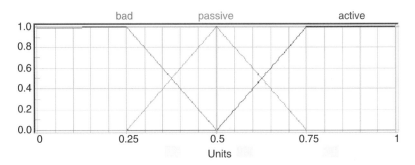

Fig. 8.101. MBF of performance

8.7.4 A Fuzzy Expert System Design for Diagnosis of Prostate Cancer

Introduction

In recent years, the methods of artificial intelligence have largely been used in the different areas including the medical applications. In the medicine area, many expert systems (ESs) were designed. ONCOCIN and ONCO-HELP are the ESs for diagnosis of the general cancer diseases. For example, ONCO-HELP is multimedia knowledge based decision support system for individual tumor entities. It makes individual and prognosis-oriented treatment of patient's tumor possible (if corresponding predictor's respective prognostic factors are known). Trough registration of individual patient data over tumor type, histology, metastatic type, methathesis localization and amount, as well as corresponding laboratory parameters together with a corresponding knowledge based on a patient individual prognosis-score can be determined. Using this score, a therapy concept is drafted. ONCO-HELP evaluates this concept by using therapy controls with regards to tumor progression/regression and side effects of the therapy.

Soft computing technology is an interdisciplinary research field in computational science. Various techniques in soft computing such as ESs, neural networks, fuzzy logic, genetic algorithms, Bayesian statistics, khaos theory, etc., have been developed and applied to solve many challenging tasks in medicine and engineering design. There are some publications in the area prostate cancer prognosis or diagnosis by aid of soft computing methods.

We have developed a rule-based FES that uses the laboratory and other data and simulates an expert-doctor's behavior. As known when the prostate cancer can be diagnosed earlier, the patient can be completely treated. If there is a biopsy for diagnosing, the cancer may spread to the other vital organs. For this reason the biopsy method is undesirable. As laboratory data, prostate specific antigen (PSA) and prostate volume (PV) and age of the patient are used. Using this data and help from an expert-doctor, the fuzzy rules to determine the necessity of biopsy and the risk factor was developed.

Table 8.17. Fuzzy rules

Rule No	PSA	Age	PV	PCR
Rule 1	VL	Very young	VS	VL
...				
Rule 43	VL	MA	H	VL
...				
Rule 77	VH	Old	VS	H
...				

The developed system gives to the user the patient possibility ratio of the prostate cancer.

Additionally, the FES is rapid, economical, without risk compared to traditional diagnostic systems, and it has also a high reliability and can be used as learning system for medical students.

Materials and Methods

The clinics and laboratory data for the developed system can be taken from the literature. For the design process PSA, age and PV are used as input parameters and prostate cancer risk (PCR) is used as output. For fuzzification of these factors the linguistic variables very small (VS), small (S), middle (M), high (H), very high (VH), very low (VL), and low (L) were used. For the inference mechanism the Mamdani max-min inference was used.

Fuzzy Expert System

The units of the used factors are: PSA ($ng\,ml^{-1}$), age (year), PV (ml) and PCR (%). Parts of the developed fuzzy rules are shown in the Table 8.17. Total of 80 rules are formed.

For example, Rule 1, Rule 43 and Rule 77 can be interpreted as follows:

Rule 1: if PSA=VL and Age=Very Young and PV=VS, then PCR=very low, i.e., if the patient's PSA is very small and patient is very young and patient's PV is very small, then patient's prostate cancer risc is very low.

Rule 43: if PSA=VL and Age=Middle Age and PV=H, then PCR=VL, i.e., if the patient's PSA is very low and patient has middle age and patient's PV is high, then patient's prostate cancer risc is very low.

Rule 77: if PSA=VH and Age=Old and PV=VS, then PCR=VH, i.e., if the patient's PSA is very high and patient is old and patient's PV is very small, then patient's PCR is high.

Fuzzfication of the used factors are made by aid of the follows functions. These formulas can be determined by aid both of the expert-doctor and literature.

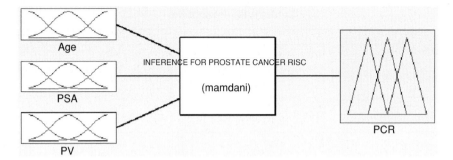

Fig. 8.102. The structure of FES

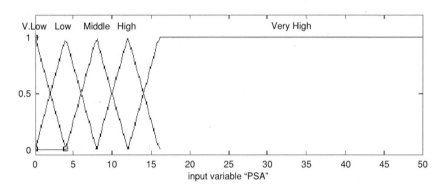

Fig. 8.103. Membership function of PSA

$$
\text{PSA}(A) = \begin{cases} a; & 0 < a < 16 \\ 1; & 50 \leq a \end{cases} \qquad \text{Age}(B) = \begin{cases} 1; & 65 \leq b \\ b; & 0 < b < 65 \end{cases}
$$

$$
\text{PV}(C) = \begin{cases} c; & 3.8 \leq c \leq 308 \\ 0; & c < 3.8 \\ 1; & c > 308 \end{cases} \qquad \text{PCR}(D) = \begin{cases} z; & 0 \leq d \leq 100 \\ 0; & d < 0 \\ 0; & d > 100 \end{cases}
$$

$$(8.12)$$

Developed FES has a structure shown as in the Fig. 8.102.

Structure of the Fuzzy Factors

The memberships of the used factors are obtained from the formulas above and shown in the Fig. 8.103–8.106.

From the developed rules and from the formulas we obtained, for example, for PSA linguistic expressions as follows:

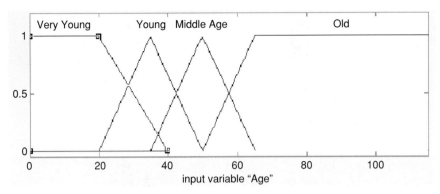

Fig. 8.104. Membership function of the age

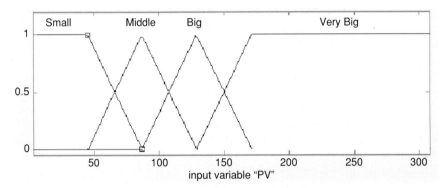

Fig. 8.105. The membership function of the PV

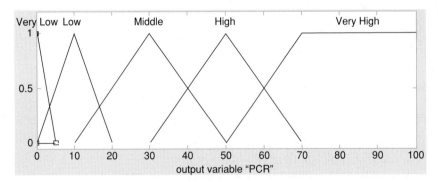

Fig. 8.106. Membership function of the PCR

$$\mu_{\text{Verylow}}(A) = \begin{cases} \frac{4-a}{4}; & 0 < a < 4 \\ 0; & \text{Diğer durumlar} \end{cases} \qquad \mu_{\text{Low}}(A) = \begin{cases} \frac{a}{4}; & 0 < a \le 4 \\ \frac{8-a}{4}; & 4 < a < 8 \\ 0; & a \ge 8 \end{cases}$$

$$\mu_{\text{Middle}}(A) = \begin{cases} 0; & a \le 4 \\ \frac{a-4}{4}; & 4 < a \le 8 \\ \frac{12-a}{4}; & 8 < a < 12 \\ 0; & a \ge 12 \end{cases} \qquad \mu_{\text{High}}(A) = \begin{cases} 0; & a \le 8 \\ \frac{a-8}{4}; & 8 < a \le 12 \\ \frac{16-a}{4}; & 12 < a < 16 \\ 0; & a \ge 16 \end{cases}$$

$$\mu_{\text{Veryhigh}}(A) = \begin{cases} 0; & a \le 12 \\ \frac{a-12}{4}; & 12 < a < 16 \\ 1; & a \ge 16 \end{cases} \tag{8.13}$$

The other linguistic expressions (Very Young, Young, Middle Age and Older) are determined similarly. For the output factor PCR the linguistic expressions are Very low, Low, Middle, High and Very high, which can be expressed as Very small, Small, Middle, High and Very high, respectively, in the formulas mentioned above. For example, for High PSA, Middle Age, and Very Big PV the membership functions will have forms, respectively:

$$\mu_{\text{high}}(\text{PSA}) = \{0/8 + 0.25/9 + 0.5/10 + 0.75/11 + 1/12 + 0.75/13 + 0.5/14$$
$$+ 0.25/15 + 0/16\},$$

$$\mu_{\text{middle}}(\text{Age}) = \{0/35 + 0.33/40 + 0.67/45 + 1/50 + 0.67/55 + 0.33/60 + 0/65\},$$
$$\mu_{\text{veryhigh}}(\text{PV}) = \{0/2129 + 01/133 + 0.2/137 + \cdots + 1/171 + 1/175 + \cdots + 1/308\}.$$

Defuzzification

In this stage, truth degrees ($\underset{\sim}{a}$) of the rules are determined for the each rule by aid of the min and then by taking max between working rules. For example, for PSA=40 ng ml^{-1}, Age=55 year, PV=230 ml the rules 60 and 80 will be fired and we will obtain:

$$\alpha_{60} = \min(\text{Very High PSA, Middle Age, Very Big PV})$$
$$= \min(1, 0.67, 1) = 0.67,$$
$$\alpha_{80} = \min(\text{Very High PSA, Old Age, Very Big PV})$$
$$= \min(1, 0.33, 1) = 0.33.$$

From Mamdani max–min inference we will obtain the membership function of our system as $\max(\underset{\sim}{a} 60, \underset{\sim}{a} 80) = 0.67$, that means Very High PCR. Then we can calculate the crisp output. The crisp value of the PCR is calculated by the method center of gravity defuzzifier by the formula:

$$D^* = \frac{\int D\mu_{\text{middle}(D)}\mathrm{d}D}{\int \mu_{\text{middle}(D)}\mathrm{d}D}$$

As also seen from the Fig. 8.107, the value of PCR=78.4. This means that the patient has the prostate cancer with a possibility 78.4%. Because this is a quite high percentage, doctor has to decide a biopsy.

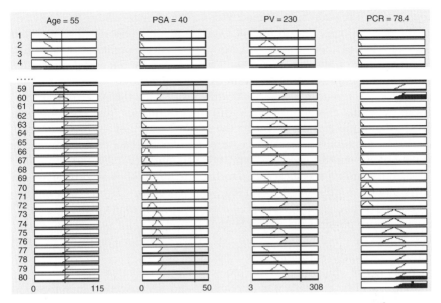

Fig. 8.107. Calculation of the value PCR for the values PSA=8 ng ml^{-1}, Age=55, PV=300 ml

Conclusion

This section describes a design of a FES for determination of the possibility of the diagnosis of the prostate cancer, which can be used by the specialist doctors for treatment and by the students for learning the scope. This system can be developed further with increasing the knowledge rules from one side and with adding the neural network to the system from the other side.

8.7.5 The Validation of a Fuzzy Expert System for Umbilical Cord Acid–Base Analysis

Introduction

The umbilical cord vein carries blood from the placenta to the fetus and the two smaller cord arteries return blood from the fetus. The blood from the placenta has been freshly oxygenated, and has a relatively high partial pressure of oxygen (pO_2) and low partial pressure of carbon dioxide (pCO_2). Oxygen in the blood fuels *aerobic* cell metabolism, with carbon dioxide produced as "waste". Thus the blood returning from the fetus has relatively low oxygen and high carbon dioxide content. Some carbon dioxide dissociates to form carbonic acid in the blood, which increases the acidity (lowers the pH).

Samples of blood may be taken from blood vessels in the umbilical cord of the neonate immediately on delivery, and a blood gas analysis machine measures the pH, partial pressure of carbon dioxide (pCO_2) and partial pressure of oxygen (pO_2). A parameter termed *base deficit of extracellular fluid* (BDec f) can be derived from the pH and pCO_2 parameters. This can distinguish the cause of a low pH between the distinct physiological conditions of *respiratory acidosis*, due to a short-term accumulation of CO_2, and a *metabolic acidosis*, due to lactic acid from a longer-term oxygen deficiency.

There are, however, a number of difficulties with the procedure. Difficulties in obtaining the samples can result in two samples from the same vessel or mixed samples, whilst blood in the syringe can alter due to exposure to air. Blood gas analysis machines require regular internal calibration and external quality control checks to ensure continuing accuracy and precision to the manufacturer's specifications, During a trial on ST-waveform monitoring in Plymouth, routine cord blood sampling on every delivery was initiated. Careful retrospective analysis of the cord blood gas results highlighted a 25% failure rate to obtain arterial and venous paired samples with all parameters. This sampling error rate is broadly in line with other studies in which the importance of paired samples was recognized.

A model of clinical expertise required for the accurate interpretation of umbilical acid-base status was developed, and encapsulated in a rule-based expert system. This expert system checks results to ensure their consistency, identifies whether the results come from arterial or venous vessels, and then produces an interpretation of their meaning.

A number of problems were identified in the implementation of conventional crisp rules used in the initial system. The interpretation section of the crisp expert system utilized a number of rules of a form similar to:

IF arterial pH $< 7{:}05$ AND arterial BDec $f \geq 12\,\mathrm{mmol\,l}^{-1}$

THEN severe arterial metabolic acidemia

Such rules feature sharp boundary cut-offs which are not representative of real decision making processes and do not employ any form of uncertainty representation in the conclusion to imply a less than certain diagnosis.

Fuzzy logic and fuzzy set theory provide a good framework for managing uncertainty and imprecision in medicine and have been successfully applied to a number of areas. It was felt that a fuzzy logic based expert system would offer more realistic and acceptable interpretation. In a fuzzy system, a rule such as above may be replaced by:

IF arterial pH is low AND arterial BDec f is high THEN arterial acidemia is metabolic

The use of fuzzy logic allows for more gradual changes between categories and allows for a representation of certainty in the rule consequence through the ability to fire rules with varying strength dependent on the antecedents. Additionally, fuzzy logic can allow the results to be presented to clinicians in a more natural form.

A preliminary investigation was performed to convert the crisp expert system directly into a FES and it was found that, after tuning, this fuzzy system improved the performance of the crisp expert system to a level effectively indistinguishable from the clinical experts. However, although this preliminary

FES utilized a set of fuzzy rules to perform the interpretation, it was characterized by a number of restrictions. It functioned with crisp input and output variables with no indication of imprecision, and it *only* interpreted results that had been already validated by the crisp expert system as comprising an error-free arterial-venous pair. An "integrated" FES was then developed, with knowledge gained from the preliminary FES, to validate and interpret *all* acid-base results. When the development of the integrated system was complete, a comprehensive validation process was undertaken to reevaluate the numeric and linguistic outputs of both the numeric and linguistic interpretations of the system.

Development of the Fuzzy Expert System

A new set of fuzzy rules was developed for both the vessel identification and the interpretation capabilities. Fresh knowledge elicitation sessions were undertaken with the same experts that had developed the crisp rules. Two sets of fuzzy rules were employed; the *vessel identification* rules and the *interpretation* rules. The sample(s) parameters are passed through the vessel identification rules to determine whether they represent an arterial-venous pair. Once vessel identification has been carried out, the sample(s) are passed through the interpretation rules. The Mamdani model of inference was used, with the min operator used for implication. Probabilistic operators were used for *and* and *or* as, in elicitation sessions with experts, the fuzzy output sets produced with the probabilistic operators were favored as they avoided the "plateau"s produced with the standard max, min operators. It was found that the probabilistic family generated smoother transition surfaces for the vessel identification rules and produced higher performance for the interpretation rules. Fuzzy sets were modeled with *sigmoid* membership functions. Center-of-gravity (centroid) defuzzification was performed on the fuzzy output variables to produce numeric outputs, and linguistic approximation was performed to produce linguistic output.

Vessel Identification Rules

As two samples may both be accidentally obtained from the vein, both from the arteries, one may be mixed arterial-venous, or both may be mixed, a "safe" vessel identification rule may be that if all parameters differ by more than a specified uncertainty, then the samples can definitely be taken as a true arterial-venous pair. The expected imprecision in each parameter may be established through a number of clinical experiments. A fuzzy rule-base was designed to produce the behavior that if all parameters differed by more than these values then the results were labeled as an arterial-venous pair – with smooth transitions between each of the categories.

Interpretation Rules

The basic principles of acid-base analysis elicited from the experts were that (1) *acidemia* is based on the absolute value of arterial pH (lower arterial pH implies worse *acidemia*), refined by the value of the venous pH; (2) *component* is based on arterial BD*ec f* (high BD*ec f* implies *metabolic* component, low BD*ec f* implies *respiratory* component), refined by venous BD*ec f*; and (3) *duration* is based on pH and BD*ec f* differences (smaller differences imply *chronic* duration, larger differences imply *acute* duration), refined by absolute arterial values. These basic principles were encapsulated in the fuzzy rules such that there was smooth transition over all input and output sets. This ensured that, as far as possible, continuous changes in input parameters resulted in continuous changes in the fuzzy output sets.

Validation of the Fuzzy Expert System

The cases for each task can be selected by the independent engineer from the database of over 10,000 results (approximately 400 abnormals), but this provided serious problems. Cases could not be selected from the entire database on a uniform random basis, as this would have resulted in approximately 75% paired arterial-venous samples, and approximately 98% *normal* interpretations. In essence it was desired to uniformally span the *target* outputs, so that a roughly even spread across the various output sets would have been obtained from the combined experts (and expert system).

Numeric Interpretation

The centroids of the integrated FES were combined into a single index by:

$$\text{condition} = \textit{acidemia} + \frac{\text{component}}{20} + \frac{\text{duration}}{10}, \qquad (8.14)$$

where the relative weighting of the three terms was determined empirically. Given that the three output variables are arranged in such a way that low scores indicate a worsening condition for the infant, to the extreme *severe, metabolic, chronic acidemia*, this index can be thought of as indicating the *health* of the infant as represented by its acid-base balance at birth. The experts were asked to rank 50 cases from "worst" to "best," in terms of likelihood that the infant may have suffered intrapartum asphyxial damage, on the basis of the acid-base information alone.

Linguistic Interpretation

The experts were given the two sets of pH and BD*ec f* parameters from each of 50 cases, and were asked to indicate their opinion of the closest linguistic interpretation for three linguistic variables; *acidemia, component*, and *duration*. For each variable they were instructed to mark *zero, one* or *two* terms to indicate the closest match. This was specifically designed to allow the expert to mark two adjacent labels if they felt a result fell in-between two labels, or to mark no label if there was insufficient information, or no label was appropriate.

Statistical Methods

Spearman rank order correlation can be used to determine the degree of association between two sets of rank-ordered data. This was used to calculate the difference between the expert system's ranking of cases, specified by the index described above, and the experts' ordering. Note that this is effectively the same as minimizing the mean square error between the desired rankings and the obtained rankings. To measure the agreement between two expert's linguistic categorization a measure of (nominal) categorical agreement was required. The χ^2 statistic can be used to measure the degree of *association* between two categorical variables, but this statistic makes no distinction between departure from chance association due to *agreement* or *disagreement*. In 1960, Cohen introduced a measure of *agreement* between two categorical variables termed the *kappa* coefficient. This plain kappa statistic measured only *exact* agreement and, to overcome this problem, Cohen later introduced *weighted kappa* to allow for partial agreement. Plain and weighted kappa was used to calculate the degree of agreement between experts and the expert system linguistic outputs.

Conclusions

Fuzzy expert system has thus been designed for umbilical cord acid-base analysis. The FES presented here, while a significant development, can be validated more thoroughly against clinical data.

8.7.6 A Fuzzy Expert System Architecture Implementing Onboard Planning and Scheduling for Autonomous Small Satellite

Introduction

Not only the improvement of technology makes the inherent characteristic of small satellite, that is "faster, better, and cheaper," to be developed forward endless, the idea of establishing "virtual presence" in space in the next century also requires the small satellite developing towards "smarter." So, several new technologies need to be demonstrated, and one of the most crucial is on board autonomy.

On Board Autonomy

For a long time, the satellite operations which include a large number of functions, such as planning mission, sequence the execution commands, tracking the spacecraft's internal hardware state, ensuring correct functioning, recovering in cases of failure, and subsequently working around faulty subsystems, or reconfiguring system, were carried out through humans intervention on the ground. This traditional approach is necessary and useful to the traditional spacecraft, but will not be viable anymore in the future due to (1) up-link and down-link communication time delay which makes driving a deep space mission impossible; (2) a desire to limit the operations team and cost; (3) ensuring high reliability through handling failures real-time. On board autonomy integrates three separate technologies: an on board planner/scheduler, a robust multithreaded executive including internal commands and telemetering, telecontrol commands, and a fault diagnosis and recovery system.

In the new model of operations, the scientists or operators will communicate high-level science goals directly to the spacecraft. The spacecraft will then perform its own science planning and scheduling, translate those schedules into commands sequence, verify that they will not damage the spacecraft, and ultimately execute them without routine human intervention. In the case of error recovery, the spacecraft will have to understand the impact of the error on its previously planned sequence and then reschedule in light of the new information and potentially degraded capabilities. The goal of the planner/scheduler is to generate a set of synchronized low level commands that once executed will achieve mission's goals.

Planning and Scheduling

Planning and scheduling is not a new subject. Many planning and scheduling methods have been proposed and analyzed since at least the 1950s. Although related and often tightly coupled, strictly speaking, planning, and scheduling are distinctly different activities. Planning is the construction of the project/process model and definition of constraints/objectives. Scheduling refers to the assignment of resources to activities (or activities to resources) at specific points in, or duration of, time. The definition of the problem is thus primarily a planning issue, whereas the execution of the plan is a scheduling issue. Yet planning and scheduling are coupled; the performance of the scheduling algorithm depends on the problem formulation, and the problem formulation may benefit from information obtained during scheduling. Because of uncertainties, and being based on incomplete data, planning and scheduling problems are dynamic. No schedule is static until the project is completed, and most plans change almost as soon as they are announced.

Depending on the duration of the project, the same may also be true for the objectives. The dynamics may be due to poor estimates, incomplete data, or unanticipated disturbances. As a result, finding an optimal schedule is often confounded not only by meeting existing constraints but also adapting to additional constraints and changes to the problem structure.

Practically speaking, finding an optimal schedule is often less important than coping with uncertainties during planning and unpredictable disturbances during schedule execution. In some cases, plans are based upon well-known processes in which resource behaviors and task requirements are all well known and can be accurately predicted. In many other cases, however, predictions are less accurate due to lack of data or predictive models. In these cases the schedule may be subject to major changes as the plan upon which it is based changes. Although methods exist for finding optimal solutions to some specific scheduling problem formulations, many methods do not work when the structure of the constraints or objectives change.

Though lots of successful applications about planning and scheduling in spacecraft operations have recently been reported, and all of them are important for the development of this field, the inherent fuzziness and uncertainty of planning and scheduling was ignored made the achieving of onboard automated planning and scheduling system become idealization and no reality. Rules-based FES not only maintain the value of based rules and the merit of using fuzzy logic control to describe uncertainty systems, and utilize the predominance of using expert systems to denote and control knowledge.

Fuzzy Expert Systems

Expert systems based conventional logic are not efficient in handling inaccurate and inexact information. Fuzzy logic based expert system is a powerful tool providing failure analysis of a complex and nonlinear dynamic system, like a final control element. To date, fuzzy expert systems are the most common use of fuzzy logic. They are used in several wide-ranging fields, including:

– Linear and nonlinear control
– Pattern recognition
– Financial systems
– Operation research
– Data analysis

As previously mentioned, this section adopts a rules-based FES architecture used to on-board planning and scheduling. The architecture also considered all kinds of limits of on board operations, such as processing capability of CPU, memory size, and the desire of real-time. In response to above consideration, this section presents an architecture, which is developed using rules-based FES. In order to adapt the requirement of on board operation, the resource restrain is also considered in the architecture, such as processing speed of CPU, the capacity of storage and the real-time requirement.

Domain and Requirements

As we known, the characteristics of small satellites make on board automated planning and scheduling function different form other project applications. Small Satellite domain places a number of requirements on the software architecture that differentiates it from domains considered by other researchers or other projects. There are some major properties of the domain that drove the architecture design as following:

1. Human could not intervene a on board small satellite real-time, but the high reliability must be ensured. Though small satellite is cheaper than large spacecraft, it is also expensive and often unique, so a high reliability must be requirement by user. However, the harsh environment of space or the inability to test in all flight conditions and still cause unexpected hardware or software failures, so that small satellite must have autonomous operations function that can rapidly react to contingencies by retrying failed actions, reconfiguring subsystems or ensuring the small satellite to prevent further, potentially irretrievable, damage.

2. The resources of small satellite is severely limited, and must be used optimized in order to achieve the missions. Small satellite uses various resources, including obvious ones like fuel and electrical power, and less obvious ones like the number of times a battery can be reliably discharged and recharged. Some of these resources are renewable but most of them are not. Hence, autonomous operations require significant emphasis on the careful utilization of nonrenewable resources and on planning for the replacement of renewable resources before they run dangerously low.

3. Small satellite operation is a complicated concurrent activity. Small satellite has a number of different subsystems, all of which operate concurrently. Hence, reasoning about the small satellite needs to reflect its concurrent nature. In particular, the planning and scheduling needs to be able to schedule concurrent activities in different parts of the small satellite, including constraints between concurrent threads active to handle concurrent commands to different parts of the small satellite.

Module and Method

Formally, resource constrained project scheduling (RCPS) is characterized by the following:

Given: A set of tasks T, a set of resources R, a capacity function $C :\rightarrow N$, a duration function $D : T \rightarrow N$, a utilization function $U : T \times R \rightarrow N$, a partial order P on T, and a deadline d.

Find: An assignment of start times $S : T \rightarrow N$, satisfying the following:

1. Precedence constraints: if t_1 precedes t_2 in the partial order P, then $S(t_1) + D(t_1) \leq S(t_2)$.

2. Resource constraints: for any time x, let $running\ (x) = \{t|S(t) \leq x < S(t)+D(t)\}$, then for all time x, and all $r \in R$, $\sum_{(running\ x)} U(t,r) \leq C(r)$.
3. Deadline: for all tasks $t : S(t) \geq 0$ and $S(t) + D(t) < d$.

Several Concepts

Activities

As a data structure, activities performs specific detail functions which space-craft must execute. An activity represents an action or step in the database. It maybe use one or more constraints. Moreover, An activity maybe include one or more subactivities.

Subactivities are activities that can be scheduled any time within the parent activity subject to resource constraints within the subactivity. Subactivities are similar to the constraint-defined activities without the exact temporal relationship between the parent and subactivities.

An activity may have multiple execution modes. Any activity may be executed in more than one manner depending upon which constrains are used to complete it. Interruption modes may depend on the resources that are applied to the activity.

Constrains

Different from typical RCPS questions, onboard autonomous planning and scheduling includes other types of constraints: temporal constraints, precedence constraints, and availability constraints, besides resource constraints. Constraints turn a relatively smooth solution space with many optimal solutions to a very nonuniform space with few feasible solutions.

A valid plan must satisfy many constraints, including ordering constraints (e.g., the catalystbed heaters must warm up for 90 min before using the reaction control thrusters), synchronization constraints (e.g., the antenna must be pointed at the Earth during up-link), safety constraints (e.g., do not point the radiators within 20° of the sun), and resource constrains (e.g., the CCD camera requires 50 W of power). These are all expressed as temporal constraints. As the most important constraints of small satellite on board planning and scheduling system, resource constraints have four types: automic, concurrency, depletable, and nondepletable. Automic resources are physical devices that can only be used (reserved) by one activity at a time, star tracker, reaction wheel, CPU are automic resources.

Concurrency resources are similar to atomic except they must be made available to the activity before they are reserved, a telecommunications downlink pass is a kind of concurrency resources.

Non-depletable resources are resources that can used more than one activity can use a different quantity of the resource, solar array power is the typical nondepletable resources.

> *CONSTRAINS_IF:*
> *Status_AttitudePrecision<0.3*
> *Status_AttitudeStability<0.001*
> *Status_StartUniversalTime=[6000,1200]*
> *Status_Latitude=[-35,14]*
> *Status_longitude=[25,70]*
> *Resource_Power>120*
> *Premise_FuzzyDegree=0.85*
> *ACTIVITIES_THEN:*
> *Activity_CCDCamera_prepare*
> *Activity_Recorder_Open*
> *Activity_CCDCamera_Work*
> *Activity_DataProcess_compress*
> *RULE_END*

Fig. 8.108. An example of constrains which is used to photograph

Depletable resources are similar to nondepletable except that their capacity is diminished after use, in some cases their capacity can be replenished (battery energy, memory capacity) and in other cases it cannot (fuel).

In many solution methods which have been proposed and implemented, heuristic methods were devised to find good solutions, or to find simply feasible solutions for the really difficult problems mostly. Most research now consists of designing better heuristics for specific instances of scheduling problems. However, heuristic solutions are typically limited to a specific set of constraints or problem formulation, and devising new heuristics is difficult at best.

Because of the severely limited on-board processing capabilities, heuristic methods, which were used frequently in other projects, are no more suitable for on-board spacecraft planning and scheduling. In order to suffice the fuzziness or uncertainty of spacecraft planning and scheduling under condition of severely limited on-board resources, we adopted onboard planning and scheduling fuzzy expert system architecture (OPSFESA) which fuzzy logic is useful in handling uncertain systems, and expert system theory is advantage of handling rule based knowledge system.

Constrains are denoted in the form of fuzzy rules as "if... then...." Figure 8.108 is the example of CCD Camera Working, in which, attitude control precision, attitude control stability, universal time, latitude position and longitude position status is required, and power resource must be sufficed, all these premises have one fuzzy degree, under the premise and fuzzy degree, the activities include CCD camera preparation, data recorder opening, CCD camera working and compressing image data. Figure 8.109 is the example of CCD camera preparation activity, the example of resources and status is in Figs. 8.110 and 8.111.

Architecture

OPSFESA is achieved through the cooperation of the following components:

ACTIVITY Activity_CCDCamera_prepare
Related_Device=[CCDCamera, heater]
Constraint= {
Status_CCDCameraTemperature>20,
Resource_Power>10,
Duration=90 }
SubActivities= {*NONE*}
Commands=14
//the procedure which achieve the activity
Running_Flag
//one of Idle, Running, Finished, Non-valid
ACTIVITY_END

Fig. 8.109. An activity example

RESOURCE Resource_Power
NO=155
Value=120
Reliability=0.97
RESOURCE_END

Fig. 8.110. An example of resource

STATUS Status_Precision
NO=22
Value=0.57
Change_Order=[]
// if Statue value changes by order
Reliability=0.97
STATUS_END

Fig. 8.111. An example of status

Goals profile contains the list of all high-level goals that have to be achieved over the entire duration of the mission, which is put into by communication from ground or initiate before being lunched;

Engine management center which attempters and coordinates every parts, has functions, activating fuzzy inference mode, retrieving mission goals and commands sequence, driving subsystems and assemblies, and so on.

Fuzzy rules base storage of all constrains and activities also include maintenance of them.

Fuzzy inference mode is the kernel of OPSFESA; in their high level goals are translated acceptable low level commands sequence which be used to control subsystems and assemblies directly.

Commands sequence The detailed sequence of low-level commands to achieve high-level goals, however, is not prestored but is generated on board by the planner.

Fuzzy planning and scheduling mode is composed of fuzzy rules base and fuzzy inference mode.

Continuous operation is achieved by repetition of the following cycle:

1. Retrieve high level mission goals from the goals profile send to fuzzy inference mode;
2. Ask the engine to activate fuzzy planning and scheduling mode, in there, mission goals are transmitted low level commands sequence;
3. Send the new generated schedule which is a set of commands sequence to the database, in there, commands are retrieved to control subsystems or assemblies. If a new schedule is generated, the engine will continue executing its current schedule and start executing the new schedule when the clock reaches the beginning of the new scheduling horizon, and making sure that the commands succeed and either retries failed commands or generates an alternate low level command sequence. Hard command execution failures may require the modification of the schedule in which case the executive will coordinate the actions needed to keep the spacecraft in a "safe state" and request the generation of a new schedule from the planning and scheduling architecture.
4. Repeat the cycle from step 1 when one of the following conditions apply
   (a) A new goals profile needs for generating a new schedule;
   (b) The engine has requested a new schedule as a result of a hard failure.

OPSFESA is the only component that is activated as a "batch process" and dies after a new schedule has been generated. This ensures the high reliability and rapidity required by the domain.

The results of OPSFESA is generation of spacecraft low level commands sequence from high level goals specifications. So, it will encoding of complex spacecraft operability constraints, flight rules, spacecraft hardware models, science experiment goals and operations procedures to allow for automated generation of low level commands sequence. Because on-board resources are severely limited, the architecture needs to generate courses of action that achieve high quality execution and cover extended periods of time. OPSFESA always has to trade off the level of goal satisfaction with respect to the long term "mission success" and within the resource limitations.

We choose to plan at infrequent intervals because of the limited on-board processing capabilities of the spacecraft. The planner must share the CPU with other critical computation tasks such as the execution engine, the real time control loops and the fault detection, isolation and recovery system. While the planner generates a plan, the spacecraft must continue to operate. More importantly, the plan often contains critical tasks whose execution cannot be interrupted in order to install newly generated plans.

There will be a copy of the on-board planner built into the ground system. This copy will be used to generate experience and rules of thumb as to what sets of goals are easily achievable and what sets are difficult to achieve for the onboard system based on these rules of thumb. The operators will define the goals for each mission phase and since the Remote agent is closing the loop around these goals, the best prediction of spacecraft behavior is that the goals will be achieved on schedule.

OPSFESA must be able to respond to unexpected events during plan execution without having to plan the response. Although it is sometimes necessary to replan, this should not be the only option. Many situations require responses that cannot be made quickly enough if the has to plan them. The executive must be able to react to events in such a way that the rest of the plan is stall valid. To support this, the plan must be flexible enough to tolerate both the unexpected events and the executive's responses without breaking. By choosing an appropriate level of abstraction for the activities, and second, by generating plans in which the activities have flexible start and end time.

Changing the start or end time of an activity may also affect other activities in the plan. Although flexibility in the activity start and end times typically occurs because the times are under-constrained, flexibility can also occur because the duration of an activity is not determined until execution time. The plan must be able to represent this kind of uncertainty. There two ways of doing this. One is to use the existing capability for flexible and end times; a second approach is to fix the end time of the activity to the latest end time, and change the semantics of the activity.

Conclusion

The most important distinction between the OPSFESA and other similar systems is that our architecture pays more attention to the characteristic of small satellite and its constrains and limitations, that makes the realized system to be smarter and terser. Besides planning and scheduling technology, the key technologies of spacecraft autonomy also include a robust multithreaded executive including internal commands and telemetering, telecontrol commands, and a model-based fault diagnosis and recovery system.

8.8 Fuzzy Logic Applications in Power Systems

8.8.1 Introduction to Power System Control

Motivation

A reliable, continuous supply of electric energy is essential for the functioning of today's complex societies. Due to a combination of increasing energy consumption and impediments of various kinds concerning the extension of existing electric transmission networks, these power systems are operated closer and closer to their limits. Deregulatory efforts will tighten the economical constraints under which utilities have to operate their own network or allow or prevent competitors from using it. This in turn will require more precise power flow control which is made possible by phase angle controllers being developed using new power electronic equipment. However, it is to be expected that these highly nonlinear components will introduce harmonics and require nonlinear control in order to prevent system destabilization. This situation requires a significantly less conservative power system operation regime, which, in turn, is possible only by monitoring and controlling the system, state in much more detail than was necessary previously.

Power System Control Tasks

In electric power systems, one can distinguish three different control levels:

- *Generating unit controls, which* consist of prime mover control and excitation control with automatic voltage control (AVR) and power system stabilization (PSS). The first controls generator speed deviation and energy supply system variable like boiler pressure or water flow. Excitation control aims at maintaining the generator terminal voltage and reactive power output within its machine-dependent limits.
- *System generation control, w*hich determines active power output such that the overall system generation meets the system load. It further controls the frequency and the tie line flows between different power system areas.
- Finally *transmission control* monitors power and voltage control devices like tap-changing transformers, synchronous condensers and static VAR compensators. In reality all controls affect both components and systems. For example the AVR is known to introduce local mode oscillations as well as inter-area oscillations, which in turn are counteracted by a well-tuned PSS.

From the viewpoint of system automation, generating unit control is a complete closed-loop system and in the last decade a lot of effort has been dedicated to improve the performance of the controllers. The main problem for example with excitation control is that the control law is based on a linearized machine model and the control parameters are tuned to some nominal operating conditions.

In case of a large disturbance, the system conditions will change in a highly nonlinear manner and the controller parameters are no longer valid. In this case the controller may even add a destabilizing effect to the disturbance by for example adding negative damping. These problems provide an important motivation to explore novel control techniques like fuzzy systems and their potential in the area of prediction, approximation, classification and control. *Power system control* consists of four steps:

1. System parametric or state-space modeling based on physical components or assumed properties

2. System parameter identification based on component data and measurements
3. System observation of inputs and outputs by filtering, prediction, state estimation, etc
4. Design of an open-loop or closed-loop system control law such that the operating conditions are met

In the case of electric power systems and electric machines, individual components are modeled in terms of resistors, inductors, capacitors, machine inertia, etc. Their interaction is modeled according to the laws of electromagnetic circuits and fields. The resulting set of differential equations then defines a state-space model whose parameters, for example the machine reactances have to be identified under steady state and transient conditions. Voltage signals on the other hand are modeled as a trignometric sum of sin and cos functions without taking the underlying physical model into account. In this case the free parameters to be identified are the signal amplitudes.

System identification may be defined as the process of determining the parameters of the dynamic system model using observed input and output data. Dynamic load modeling attempts to model individual as well as composite loads of the system. Identification of machine parameters is another identification task.

System observation in power systems concerns off- and online monitoring of directly or indirectly observable system variables. Load forecasting, for example is an offline monitoring task, power quality monitoring an online monitoring task. *State estimation* calculates the most likely values for power system parameters like bus voltage and line flow by giving a least-squares estimate of a set of redundant measurements.

Closed-loop control attempts to counter-balance undesirable effects like undamped frequency oscillations or voltage deviations in a closed-loop feedback environment. Excitation control, automatic voltage regulation and PSS fall in this category.

Assume that the plant has been modeled by the following single-input, single-output noise-free continuous system and the parameters of the nonlinear plant model f and observer model h have been identified.

$$\mathbf{dx}(t)/\mathrm{d}t = f(\mathbf{x}, u, t),$$
$$y = h(\mathbf{x}, u, t),$$

where $x \in \Re^m$ is the state vector of the modeled plant, $u \in \Re$ is the control input of the system, $y \in \Re$ is the observed output of the system, f is a nonlinear state-space model of the plant, and h is the nonlinear observer model.

r is a reference signal, $f(e, t)$ denotes the controller and $e = r - y$ is the feedback error to be minimized.

In the case of a power system stabilizer the reference signal is the reference voltage V_{ref} and the terminal voltage V_{t}, the plant output is the angular

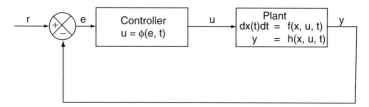

Fig. 8.112. Closed loop control system

speed deviation Dw, and the control signal is the stabilizing voltage vs. The plant model comprises excitation system, AVR and generator model. So far the continuous system model has been considered. If the system in Fig. 8.112 is discretized taking into account $k = 1 \ldots n$ time-steps $t = kT$, the control, output and error signals will become higher-dimensional vectors, as for example

$$\mathbf{u} = [u(k), u(k-1), \ldots, \mathbf{u}(k-m)]^{\mathrm{T}}$$
$$\mathbf{e} = [e(k), e(k-1), \ldots, \mathbf{e}(k-n)]^{\mathrm{T}}$$
$$y(k) = h_{\mathrm{d}}(\mathbf{x}(k), \mathbf{x}(k-1), \ldots, u(k), u(k-1) \ldots)$$

In the case of feedback error minimization, task 4 of the controller design consists in finding a function

$$\phi : \Re^n - > \Re \text{ such that } \phi(\mathbf{e}) = u(k+1) \text{ and } |e(k+1)| = \min.$$

In the case of linear controllable and observable systems, a controller f can be found through inversion of the system transfer function and pole placement. In the case of nonlinear systems there is no general closed form of f. How fuzzy systems can provide an approximation of the controller f is explored in the following sections.

State of the Art of Fuzzy Control for Power Systems

In the case of power systems, control measurement data can be obtained for the discretized plant output $\mathbf{y}$, the reference signal $\mathbf{r}$ and the control input $\mathbf{u}$. In analogy to neural networks, let this data be referred to as the training data. A short overview of the studies of fuzzy systems of type $\mathbf{u} = F(\mathbf{e})$ as controllers $f(e)$ in the area of power systems or generation control is now given. The majority of fuzzy controllers can be found in the area of excitation control, especially power system stabilizers (PSS). An upcoming important area is control of FACTS devices like thyristors and GTOs. Table 8.18 gives an snapshot of the state-of-the-art. Given the considerable number of publications Table 8.18 cannot claim to give a complete overview. Instead this short

Table 8.18. Overview of fuzzy systems applications to power system control

Application	Fuzzy approach	Membership functions	comments
PSS	Self-adaptive	B-Spline	1-machine, 2-line-inf. Bus; Lab experiment, micromachine system DSP controller
PSS	Rule-based	Triangular	1-machine, 2-line-inf. Bus; Lab experiment, micromachine system DSP controller
PSS	Self-adaptive	Trapezoidal	Simulation on analog power system simulator (12 machine max); Prototype field test on two hydro units, frequency response study, capacitor bank switching
PSS	Self-adaptive	Gaussian	Computer simulation, utility power system
AVR & PSS	Rule-based	Trapezoidal	Computer simulation, 3-machine test system
PSS	Rule-based	Triangular	Computer simulation, 2-machine, 4-line-inf. bus
PSS	Self-adaptive	Triangular	Computer simulation, 1-machine, 2-line-inf. bus
PSS	Rule-based	Triangular	Computer simulation, 3-machine, 7-line-inf. bus
FACTS	Self-adaptive	Trapezoidal	Computer simulation, 5-macine, 13-line-inf. Bus, capacitor bank switching, thyristor controlled braking resistor, static VAR compensator
Induction motor	Rule-based	Trapezoidal	Lab inverter/3 hp induction motor, PC-based microcontroller
Variable speed drive	Rule-based	Triangular	Lab inverter/reduced ratings, induction motor, DSP-based microcontroller
PWM inverter	Rule-based	Triangular	Lab prototype of wind energy conversion system

summary intends to provide the reader with some information on typical approaches and project states. The implementation of the fuzzy controller on a PC or DSP in order to control actual small generators or motors in an experimental laboratory environment. In most cases the membership functions are established based on data samples.

Those approaches listed as rule-based attempt to justify the fuzzy sets in terms of linguistic descriptions like "if angle is small then deviation should be small." The comparison of fuzzy controllers and conventional controllers stresses advantages of fuzzy controllers as being "generic" parametric models instead of circuit based state space models. The self-adaptive controllers can be easily tuned to different operating conditions and all projects report better tracking capabilities of the fuzzy controllers when compared to conventional controllers. However, the sensitivity issues concerning the range of validity of the tuning and the detection of changes of operating conditions still need to be investigated for conventional as well as for fuzzy controllers. This is especially important for power system control where topology, load and generation can change stochastically and discontinuously.

Application of a Fuzzy Logic Controller as a Power System Stabilizer

The design process of the fuzzy logic controller (FLC) has five steps:

 – Selection of the fuzzy control variables
 – Membership function definition
 – Rule creation
 – Inference engine
 – Defuzzification strategies

To design the FLC, variables which can represent the dynamic performance of the plant to be controlled should be chosen as the inputs to the controller. In addition to the proper input signals, signal gains and fuzzy subsets should be defined. It is common to use the output error (e) and the rate or derivative of the output (e') as controller inputs. In the case of the fuzzy logic based power system stabilizer (FPSS), the generator speed deviation (Dw) and its derivative (Dw), the acceleration, are considered as the inputs of the FPSS. After sampling, two appropriate gains, SG and AG are applied to speed deviation and acceleration, respectively, and then fed to the FPSS. The output of the controller is also scaled by an output gain, UG, and added to the AVR input signal.

The measured input variables are converted into suitable linguistic variables. In this case, seven fuzzy subsets, NB (Negative Big), NM (Negative Medium), NS (Negative Small), Z (Zero), PS (Positive Small), PM (Positive Medium) and PB (Positive Big) have been chosen. Membership functions for the input variables used here are shown in Fig. 8.113. These membership functions are symmetrical and each one overlaps with the adjacent functions by 50%.

In practice, the membership functions are normalized in the interval $[-L, L]$, which is symmetrical around zero. Thus, control signal amplitudes (fuzzy variables) are expressed in terms of controller parameters (gains). These parameters can be defined as:

$$Kj = 2L/X_{\text{range}j},$$

where $X_{\text{range}j}$ defines the full range of the control variable Xj.

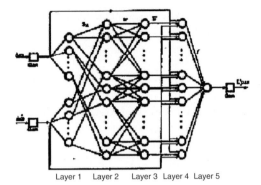

Layer 1   Layer 2   Layer 3   Layer 4 Layer 5

Fig. 8.113. Architecture of ANF PSS

Table 8.19. FPSS control rule table

		$\Delta\omega$						
		NB	NM	NS	Z	PS	PM	PB
	NB	NB	NB	NB	NM	NM	NS	Z
	NM	NB	NB	NM	NM	NS	Z	PS
	NS	NB	NM	NS	NS	Z	PS	PM
$\Delta\dot{\omega}$	Z	NM	NM	NS	Z	PS	PM	PM
	PS	NM	NS	Z	PS	PS	PM	PB
	PM	NS	Z	PS	PM	PM	PB	PB
	PB	Z	PS	PM	PM	PB	PB	PB

In this case, both inputs of the FPSS have seven subsets. Thus, a fuzzy rule table with 49 rules should be constructed. A rule table that is formulated based on the past experience of manual tuning of a conventional PSS (CPSS) is shown in Table 8.19. In the next step, the controller output is computed by the inference mechanism. As an example, consider a pair of D w and D &w inputs to the controller. In fuzzification stage these inputs are converted to membership grades for each of the seven subsets, e.g., mw(PB), mw(PM), etc., and mw& (PB), mw& (PM), etc. Thus, there is a set of 49 pairs of membership grades for each of input pair. The smaller element in each pair would be the grade of membership for any of the possible control actions. For example, the FPSS output membership grade for the first rule in Table 8.19 is given by:

$$\mu_{out}^{1} = \min[\mu_{\omega}(NB), \mu_{\dot{\omega}}(NB)].$$

The output of the FPSS is limited to 0.1 pu and is divided in seven subsets. Also, the output membership functions are chosen as singleton functions as indicated in Table 8.20.

The output of the inference process at this stage is a fuzzy set. In order to take a nonfuzzy (crisp) control action, the fuzzy control action inferred from the fuzzy control algorithm must be defuzzified. Three different defuzzification

Table 8.20. Output membership functions

Output subsets	NB	NM	NS	Z	PS	PM	PB
u_{out} pu	-0.10	-0.06	-0.03	0.0	0.03	0.06	0.10

methods, the Max criterion method, the MoM method and the centre of gravity method are commonly used. To ensure that all of the fired rules have some contribution in the output control action, the centre of gravity method, using the following equation, is employed in this study:

$$u_{pss} = \frac{\sum_{i=1}^{Rules} u_{out}(Z)\mu_{out}^i(Z)}{\sum_{i=1}^{Rules} \mu_{out}^i(Z)},$$

where $\mu_{out}^i(Z)$ denotes the output membership grade for ith rule with the output subset of Z. To achieve the best performance with FPSS, the input and output gains need to be selected properly. For this purpose, the speed deviation and its derivative were measured for a variety of small and large disturbances applied to the system. The universe of discourse for both inputs of the FPSS is normalized.

Therefore, appropriate gains should be chosen such that they map the measured inputs of the FPSS to their suitable linguistic variables. For example, for a small disturbance the measured inputs should be mapped to the "Small" domain, whereas for a large disturbance they should be mapped to the saturated region of the "Large" domain. It was found for the system under study that for different applied disturbances on the system, the magnitude of D &w was about ten times that of D w. As the same membership functions are used here for both inputs, the input gain for D w should be about ten times the input gain for D &w. After fixing the input gains, the output gain should be selected such that the controller is sensitive to the errors in the lower region of the universe of discourse. At the same time, to minimize the time the control stays in the saturated region of the controller, the output gain selected should not be very high.

A Self Learning Fuzzy Power System Stabilizer

A lot of effort is required in the creation and tuning of the fuzzy rules for an FLC which can be time consuming and nontrivial. A self-learning adaptive network can be used to reduce this effort. A class of adaptive networks, which are functionally equivalent to FLC, combines the idea of the FLC and adaptive network structures.

Essentially, an adaptive network is a superset of multilayer feed forward network with supervised learning capability. The network consists of nodes and directional links through which the nodes are connected. Each node performs a particular function, which may vary from node to node. In this network, the links between the nodes only indicate the direction of flow of signals and a part or all of the nodes contain the adaptive parameters. These parameters are specified by the learning algorithm and should be updated to achieve the desired input–output mapping.

An adaptive network based FLC employed as a fuzzy logic PSS (ANFPSS), Fig. 8.113, has two inputs, the generator speed deviation and its derivative, and one control output. The node functions in each layer are:

- Layer 1 performs a membership function
- Layer 2 represents the firing strength of each rule
- Layer 3 calculates the normalized firing strength of each rule
- Layer 4 output is the weighted consequent part of the rule table
- Layer 5 computes the overall output as the summation of all incoming nodes.

This adaptive network is functionally equivalent to a fuzzy logic PSS. Because the adaptive network has the property of learning, the learning algorithm can tune fuzzy rules and membership functions of the controller automatically. Learning is based on the error evaluated by comparing the output of the ANFPSS and a desired PSS.

In a typical situation, a desired controller may either not be available, or the extensive input–output data required for training may not be easy to procure. A self-learning FLC does not require another desired controller to obtain the training data. It is trained from the controlled plant output, which in the case of the self-learning ANFPSS has been taken as the generator speed deviation.

In this approach, first a function approximator (or model) is required to represent the input–output behavior of the plant. An adaptive network based fuzzy logic model, which has the same structure as the controller, is employed to model the plant. The function of this model is to compute the derivative of the model output with respect to its input by means of the back propagation process. Consequently, by propagating errors between the actual and the desired plant outputs, back through the model, error in the control signal can be calculated. The error in the control signal can be used to train the controller. A block diagram of the self-learning FLC, showing an adaptive network containing two subnetworks, the fuzzy controller and the plant model, is shown in Fig. 8.114.

The training process for the controller starts from an initial state at $t = 0$. Then the FLC and the plant model generate the next states of control and D w at time $t = h$. The process continues till the plant state trajectory is determined based on the minimization of a performance index.

A number of studies have been performed to investigate the performance of the self-learning FLC structure employed as a self-learning ANF PSS on a single-machine infinite-bus system. One such for a 0.2 pu step increase in torque under leading power factor conditions is shown in Fig. 8.115.

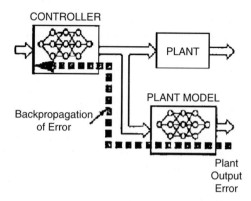

Fig. 8.114. Error propagation through plant model

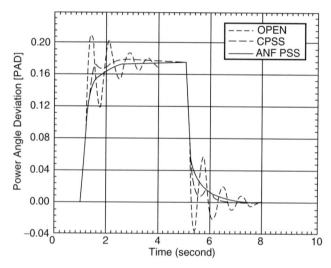

Fig. 8.115. Response curve

Control Design Techniques

When fuzzy systems are used as controllers, they are called *fuzzy controllers*. If fuzzy systems are used to model the process and controllers are designed based on the model, then resulting controllers also are called fuzzy controllers. Therefore, fuzzy controllers are nonlinear controllers with a special structure. Fuzzy control has represented the most successful applications of fuzzy theory to practical problems. Fuzzy control can be classified into static fuzzy control and adaptive fuzzy control. In static fuzzy control, the structure and parameters of the fuzzy controller are fixed and do not change during real-time operation. On the other hand in adaptive fuzzy control, the structure and/or parameters of the fuzzy controller change during real-time operation. Fixed fuzzy control is simpler than adaptive fuzzy control, but requires more knowledge of the process model or heuristic rules. Adaptive fuzzy control, on the other hand, is more expensive to implement, but requires less information and may perform better.

Fixed Fuzzy Controller Design

Fuzzy control and conventional control have similarities and differences. They are similar in the sense that they must address the same issues that are common to any control problem, such as stability and performance. However, there is a fundamental difference between fuzzy control and conventional control. Conventional control starts with a mathematical model of the process and controllers are designed based on the model. Fuzzy control, on the other hand, starts with heuristics and human expertise (in terms of fuzzy IF–THEN rules) and controllers are designed by synthesizing these rules. That is, the information used to construct the two types of controllers is different; see Fig. 8.116. Advanced fuzzy controllers, however, can make use of both heuristics and mathematical models.

For many practical problems, it is difficult to obtain an accurate yet simple mathematical model, but there are human experts who can provide heuristics and rule-of-thumb that are very useful for controlling the process. Fuzzy control is most useful for these kinds of problems. If the mathematical model of the process is unknown, we can design fuzzy controllers in a systematic manner that guarantees certain key performance criteria.

The design techniques for fuzzy controllers can be classified into the trial-and-error approach and the theoretical approach. In the trial-and-error approach, a set of fuzzy IF–THEN rules are collected from human experts or documented knowledge base, and the fuzzy controllers are constructed from these fuzzy IF–THEN rules. The fuzzy controllers are tested in the real system and if the performance is not satisfactory, the rules are fine-tuned or redesigned in a number of trial-and-error cycles until the performance is satisfactory. In theoretical approach, the structure and parameters of the fuzzy controller are designed in such a way that certain performance criteria are

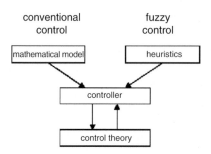

Fig. 8.116. Fuzzy control and conventional control

guaranteed. Both approaches, of course, can be combined to give the best fuzzy controllers.

Trial-and-Error Approach

The trial-and-error approach to fuzzy controller design can be summarized in the following steps:

1. Select state and control variables. The state variables should characterize the key features of the system and the control variables should be able to influence the states of the system. The state variables are the inputs to the fuzzy controller and the control variables are the output of the fuzzy controller.
2. Construct IF–THEN rules between the state and control variables. The formulation of these rules can be achieved in two different heuristic approaches. The most common approach is the linguistic verbalization of human experts. Another approach is to interrogate experienced experts or operators using a carefully organized questionnaire.
3. Test the fuzzy IF–THEN rules in the system. The closed-loop system with the fuzzy controller is run and if the performance is not satisfactory, fine tune or redesign the fuzzy controller and repeat the procedure until the performance is satisfactory.

The resulting fuzzy IF–THEN rule can be in the following two types:

$$\text{Type I}: \quad \text{IF } x_1 \text{ is } A_1^t \text{ AND} \ldots \text{AND } x_n \text{ is } A_N^t,$$
$$\text{THEN } u \text{ is } B^f.$$

$$\text{Type II}: \quad \text{IF } x_1 \text{ is } A_1^t \text{ AND} \ldots \text{AND } x_n \text{ is } A_n^t,$$
$$\text{THEN } u \text{ is } c_0^t + c_1^t x_1 + \ldots + c_n^t x_n.$$

In Type I, both the antecedent and consequent have linguistic variables, On the other hand in Type II, the consequent is parameterized function of the input to the fuzzy controller, or the state variables. Comparing the two types, the THEN part of the rule is changed from a linguistic description to a simple mathematical formula. This change makes it easier to combine the rules. In fact, Type II, the Takagi-Sugeno system, is a weighted average of the rules in the THEN parts of the rules. This framework is useful in tuning the rules mathematically. Type II, on the other hand has drawbacks (1) its THEN part is a mathematical formula and therefore may not provide a natural framework to represent human knowledge and (2) there is not much freedom left to apply different principles in fuzzy logic, so that the versatility of fuzzy systems is not fully represented in this framework.

Theoretical Approach

Knowing the mathematical model of a system is not a necessary condition for designing fuzzy controllers. However, in order to analyze the performance of the closed loop fuzzy control system theoretically, we need to have some knowledge on the model of the system. This approach assumes a mathematical model for the system, so that mathematical analysis can be performed to establish the properties of the designed system.

Theoretical approach can be classified into the following categories:

1. Stable controller design
2. Optimal controller design
3. Sliding mode controller design
4. Supervisory controller design
5. Fuzzy system model-based controller design

Expert System Applications

The major use of artificial intelligence today is in expert systems, AI programs that act as intelligent advisors or consultants. Drawing on stored knowledge in a specific domain, an inexperienced user applies inferencing capability to tap the knowledge base. As a result, almost anyone can solve problems and make decisions in a subject area nearly as well as an expert. It is not easy to give a precise definition of an expert system, because the concept of expert system itself is changing as technological advances in computer systems take place and new tasks are incorporated into the old ones. In simple words, it can be defined as a computer program that models the reasoning and action processes of a human expert in a given problem area. Expert systems, like human experts, attempt to reason within specific knowledge domains.

An expert system permits the knowledge and experience of one or more experts to be captured and stored in a computer. This knowledge can then be used by anyone requiring it. The purpose of an expert system is not to replace the experts, but simply to make their knowledge and experience more idely available. Typically there are more problems to solve than there are experts available to handle them. The expert system permits others to increase their productivity, improve the quality of their decisions, or simply to solve problems when an expert is not available. Valuable knowledge is a major resource and it often lies with only a few experts. It is important to capture that knowledge so others can use it. Experts retire, get sick, move on to other fields, and otherwise become unavailable. Thus the knowledge is lost. Books can capture some knowledge, but they leave the problem of application up to the reader. Expert systems provide a direct means of applying expertise.

An expert system has three main components: a *knowledge base*, an *inference engine*, and a *man-machine interface*. The knowledge base is the set of rules describing the domain knowledge for use in problem solving. The prime element of the man-machine interface is a working memory, which serves to store information from the user of the system and the intermediate results of knowledge processing. The inference engine uses the domain knowledge together with the acquired information about the problem to reason and provide expert solution.

A Working Definition

An expert system is an artificial intelligence (AI) program incorporating a knowledge base and an inferencing system. It is a highly specialized piece of software that attempts to duplicate the function of an expert in some field of expertise. The program acts as an intelligent consultant or advisor in the domain of interest, capturing the knowledge of one or more experts. Nonexperts can then tap the expert system to answer questions, solve problems, and make decisions in the domain.

The expert system is a fresh new, innovative way to capture and package knowledge. Its strength lies in its ability to be put to practical use when an expert is not available. Expert systems make knowledge more widely available and help overcome the age-old problem of translating knowledge into practical, useful results. It is one more way that technology is helping us get a hand on the oversupply of information. All AI software is knowledge-based as it contains useful facts, data, and relationships that are applied to a problem.

Expert systems, however, are a special type of knowledge based system, they contain heuristic knowledge. Heuristics are primarily from real world experience, not from textbooks. It is knowledge that directly from those people the experts – who have worked for years within the domain. It is knowledge derived from learning by doing. It is perhaps the most useful kind of knowledge, specifically related to everyday problems. It has been said that knowledge is power.

Certainly there is truth in that but in a more practical sense, knowledge becomes power only when it is applied. The bottom line in any field of endeavor is RESULTS, some positive benefit or outcome. Expert systems are one more way to achieve results faster and easier.

Desirable Expert System Features

Expert systems are far more useful if they have some additional features. These include an explanation facility, ease of modification, transportability, and adaptive learning ability. Let us take a look at each of these key features.

Expert systems are very impersonal and get right to the point. Many first time users are surprised at how quickly the expert system comes up with a recommendation, conclusion, or selection. The result is usually stated concisely, and sometimes very curtly, using rule clauses. A natural language interface will help improve this situation, but that is not the main problem. A more important issue is that often users have difficulty in "buying" the output decision. They question it or perhaps do not believe it. Users frequently want to know how the expert system arrived at that answer. Most of the better expert systems have a means for explaining their conclusion. Typically, this takes the form of showing the rules involved in the decision and the sequence in which they were fired. All of the information is retained in the database for that purpose. When users want to know the expert system's line of reasoning, they can read the rules and follow the logic themselves. Some rule formats permit the inclusion of an explanation statement that justifies or elaborates on the need for or importance of the rule.

The explanation facility is important because it helps the user feel comfortable with the outcome. Sometimes the outcome is a surprise or somewhat different than expected. It is difficult for an individual to follow the advice of the expert in these cases. However, once the expert system explains itself, the user better understands the decision and feels more at ease in making a decision based upon it.

Ease of Modification

As indicated earlier, the integrity of the knowledge base depends upon how accurate and up to date it is. In domains where rapid changes take place, it is important that some means be provided for quickly and easily incorporating this knowledge. When the expert system was developed using one of the newer development tools, it is usually a simple matter to modify the knowledge base by writing new rules, modifying existing rules, or removing rules. The better systems have special software subsystems, which allow these changes to be made without difficulty. If the system has been programed in LISP or Prolog, changes are much more difficult to make. In examining or evaluating an expert system, this feature should be considered seriously in context of the modification.

Transportability

The wider the availability of an expert system the more useful the system will be. An expert system is usually designed to operate on one particular type of computer, and the software development tools used to create the expert system usually dictate this. If the expert system will operate on only one type of computer, its potential exposure is reduced. The more different types of computers for which the expert system is available, the more widely the expertise can be used. If possible, when the expert system is to be developed, it should be done in such a way that it is readily transportable to different types of machines. This may mean choosing a programing language or software development tool that is available on more than one target machine.

Adaptive Learning Ability

This is an advanced feature of some expert systems that allows them to learn their own use or experience. As the expert system is being operated, the engine will draw conclusions that can, in fact, produce new knowledge. New functions stored temporarily in the database, but in some systems they can lead to the development of a new rule which can be stored in the knowledge base and used again in the problem. The more the system is used, the more it learns about the domain and more valuable it becomes.

The term learning as applied to expert systems refers to the process of the expert system new things by adding additional rules or modifying existing rules. On the other hand, if the system incorporates the ability to learn it becomes a much more powerful and effective problem solver. Today few expert systems have this capability, but it is a feature that is sure to be further developed into future systems.

Suitable Application Areas for Expert Systems

Expert systems are best suited for problems with limited domains and well-defined expertise. Application areas involving common sense and analogical reasoning do not lend themselves well to expert system development. The suitability of expert system-based approaches can be determined by taking into consideration some criterion based on general experience in this field. Expert systems are found to be suitable for those problems for which the solution steps are not clearly defined. The action taken depends not only on the present values of data but on the outcome of previous decisions, historical data, past experience and trends. In power systems, many promising applications have been reported in the broad fields of system control, alarm processing and fault diagnosis, system monitoring, decision support, system analysis and planning.

Expert System Applications

Expert systems are ideal when it is necessary for an individual to select the best alternative from a long list of choices. Based on the criteria supplied to it, the expert system can choose the best option. For example, there are expert systems that will help you select one of the many places to invest your money based on your own financial condition, goals, and personality traits.

An expert system can be created to help an individual troubleshoot and repair a complex piece of equipment. The various troubles and symptoms can be given to an expert system, which then identifies the problem and suggests courses of action for repair. Expert systems also can be used to aid in diagnosing medical cases. Symptoms and test results can be given to the expert system, which then searches its knowledge base in an attempt to match these input conditions with a particular malady or disease. This results in a conclusion about the illness and some possible suggestions on how to treat it. Such an expert system can greatly aid a doctor in diagnosing an illness and prescribing treatment. It does not replace doctors, but helps them confirm their own decisions and may provide alternative conclusions.

Expert systems perform financial analysis. Some expert systems evaluate stocks and recommend buy, sell, or hold positions. Other expert systems can be used in tax planning and budgeting. Expert systems have been used to help

Table 8.21. Generic expert system categories

Control – intelligent automation
Debugging– renovation corrections to faults
Design – development products to specification
Diagnosis – estimated defects
Instruction – optimized computer instruction
Interpretation – clarification of situations
Planning – developing goal-oriented scheme
Prediction – intelligent guessing of outcome
Repair – automatic diagnosis, debugging, planning, and fixing

locate oil and mineral deposits or to configure complex computer systems and recommend a specific policy in a variety of insurance applications. Expert systems also have been used to locate oil spills and provide speedy critical advice to commanders in battlefield situations. The variety of potential applications is enormous. If one or more experts exist in the domain of interest, and the knowledge can be codified and represented in symbolic form, then an expert system can be created.

In power systems, many promising applications have been reported in the broad field of system control, alarm processing and fault diagnosis, system monitoring, decision support, system analysis and planning. An excellent review of the popular application areas can be found. Table 8.21 shows the main categories of applications suitable for expert systems. If the problem to be solved falls into one of these categories, it is a candidate for expert systems solution. This is not to imply that an expert system is the only answer. There may very well be a more conventional algorithmic program that will do the job. In any case, assuming the problem is one of these types, an expert system should most certainly be considered as an alternative. Now let us take a look at each category in more detail.

Reasoning with Uncertainty in Rule Based Expert Systems

One of the important feature in expert systems is their ability to deal with incorrect or uncertain information. There will be times when an expert system, in gathering initial inputs, will ask you a question for which you do not have the answer. In such a case, you simply say that you do not know. Expert systems are designed to deal with cases such as this. Because you may not have a particular fact, the search process will undoubtedly take a different path. It may take longer to come up with an answer, but the expert system will give you an answer.

Traditional algorithmic software simply cannot deal with incomplete information. If you leave out a piece of data, you may not receive an answer at all. If the data is incorrect, the answer will be incorrect. This is where artificial intelligence programs, particularly expert systems, are particularly useful.

When the inputs are ambiguous or completely missing, the program may still find a solution to your problem. The system may qualify that solution, but at least it is an answer that can in many cases be put to practical use. This is consistent with expert level problem solving where one rarely has all the facts before making a decision. Our common sense or knowledge of the problem tells us what is important to know and what is less important. Experts almost always work with incomplete or questionable information, but that it does not prevent them from solving the problem.

Thus, increasingly in the design of expert systems, there has been a focus on methods of obtaining approximate solutions to a problem when there is no clear conclusion from the given data. Logically, as expert system problems become more complex, the difficulty of reaching a conclusion with complete certainty increases, so in some cases, there must be a method of handing uncertainty. In, researchers report that a classical expert system gave incorrect results due to the sharpness of the boundaries created by the if–then rules of the system; however, once a method for dealing with uncertainty (in these two cases fuzzy set theory) was used, the expert system reached the desired conclusions. The successful performance of expert systems relies heavily on human expert knowledge derived from domain experts based on their experience. The other forms of knowledge include causal knowledge and information from case studies, databases, etc. Knowledge is typically expressed in the form of high-level rules. The expert knowledge takes the form of heuristics, procedural rules and strategies in nature.

It inherently contains vagueness and imprecision because an expert is not able to explicitly express their knowledge. The process of acquiring knowledge is also quite imprecise, because the expert is usually not aware of all the tools used in the reasoning process. The knowledge that one reasons with may itself contains uncertainty. Uncertain data and incomplete information are other sources of uncertainty in expert systems.

Uncertainty in rule-based expert systems occurs in two forms. The first form is linguistic uncertainty which occurs if an antecedent contains vague statements such as the "level is high" or "the value is near 20." The other form of uncertainty, called evidential uncertainty, occurs if the relationship between an observation and a conclusion is not entirely certain. This type of uncertainty is most commonly handled using conditional probability that indicates the likelihood that a particular observation leads to a specific conclusion. The study of making decisions under either of these types of uncertainty will be referred to as plausible or approximate reasoning in this work. Several methods of dealing with uncertainty in expert systems have been proposed, including

- Subjective probability
- Certainty factors
- Fuzzy measures
- Fuzzy set theory

The first three methods are generally used to handle evidential uncertainty, while the last method, fuzzy set theory is used to incorporate linguistic uncertainty. These methods of reasoning with uncertainty will be discussed in the following sections. For a comprehensive list of methods used in reasoning with uncertainty including a discussion about their application.

As expert assessments of the indicators of the problem may be imprecise, fuzzy sets may be used for determining the degree to which a rule from the expert system applies to the data that is analyzed. When applying a method of reasoning with uncertainty to a rule based expert system, there must be a method of combining or propagating uncertainty between rules. A method of propagating uncertainty for the method of reasoning with uncertainty will be discussed in the next section.

Example Application: Fault Diagnosis

In the past few years, great emphasis has been put in applying the expert systems for transmission system fault diagnosis. However, very few papers deal with the unavoidable uncertainties that occur during operation involving the fault location and other available information. This example shows a method using fuzzy sets to cope with such uncertainties.

Problem Statement

To reduce the outage time and enhance service reliability, it is essential for dispatchers to locate fault sections in a power system as soon as possible. Currently, heuristic rules from dispatchers' past experiences are extensively used in fault diagnosis. The important role of such experience has motivated extensive recent work on the application of expert system in this field. The uncertainties occur due to failures of protective relays and breakers, errors of local acquisition and transmission, and inaccurate occurrence time, etc. An effective approach is thus necessary to deal with uncertainties in these expert systems. Fault diagnosis in electric power system is a facet operation. Every signal and step contains some uncertainties, which can be modeled by membership functions (Fig. 8.117).

Fuzzy set theory is used to determine the most likely fault sections in the approach presented here. Membership functions of the possible fault sections are the most important factors in the inference procedures and decision making. In this example, the membership function of a hypothesis is used to describe the extent to which the available information and the system knowledge match the hypothesis. They are manipulated during inference based on rules concerning fault sections.

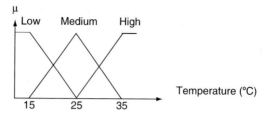

Fig. 8.117. Fuzzy sets for representation of uncertainty

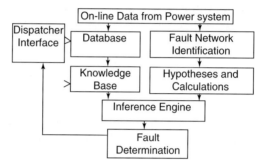

Fig. 8.118. Fuzzy expert system structure

Structure of the Fault Diagnosis System

The FES structure is shown in Fig. 8.118. Its database contains the power system topology, and the status of all breakers and protective relays after the fault.

The knowledge base of the FES contains all the data of the protection system. The information is based on known statistics of protection performance used in the system. If these data are not available when a fault occurs, the FES asks the dispatcher to provide them and then saves them in the database for future use. Models for estimation of possible faults, and heuristic rules about the relay characteristics for actual fault determination are also included here.

Island Identification

When a fault occurs in a power system, the relays corresponding to the fault sections should trip the circuit breakers to isolate the fault sections from being extended. Thus the power system is separated into several parts named subnetworks after the operation of protective relays and circuit breakers (Fig. 8.119). Generally, only a few subsections are formed from the faults. Since the fault sections are confined to these subnetworks, the magnitude of the problem can be reduced greatly. An expert system is developed to identify the island by using the real-time information of circuit breakers and adopting the real-time network topology determination method.

The framework of this efficient method is described as follows:

- Initializing the network: the expert system identifies the power system pre-fault status as the normal operation state by using the real-time network topology determination method. When a fault occurs, the power system status would be changed by the operation of relays and circuit breakers.
- Healthy subnetwork identification: the next step is to identify the network topology of the healthy part of the postfault power system by using the real-time network topology determination method. The healthy subnetwork is called set Shealthy.
- Island identification: by comparing the initial network topology with the healthy subnetwork topology, the differences between them are identified as the island. This subnetwork is called Sisland.

Fault Section Identification

When a fault occurs, the change in breaker status activates the FES. It then classifies the breakers into two sets: no-trip status set and tripped status set. According to the procedures described in previous section, a fault hypothesis F_i is formed as follows:

$$F_i = Fi(\text{CB}) \cup F_i(\text{RL})$$
$$= \{(C_i, \mu_{p_{\text{fault}}}(C_i)|C_i) \in S_{\text{island}}\} \quad (8.15)$$
$$P_{\text{fault}} = \{F_t\} \quad (8.16)$$
$$F_t(\text{CB}) = \{C_t, \mu_{P_{\text{fault}}}^{\text{CB}}(C_t)\} \quad (8.17)$$
$$F_t(\text{RL}) = \{C_t, \mu_{P_{\text{fault}}}^{\text{RL}}(C_t)\} \quad (8.18)$$

where C_i is one of the possible fault sections being considered; P_{fault} is the fuzzy set which contain all the possible fault sections and their membership functions; $F_i(\text{CB})$ is the fuzzy subset by considering only the tripped circuit breakers; $F_i(\text{RL})$ is the fuzzy subset by considering only the operated relays.

The following rules are used to determine the overall grade of the results:

Rule 1: if (first stage protection has operated) then (ignore the signals in second and third stage protections)

Rule 2: if (first and second protection have isolated the suspected fault section) then (ignore the signals of third stage protection)

Rule 3: if (all three stage protections have not isolated the suspected fault section) then (no fault at this section)

During decision making, the most likely fault sections are determined by comparing the above membership grades for each possible fault section using either or both of the following methods

(1) *Maximum selection*: the most likely fault section is the one with the highest membership grade

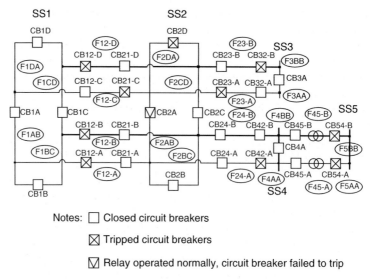

Notes: □ Closed circuit breakers

☒ Tripped circuit breakers

☑ Relay operated normally, circuit breaker failed to trip

Fig. 8.119. Operating relays and tripped circuit breakers

(2) *a-Level selection*: the a-level set includes all fault sections with a membership grade

According to the methods described above, two additional rules are formed for decision making:

Rule 4: (maximum selection): if (section M has the greatest membership function compared with all the other possible fault sections) then (select M as the fault section)

Rule 5: (a-level selection): if (the membership grade of a fault section is greater than the constant a) then (add this section to the fault section set) case study

Conclusion

A fuzzy system can be used to approximate a controller. The determination of the membership functions and the fuzzy rule base is illustrated in two ways:

– The empirical way using linguistic sets and rules and human knowledge
– The self-organizing way using data samples and analysis

Both approaches do not necessarily need a detailed state space model of the plant. The advantage of the first approach is the use of heuristics and human knowledge. However the demonstration of stability for this type of controller is very tedious if not impossible.

A lot of progress has been made concerning the application of fuzzy systems to power system control problems. Recently, fuzzy controllers have achieved commercialization. Generally, a conventional rule-based expert system for bulk power system needs several hundreds of rules. It is time consuming in inference procedures to search for suitable rules during inferencing. On the other hand, fuzzy set based expert systems tend to be much faster compared to traditional rule-based expert systems for most of the rules are replaced by the calculation of the membership functions of the applicable rules. Only a few rules or functions are used in the inference engine.

The fuzzy set approach for uncertainty processing in expert systems offers many advantages to compared other approaches to deal with uncertainty.

- *Small memory space and computer time*: the knowledge base is very small because there are only a few rules needed during inference. The computation time is therefore also small.
- *Small number of rules*: with properly designed linguistic variables and level of granularity, only a few fuzzy rules are needed for each situation.
- *Flexibility of the system*: membership functions representing the parameters can be changed dynamically according to the situation. It is also possible to develop a self-learning module that modifies the grades of membership automatically according to changing situations.

8.9 Fuzzy Logic in Control

8.9.1 Fuzzy Logic Controller

Control systems are an arrangement of physical components designed to alter or regulate the system based on control action

Control system may be of two systems:

(1) Open loop control system – control action is independent of system output
(2) Closed loop control system – control action depends on system output

Controllers designed may be of two types:

(1) Regulatory type of control
(2) Tracking controllers

Regulatory type of control or regulator: object of the control system is to maintain the physical variable at some constant value in the presence of disturbances, it is a regulator.

Ex: Room temperature control, autopilot.
Tracking controllers: a physical variable is required to follow or track some desired time function.

Ex: Automatic aircraft landing function.

Control Problem: It is stated as:

(1) The output or response of the physical system under control is adjusted as required by the error signal.
(2) The error signal is the difference between the actual (sensor) response of the plant and the desired response, as specified by a reference input.

General Control System Design Stages

The following steps are followed in designing a controller for a complex system:

(1) Large scale systems are decomposed into a collection of decoupled subsystems.
(2) The temporal variations of the systems are made to be "slowly varying."
(3) The nonlinear plant dynamics are locally linearized about a set of operating points.
(4) A set of state variables, control variables, or output features is made available.
(5) A simple P, PD, and PID controllers are designed for each decoupled system.
(6) There may be uncertainties due to external environment, the controller design should be such that to meet all the requirements.
(7) A supervisory control system, either automatic or human expert operator, forms a feedback control loop and helps in adjusting the controllers parameters.

Assumptions Made in Fuzzy Control System Design

Whenever a fuzzy logic based control policy is selected, the following assumptions are made:

(1) The plant is observable and controllable: state, input and output variables are usually available for observation and measurement or computation
(2) There exists knowledge about
    (a) Set of expert production linguistic rules
    (b) Engineering common sense
    (c) Intuition
    (d) A set of input/output measurements data
    (e) Analytic model
(3) A solution exists
(4) The control engineer is looking for a "good enough" solution, not necessarily the optimum one
(5) A controller is to be designed with the best of our knowledge and within an acceptable range of precision
(6) The problems of stability and optimality are still open problems in fuzzy controller design

Control Surface (Decision Surface)

The concept of control surface or decision surface is important in fuzzy control systems methodology.

Consider a nonlinear system,

$$u(t) = h[t, x(t), r(t)]$$

The function h is a *nonlinear hyper surface* in an n-dimensional space.

Consider a linear system, with output feedback or state feedback, it generally is a *hyper plane* in an n-dimensional space.

This surface is known as *control or decision surface*. The control surface describes the dynamics of the controller.

How Control Surface is Obtained?

The control surface, which relates the control action to measure the state and output variables is obtained using four structure.

A fuzzy production rule system consists of four structures:

(1) A set of rules that represents the policies and strategies of the expert decision maker
(2) A set of input data assessed prior to actual decision
(3) A method for evaluating any proposed action, given the available data
(4) A method for determining when to stop searching for better ones

The input data, rules and output data are generally fuzzy sets expressed as membership function defined on a proper space. The method used for the evaluation of rules is known as approximate reasoning or interpolative reasoning.

The above reasoning is commonly represented by composition of fuzzy relations applied to a fuzzy relational equation.

Thus the control surface is obtained from the four structured rules. It is then sampled at a finite number of points, depending on the required resolutions and a lookup table is constructed. The look up table could be downloaded onto a real memory chip and would constitute a fixed controller for the plant.

Simple Fuzzy Logic Controller

The block diagram for simple fuzzy logic controller is shown in Fig. 8.120.
Knowledge base: this is a module which contains the knowledge about all the input and output fuzzy partitions. It will include the term set and the corresponding membership functions defining the input variables to the fuzzy rule base system and the output variables, or control actions, to the plant under control.

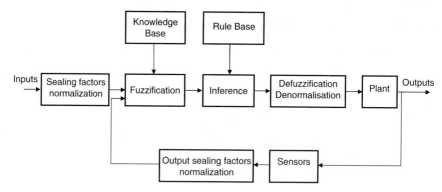

Fig. 8.120. Simple fuzzy logic controller

Design of a Simple Fuzzy Logic Control

It involves the following steps:

(1) Identify the variables of the plant, i.e., inputs, states, and the outputs
(2) Partition the universe of discourse or the interval spanned by each variable into a number of fuzzy subsets, assigning each a linguistic label
(3) Assign or determine a membership function for each fuzzy subset
(4) Assign the fuzzy relationships between the inputs', states' fuzzy subsets on the one hand and the outputs' fuzzy subsets on the other hand, to form the rule base
(5) Choose appropriate scaling factors for the input and output variables in order to normalize the variables to the $[0, 1]$ or the $[-1, 1]$ interval
(6) Fuzzify the inputs to the controller
(7) Use fuzzy approximate reasoning to infer the output contributed from each rule
(8) Aggregate the fuzzy outputs recommended by each rule
(9) Apply defuzzification to form a crisp output

Feature of Simple Fuzzy Logic Control

(1) Fixed and uniform input and output scaling factors
(2) Flat, single partition rule-base with fixed and noninteractive rules
(3) Fixed membership functions
(4) Limited number of rules, which increases exponentially with the number of input variables
(5) Fixed metaknowledge, including the methodology for approximate reasoning, rules aggregation and output defuzzification
(6) Low level control and no hierarchial rule structure

This is a nonadaptive simple fuzzy logic controller, whereas for adaptive fuzzy logic controller, they are adaptively based on some adaptation law, to optimize the controller.

General Fuzzy Logic Controllers

The principal design elements in this case are:

(1) Fuzzification strategies and the interpretation of fuzzification operator or fuzzifier
(2) Knowledge base
    (a) Normalization of the universe of discourse
    (b) Fuzzy partitions of input and output spaces
    (c) Completeness of the partitions
    (d) Choice of the membership function
(3) Rule base
    (a) Choice of process of state and control variables (input and output)
    (b) Source of derivation of fuzzy control rules
    (c) Types of fuzzy control rules
    (d) Completeness of fuzzy control rules
(4) Decision making logic
    (a) Definition of fuzzy implication
    (b) Interpretation of sentence connective
    (c) Interpretation of sentence connective
    (d) Inference mechanism
(5) Defuzzification strategies and the interpretation of a defuzzification operator (defuzzifier)

If all the five above are fixed, the fuzzy logic control system is simple and nonadaptive. Adaptation or change in any of the five design parameters above creates an adaptive fuzzy control system.

Fuzzy Logic Control System Models

The fuzzy logic control system models are expressed in two different forms:

(1) Fuzzy rule based structure
(2) Fuzzy relational equations

(1) *Fuzzy Rule Based Structure*

There are five types of fuzzy rule based system models:

(1) If the input and the output restrictions are given in the form of singletons

$$\text{IF } A^i : x = x_i \text{ THEN } B^i : y = y_i$$

(2) If the input restrictions are in the form of crisp sets and output are given by singleton

$$\text{IF } A^i : x_{i-1} < x < x_i \text{ THEN } B^i : y = y_i$$

(3) If the input conditions are crisp sets and the output is a fuzzy set

$$\text{IF } A^i : x_{i-1} < x < x_i \text{ THEN } y = B^i$$

(4) If the input conditions are fuzzy sets and the outputs are crisp functions,

$$\text{IF } x = A^i \text{ THEN } B^i : y = f_i(x)$$

(5) If both the input and output restrictions are given by fuzzy sets

$$\text{IF } A^i \text{ THEN } B^i$$

(2) *Fuzzy Relational Equations*

The following are the fuzzy relational equations describing a number of commonly used fuzzy control system models,

(1) For a discrete first-order system with input U, the fuzzy model is

$$x_{k+1} = x_k \text{ o } u_k \text{ o } R \quad \text{for } k = 1, 2, \ldots$$

(2) For a discrete Pth order system with single input U, the fuzzy system equation is,

$$
\begin{aligned}
x_{k+p} &= x_k \text{ o } x_{k+1} \text{ o} \cdots \text{o } x_{k+p-1} \text{ o } u_{k+p-1} \text{ o } R \\
y_{k+p} &= x_{k+p}
\end{aligned}
$$

(3) A second-order system with full state feedback is described as

$$
\begin{aligned}
u_k &= x_k \text{ o } x_{k-1} \text{ o } R \\
y_k &= x_k
\end{aligned}
$$

(4) A discrete Pth order SISO system with full state feedback is represented by the following fuzzy relational equation,

$$u_{k+p} = y_k \text{ o } y_{k+1} \text{ o} \cdots \text{o } y_{k+p-1} \text{ o } R$$

In all the cases,

$$
\begin{aligned}
R &= R^1 \text{ U } R^2 \text{ U} \cdots, \text{U } R^r, \\
R &= \{R^1, R^2, \ldots, R^r\},
\end{aligned}
$$

where R is a system transfer function.

$$R^1 : \text{IF } x \text{ is } A^1, \text{ THEN } y \text{ is } B^1$$
$$R^2 : \text{IF } x \text{ is } A^2, \text{ THEN } y \text{ is } B^2$$

.

.

.

$$R^r : \text{IF } x \text{ is } A^r, \text{ THEN } y \text{ is } B^r$$

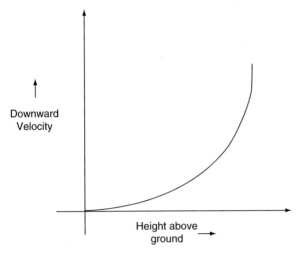

Fig. 8.121. Landing approach of an aircraft

Aircraft Landing Control Problem

This example shows the flexibility and reasonable accuracy for the fuzzy control.

Simulation of the Final Descent and Landing Approach of an Aircraft: the desired downward velocity is proportional to the square of the height (Fig. 8.121):

(1) At higher altitudes, a large downward velocity is desired
(2) As altitude (height) diminishes, the desired downward velocity gets smaller and smaller
(3) As the height becomes vanishingly small, the downward velocity also goes to zero
   In this way, the aircraft will descend from altitude promptly, but will touch down very gently to avoid damage.

State Variables Assumed: The two state variables for this simulation will be:

(1) The height above the ground h
(2) The vertical velocity of the aircraft v

The control output will be a force, that, when applied to the aircraft, will alter its height h and velocity v. The differential control equations are,
   Mass m moving with velocity v has momentum,

$$p = mv$$

(1) If no external forces is applied, the mass will continue in the same direction at the same velocity v

(2) If a force f is applied over a time interval Δt, a change in velocity of $\Delta v = f \cdot \Delta t / m$ is obtained.

$$\text{If } \Delta t = 1.0\,\text{s and } m = 1.0 (\text{lb s}^2\,\text{ft}^{-1})$$
$$\Delta v = f(\text{lb})$$

(or) change in velocity is proportional to the applied force.
In different notation,

$$V_{i+1} = V_i + f_i$$
$$h_{i+1} = h_i + V_i$$

V_{i+1} – new velocity, V_i – old velocity, h_{i+1} – new height, and h_i – old height.

The above "control equations" define the new value of the state variables v and h in response to control input. The following procedures are done:

STEP 1: Construct membership functions for the height h, the velocity v, and the control force f. Define membership functions for state variable.

STEP 2: Define membership function for the control output.

STEP 3: Define the rules and form FAM table. The values in FAM (Fuzzy Associative Memory) table, are the control outputs.

STEP 4: Define the initial conditions and perform simulation.

Here the simulation is done for four cycles, to obtain the accuracy.

Since the task at the hand is to control the aircrafts vertical descent during approach and landing, we will start with the aircraft at an altitude of 1,000 ft, with a downward velocity of $-20\,\text{ft s}^{-1}$.

The state variables are updated for each cycle using,

$$V_{i+1} = V_i + f_i$$
$$h_{i+1} = h_i + V_i$$

Membership values for height (Fig. 8.122)

Linguistic variables	Height (ft)										
	0	100	200	300	400	500	600	700	800	900	1,000
Large (L)	0	0	0	0	0	0	0.2	0.4	0.6	0.8	1.0
Medium (M)	0	0	0	0	0.2	0.4	0.6	0.8	1.0	0.8	0.6
Small (S)	0.4	0.6	0.8	1.0	0.8	0.6	0.4	0.2	0	0	0
Near zero (NZ)	1.0	0.8	0.6	0.4	0.2	0	0	0	0	0	0

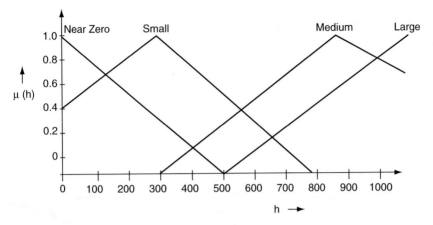

Fig. 8.122. Membership values of height

Membership values for velocity (Fig. 8.123)

Linguistic	Velocity (ft s^{-1})												
variables	−30	−25	−20	−15	−10	−5	0	5	10	15	20	25	30
Up large (UL)	0	0	0	0	0	0	0	0	0	0.5	1.0	1.0	1.0
Up small (US)	0	0	0	0	0	0	0	0.5	1.0	0.5	0	0	0
Zero (Z)	0	0	0	0	0	0.5	1	0.5	0	0	0	0	0
Down small (DS)	0	0	0	0.5	1	0.5	0	0	0	0	0	0	0
Down large (DL)	1	1	1	0.5	0	0	0	0	0	0	0	0	0

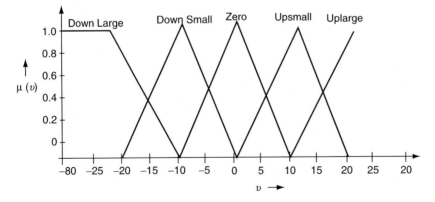

Fig. 8.123. Membership value of velocity

Membership values for control force (Fig. 8.124)

Linguistic variables	Control force (lbs)												
	−30	−25	−20	−15	−10	−5	0	5	10	15	20	25	30
Up large (UL)	0	0	0	0	0	0	0	0	0	0.5	1.0	1.0	1.0
Up small (US)	0	0	0	0	0	0	0	0.5	1.0	0.5	0	0	0
Zero (Z)	0	0	0	0	0	0.5	1	0.5	0	0	0	0	0
Down small (DS)	0	0	0	0.5	1	0.5	0	0	0	0	0	0	0
Down large (DL)	1	1	1	0.5	0	0	0	0	0	0	0	0	0

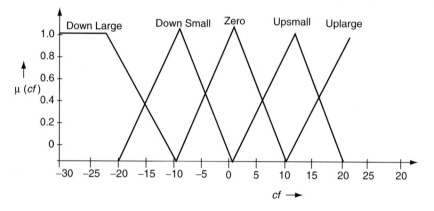

Fig. 8.124. Membership values for control force

Based on the rule base formed, FAM table is formed,

	FAM Table				
			Velocity		
Height	DL	DS	Z	US	UL
L	Z	DS	DL	DL	DL
M	US	Z	DS	DL	DL
S	UL	US	Z	DS	DL
NZ	UL	UL	Z	DS	DS

SIMULATION

Simulation is done assuming the initial conditions,
Cycle 1:

Initial height h_0: 1,000 ft
Initial velocity V_0: $-20\,\text{ft s}^{-1}$
Control f_0: has to be computed

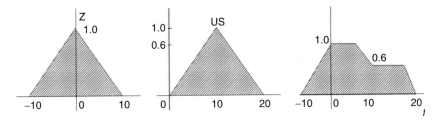

Fig. 8.125. Defuzzification (centroid method)

From the membership values,
    Height h fires L at 1.0 and M at 0.6
    Velocity v fires only DL at 1.0

From the FAM table,
    Rule: IF $h = L$ and $v = DL$ THEN $f = Z$ min$(1.0, 1.0) = 1.0$

Height		Velocity	Output
L(1.0)	AND	DL(1.0) –	Z(1.0)
M(0.6)	AND	DL(1.0) –	US(0.6)

Defuzzification using centroid method (as shown in Fig. 8.125)
The new values of state variables,

$$h_1 = h_0 + v_0 = 1,000 + (-20) = 980\,\text{ft}$$
$$V_1 = V_0 + f_0 = -20 + 5.8 = -14.2\,\text{ft s}^{-1}$$

Cycle 2:
    The new h_1 and V_1 are used for computation.
From the membership values,
    Height $h = 980\,\text{ft}$ fires L at 0.96 and M at 0.64
    Velocity $v = -14.2\,\text{ft s}^{-1}$ fires DS at 0.58 and DL at 0.42
From the FAM table,

Height	Velocity		Output
L(0.96)	AND	DS(0.58) –	DS(0.58)
L(0.96)	AND	DL(0.42) –	Z(0.42)
M(0.64)	AND	DS(0.58) –	Z(0.58)
M(0.64)	AND	DL(0.42) –	US(0.42)

Defuzzification using centroid method (as shown in Fig. 8.126),
we get, $f_1 = -0.5\,\text{lbs}$ – the output force computed from results of cycle 1
The new values of state variables,

$$h_2 = h_1 + V_1 \quad = 980 + (-14.2) = 965.8\,\text{ft}$$
$$V_2 = V_1 + f_1 \quad = -14.2 + (-0.5) = 14.7\,\text{ft s}^{-1}$$

Cycle 3:
    The new h_2 and V_2 are used for computation.

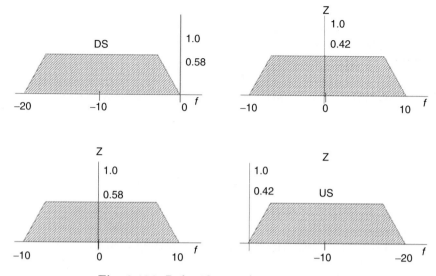

Fig. 8.126. Defuzzification (centroid method)

From the membership values,

Height $h = 965.8$ ft fires L at 0.93 and M at 0.67

Velocity $v = -14.7$ ft s^{-1} fires DL at 0.43 and DS at 0.57

From the FAM table,

Height		Velocity		Output
L(0.93)	AND	DL(0.43)	–	Z(0.43)
L(0.93)	AND	DS(0.57)	–	DS(0.57)
M(0.67)	AND	DL(0.43)	–	US(0.43)
M(0.67)	AND	DS(0.57)	–	Z(0.57)

Defuzzification using centroid method (as shown in Fig. 8.127),
we get, $f_2 = -0.4$ lbs – the output force computed from results of cycle 2
The new values of state variables,

$$h_3 = h_2 + v_2 = 965.8 + (-14.7) = 951.1 \text{ ft}$$
$$V_3 = V_2 + f_2 = -14.7 + (-0.4) = -15.1 \text{ ft s}^{-1}$$

Cycle 4:

The new h_3 and V_3 are used for computation.

From the membership values,

Height $h = 951.1$ ft fires L at 0.9 and M at 0.7

Velocity $v = -15.1$ ft s^{-1} fires DS at 0.49 and DL at 0.51

From the FAM table,

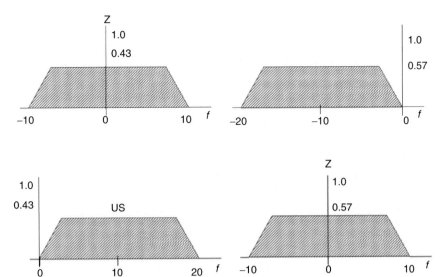

Fig. 8.127. Defuzzification (centroid method)

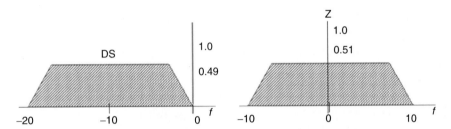

Fig. 8.128. Defuzzification (centroid method)

Height		Velocity	Output
L(0.9)	AND	DS(0.49) −	DS(0.49)
L(0.9)	AND	DL(0.51) −	Z(0.51)
M(0.7)	AND	DS(0.49) −	Z(0.49)
M(0.7)	AND	DL(0.51) −	US(0.51)

Defuzzification using centroid method (as shown in Fig. 8.128),
we get, $f_3 = 0.3$ lbs − the output force computed from results of cycle 4
The new final values of state variables,

$$h_4 = h_3 + v_3 = 951.1 + (-15.1) = 936.0 \,\text{ft}$$
$$V_4 = V_3 + f_3 = -15.1 + 0.3 = -14.8 \,\text{ft s}^{-1}$$

Thus we may obtain the curve of downward velocity vs. the altitude as
discussed at the starting.

	Cycle 0	Cycle 1	Cycle 2	Cycle 3	Cycle 4
Height (ft)	1000	980	965.8	951.1	936
Velocity (ft s^{-1})	−20	−14.2	−14.7	−15.1	−14.8
Control Force	5.8	0.5	−0.4	0.3	−

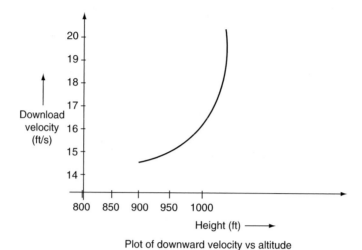

Plot of downward velocity vs altitude

Advantages of Using This Control

(1) It is accurate and flexible
(2) The trend towards the compilation and fusion of different forms of knowledge representation for the best possible identification and control of ill-defined complex systems

8.9.2 Automatic Generation Control Using Fuzzy Logic Controllers

Objective: The objective of the controller is to generate and deliver power in an interconnected system as economically and reliably as possible while maintaining the voltage and frequency within permissible limits:

- AGC is used in real time control to match the generation with demand
- The rising and falling of demand, alters the system voltage and frequency
- Real power and reactive power is changed due to change in frequency and voltage
- *LFC* controls real power and frequency
- *AVR* controls the reactive power and voltage
- *AGC* tracks system load and generation level of each committed unit

  Now it is dealt only with load frequency control (LFC):

- LFC is of importance in electric power system design and operation

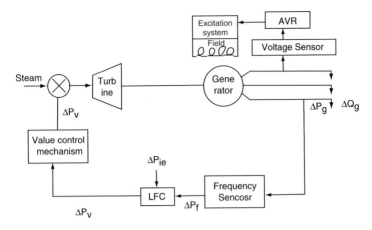

Fig. 8.129. Modeling of load frequency control

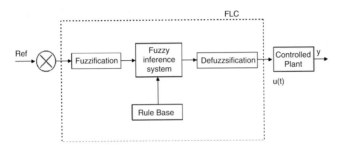

Fig. 8.130. Block diagram of fuzzy logic controller

– Conventionally PI (Proportional Integral) controllers, is applied to speed
  governor system for load frequency control Modeling of LFC is shown in
  Fig. 8.129

PI control approach achieves zero steady state error in frequency but it
exhibits relatively poor dynamic performance by large overshoots and large
settling time.

Using FLC, applying to LFC problems, gives a better dynamic response.

Design of Fuzzy Logic Controllers

The block diagram of fuzzy logic controller is shown in Fig. 8.130. It involves:

(1) Fuzzification of input parameters
(2) Rule definition
(3) Inference mechanism
(4) Defuzzification

Fuzzification of Input Parameters

(1) Process of converting crisp input variables into fuzzy variables
(2) Fuzzy variables depends on system nature
(3) The error and derivative of error are selected as controlled inputs
(4) Del f and Del Ptie are inputs
(5) Input and output signals are expressed as linguistic variables
(6) Seven linguistic variables are taken – NL, NM, NS, Z, PS, PM, PL
(7) Each linguistic variable has a fuzzy membership value.

The membership functions of Del f, Del Ptie and output are shown in Figs. 8.131–8.133.

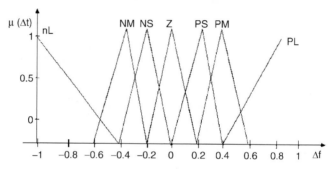

Fig. 8.131. Membership function of Del f

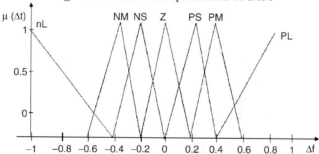

Fig. 8.132. Membership function of Del Ptie

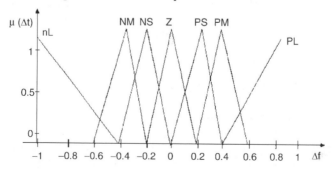

Fig. 8.133. Membership function of output

Rule Definition

Knowledge based rules are defined and based on that FAM table is formed

Def f/ Del Ptie	NL	NM	NS	Z	PS	PM	PL
NL	NL	NL	NL	NL	NM	NS	Z
NM	NL	NL	NM	NM	NS	Z	PS
NS	NL	NM	NS	NS	Z	PS	PM
Z	NL	NM	NS	Z	PS	PM	PL
PS	NM	NS	Z	PS	PS	PM	PL
PM	NS	Z	PS	PM	PL	PL	PL
PL	Z	PS	PM	PL	PL	PL	PL

Inference Mechanism

(1) It takes fuzzy input from fuzzifier in form of membership value matrix and uses fuzzy rule base to decide the fuzzy value of output

$$\text{IF (Del } f = \text{NS) AND (Del Ptie} = \text{PS) THEN (O/P} = \text{Z)}$$
$$\min (\mu(\text{NS}), \ \mu(\text{PS})) = \text{output}$$

(2) Product inference method may be used

Defuzzification

(1) This stage converts fuzzy output into a crisp value
(2) Crisp control action is required in practical application
(3) Centroid method is employed here

Thus by using the above, FLC is implemented for the AGC

Simulation Results

Thus in the case of PI the settling time is high and large overshoots whereas using fuzzy logic controller, the convergence has reached faster and no large overshoots. Figure 8.134 shows the comparison of output of fuzzy logic controller and proportional integral controller.

8.10 Fuzzy Pattern Recognition

Pattern recognition means determining the structure in the data by comparison to known structures. The known structures are obtained from different methods of classifications. These classification methods may be classification using equivalence relation and classification using fuzzy c means. In statistical

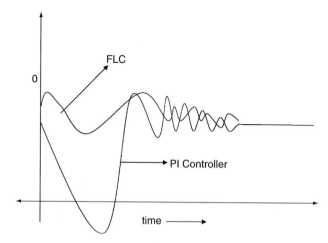

Fig. 8.134. Comparison of output of FLC and PI controller

pattern recognition, each input observation is stored as a multidimensional data vector (also know as feature vector) and each component of it is known as feature. Feature can be defined as the measurement done on the input data that is to be classified. The purpose of the pattern recognition process is to assign each input data to one of the "c" possible classes. The fundamental objective of pattern recognition is classification. Different input observations should be assigned the same class if they have similar features and to different classes if they have dissimilar features.

Typical pattern recognition as system is as shown in Fig. 8.135:

The data used to design a pattern can be divided into two categories:

Design data (training data)

Test data

1. Design data: design samples are used to establish the algorithmic parameters of the pattern recognition system. The design samples may be labeled (the class to which they belong is known) or unlabeled (the class to which they belong is unknown)
2. Test data: test samples are used to test the overall performance of the system.

Obviously, the test samples are labeled.

Feature analysis:

This is a method of reconditioning the raw data so that the information that is most relevant and for classification and interpretation (recognition) is enhanced and represented in a minimal number of features.

It consists of three components:

1. Feature nomination
2. Feature selection

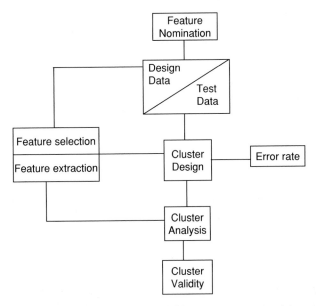

Fig. 8.135. Block diagram of fuzzy pattern recognition

3. Feature extraction

Feature nomination: it is the process of proposing the original "p" features. It is usually done by the workers close to the physical process. These features can be characteristics of various sensors. These are influenced by the physical constraints.

Feature selection: it is the process of choosing the best subset of "s" features from the original "p" features.

Feature extraction: it describes the process of transforming the original "p" dimensional feature space into an "s" dimensional feature space in some manner that the best preserves the information available in the original "p" space.

Cluster analysis:

It refers to the process of identifying the number of subclasses of "c" clusters in a data universe X comprised of "n" data samples, and partitioning X into "c" clusters in the data.

Cluster validity:

In many cases, the number of clusters in the data is known. However, it may be reasonable to expect cluster substructure at more than one value of "c". In this situation it is necessary to identify the value of "c" that gives the most plausible number of clusters in the data for the analysis. This problem is known as cluster validity. If the data used are labeled, there is unique and absolute measures of cluster validity. For unlabeled data, no absolute measure of clustering validity exits.

Classifier design:

Partitioning the feature space into "c" regions and for each subclass in the data, is usually in the domain of the classifier design. Crisp classifier partitions the feature space into disjoint sets. Fuzzy classifiers assign a fuzzy labeled vector to each vector in feature space. The classifier maps the input features onto a classification state. There are many similarities between the classification and pattern recognition. Basically, classification determines the structure in the data whereas pattern recognition attempts to take a new data and assigns them to one of the classes defined in the classification process. Classification defines the "patterns" and pattern recognition assigns a data to the class.

In both classification and pattern recognition process there are feed back loops as shown in Fig 8.136.

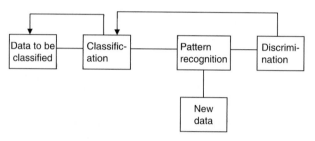

Fig. 8.136. Classification and pattern recognition process

The loop for classification is required when one needs better segmentation. Second loop is required when pattern matching fails.

Single sample identification:

Typical problem in pattern recognition is data collection from the physical process and classify them into known patterns. The known patterns are represented as class structure and each class structure is described by number of features.

Suppose we have several typical patterns stored in our knowledge base and we are given a new data sample that has not yet been classified. We want to determine which pattern the sample most closely resembles.

Express typical patterns as fuzzy sets $A_1, A_2, A_3, \ldots, A_m$.

Now suppose we are given a new data sample, which is characterized by crisp singleton, x_0. Using the simple criterion of maximum membership, the typical pattern that the data sample most closely resembles is found by the following expression:

$$\mu_{Ai}(x_0) = \max \{\mu_{A1}(x_0), \mu_{A2}(x_0), \cdots, \mu_{Am}(x_0)\}, \qquad (8.19)$$

where x_0 belongs to the fuzzy set A_i, which is the set indication for the set with highest membership at point x_0.

The below Fig. 8.137 shows this expressed by above (8.19), where clearly the new data sample defined by the singleton expressed by x_0 most closely resembles the pattern described by fuzzy set A_2.

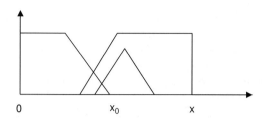

Fig. 8.137. Data sample resemblance

Example: suppose the single data sample is described by a data trilet, where the three coordinates are the angles of a specific triangle:

(1) *New data is crisp singleton*
    For example $x_0 = \{A = 85, B = 50, C = 45\}$; the five known patterns are isosceles, right, right isosceles, equilateral and all other triangles.

$$\mu_I(A, B, C) = 1 - 1/60 \; \min(A - B, B - C) \to \mu_I(85, 50, 45) = 0.916$$
$$\mu_R(A, B, C) = 1 - 1/90|A - 90| \to \mu_R(85, 50, 45) = 0.94$$
$$\mu_{IR}(A, B, C) = \mu_{IV}\mu_R \to \mu_{IR}(85, 50, 45) = 0.916$$
$$\mu_E(A, B, C) = 1 - 1/180(A - C) \to \mu_E(85, 50, 45) = 0.8$$
$$\mu_I(A, B, C) = 1/180 \; \min\{3(A - B), 3(B - C), 2|A - 90|, |A - C|\}$$
$$\to \mu_T(85, 50, 45) = 0.05$$

using the criterion of maximum membership, we see from these values that x_0 most closely resembles the right triangle pattern R.

(2) *New data is fuzzy set*
    Consider the case where the new data sample is not crisp, but rather a fuzzy set itself. Suppose we have m typical patterns represented as fuzzy sets A_I on $X(I = 1, 2 \ldots m)$ and a new piece of data, perhaps consisting for group of observations, is represented by a fuzzy set B on X. The task is to find which A_i the sample B most closely matches.

If we define two identical fuzzy vectors (they are of same length and they contain the same elements in an ordered sense) $\mathbf{a}$ and $\mathbf{b}$, then the *inner product* $\mathbf{a} \cdot \mathbf{b}^{\mathbf{T}}$ reaches a *maximum* value as the *outer product* $\mathbf{a} \oplus \mathbf{b}^{\mathbf{T}}$ reaches a *minimum*. These two norms, the inner product and outer product can be used simultaneously in pattern recognition studies because they measure closeness or similarity.

We can extend the fuzzy vectors to the case of fuzzy sets whereas vectors are defined on a finite countable universe, sets can be used to address infinite-valued universes.

Let A and B be two fuzzy sets then either of the expressions

$$(A, B)_1 = [(A \cdot B) \cap (\overline{A \oplus B}), \tag{8.20}$$
$$(A, B)_2 = 1/2[(A \cdot B) + (\overline{A \oplus B}). \tag{8.21}$$

Describes the two metrics to assess the degree of similarity of the two sets A and B:

$$(A, B) = (A, B)_1 \text{ or } (A, B) = (A, B)_2 \tag{8.22}$$

In particular, when either of the values of (A, B) approaches 1, then the two fuzzy sets A and B are "more closely similar" when either of the values (A, B) approaches a value 0, the two fuzzy sets are " far more apart" (dissimilar). The metric 1 uses minimum property to describe similarity and the metric 2 uses arithmetic property to describe similarity.

The first metric always gives a value lesser than the value obtained from the second metric both of these represent a concept called "*APPROACHING DEGREE.*"

Example: suppose we have a universe of five discrete elements, $X = \{x_1, x_2, x_3, x_4, x_5\}$, and we have two fuzzy sets A and B, on this universe. Note that the two fuzzy sets are actually crisp sets and complements to each other

$$A = \{1/X_1 + 1/X_2 + 0/X_3 + 0/X_4 + 0/X_5\}$$
$$B = \{0/X_1 + 0/X_2 + 1/X_3 + 1/X_4 + 1/X_5\}$$

If we calculate the inner and outer products we get

$$\mathbf{A \cdot B = 0}$$
$$\mathbf{A \oplus B = 1}$$
$$(A, B)_1 = (A, B)_2$$

The conclusion is that a crisp set and its complement are completely dissimilar.

Example: suppose we have a one-dimensional universe on the real line, $X = [-\infty, \infty]$; and we define two fuzzy sets A and B having normal Gaussian membership functions which are defined mathematically as

$$\mu_B(x) = \exp[-(x - b)^2/\sigma_b^2],$$
$$\mu_A(x) = \exp[-(x - a)^2/\sigma_a^2]$$
$$A \cdot B = \exp[-(a - b)^2/(\sigma_a + \sigma_b)^2] = \mu_A(x_o) = \mu_B(x_o).$$

It can be shown that the inner product of the two fuzzy sets is equal to

$$x_o = (\sigma_a \cdot b + \sigma_b \cdot a)/(\sigma_a + \sigma_b)$$

and the outer product is calculated to be

$$A \oplus B = 0.$$

Hence the values are

$$(A, B)_1 = \exp[-(a - b)^2/(\sigma_a + \sigma_b)^2] \wedge 1$$
$$(A, B)_2 = 1/2[\exp[-((a - b)^2/(\sigma_a + \sigma_b)^2) + 1]$$

In usual pattern recognition problems we are interested in comparing a data sample to a number of known patterns.

Suppose we have a collection of M patterns, each represented by a fuzzy set, A_i where $i = 1, 2, \cdots, m$ and a sample pattern B, all defined on the universe. Now the question of which pattern most closely resembles the pattern B. A useful metric that has appeared in the literature is to compare the data sample to each of the known patterns in a pairwise comparisons, then select the pair with largest approaching degree values as the one governing the pattern recognition process. The known pattern that is involved in the maximum approaching degree value is then the pattern the data sample most closely resembles in a maximal sense. This concept has been termed the "maximum approaching degree."

The below equation shows this concept for m known patterns:

$$(B, A_i) = \max\{(B, A_1), (B, A_2), (B, A_3) \cdots (B, A_m)\} \qquad (8.23)$$

Example: suppose an earthquake engineering consultant have to assess earthquake damage in a region just hit by a large earthquake. The assessment of damage will base very important to residents of the area because the insurance companies will base their claim payouts on the assessment. From previous historical records you determine that the six categories of the modified Mercalli Intensity (I) scale (VI) to (XI) are most appropriate for the range of damage to the buildings in the region. These damage patterns can all be represented by Gaussian membership functions $A_i i = 1, 2, \ldots, 6$ of the following form.

$$\mu_{A_i}(x) = \exp[-(x - a_i)^2/\sigma_{a_i}^{2}]$$

Where parameters a_I and σ_I define the shape of each membership function. Historical database provides the information shown in Table 8.22 for the parameters for the six regions.

Table 8.22. Parameters for Gaussian membership functions

	A_1	A_2	A_3	A_4	A_5	A_6
	VI	VII	VIII	IX	X	XI
a_i	5	20	35	49	71	92
σ_{ai}	5	10	13	26	18	4

The pattern of damage to buildings in a given location is determined by inspection and is represented by a fuzzy set B, with the following characteristics:

$$\mu_B(x) = \exp[-(x-b)^2/\sigma_b{}^2]; \qquad b = 41 \quad \text{and} \quad \sigma_b = 10$$

$$\mu_B(41) = 1; \quad \mu_B(x \neq 41) = 0; \quad x_o = 41$$

Now conducting the following calculations, using the similarity metric to determine the maximum approaching degree:

$$(B, A_1) = 1/2(0.004 + 1) \approx 0.5$$

$$(B, A_2) = 0.67$$

$$(B, A_3) = 0.97$$

$$(B, A_4) = 0.98$$

$$(B, A_5) = 0.65$$

$$(B, A_6) = 0.5$$

from this list we can see that Mercalli Intensity IX (A_4) most closely resembles the damaged area because of the maximum membership value of 0.98.

Suppose if the membership function of the damaged region to be a simple singleton with the following characteristics:

This example reduces to the single data sample problem posed earlier, i.e.,

$$(B, A_i) = \mu_{A_i}(x_o) \wedge 1 = \mu_{A_i}(x_o)$$

the calculations produce the following results:

$$\mu_{A1}(41) \approx 0$$

$$\mu_{A2}(41) = 0.01$$

$$\mu_{A3}(41) = 0.81$$

$$\mu_{A4}(41) = 0.91$$

$$\mu_{A5}(41) = 0.06$$

$$\mu_{A6}(41) \approx 0$$

Again, Mercalii scale IX (A_4) would be chosen on the basis of maximum membership (0.91). If we were to make selection without regard to the shapes of the membership values, we would be inclined erroneously to select region VIII because its mean value of 35 closer to the singleton 41 then it is to the mean value of region IX, i.e., to 49.

8.10.1 Multifeature Pattern Recognition

So far we have considered only one-dimensional pattern recognition; that is, the patterns here have been constructed only on a single feature. If we have to consider many features in the pattern recognition process there are three popular methods:

- Nearest neighbor classifier
- Nearest center classifier
- Weighted approaching degree

The first two methods are restricted to the recognition of crisp singleton data samples.

In the nearest neighbor classifier method we can consider m features for each data sample, so each sample (x_I) is a vector of features,

$$X_i = \{x_{i1}, x_{i2}, x_{i3}, \ldots, x_{i1}\}.$$

Now suppose we have n data samples in a universe, or $X = \{x_1, x_2, x_3, x_4, \ldots, x_m\}$.

Using a conventional fuzzy classification approach, we can cluster the samples into c-fuzzy partitions, then get c-hard partitions from these by using the equivalent relations idea or by hardening soft partitions U.

Now if we have a new singleton data sample, say x, then the nearest neighbor classifier is given by the following distance measure d:

$$d(X, X_i) = \min_{1 \leq k \leq n} \{d(X, X_k)\}$$

for each of the n data samples where $X_I \in A_j$, that is, points X and X_i are nearest neighbors, hence both would belong to the same class.

In another method for singleton recognition, the nearest classifier method works as follows.

We again start with n known data samples, $X = \{x_1, x_2, x_3, x_4, \ldots, x_n\}$ and each data sample is m-dimensional (characterized by m features). We can cluster these samples into fuzzy classes and each have a centre class, so

$$V = \{v_1, v_2, v_3, \ldots, V_c\}$$

Is a vector of the "c" class centers.

If we have a new singleton data sample, say x, the nearest center classifier is then given by

And now the data singleton, x, is classified as belonging to the fuzzy partition A_i.

$$d(X, V_i) = \min_{1 \leq k \leq c} \{d(X, V_k)\}$$

In the third method for addressing multifeature pattern recognition for a sample with several (m) fuzzy features, we will use the approaching degree concept again to compare the new data pattern with some known data patterns. Define a new data sample characterized by m features as a collection of noninteractive fuzzy sets, $B = \{B_1, B_2, B_3, \ldots, B_m\}$. Because the new data sample is characterized by m features, each of the known patterns, A_i, is also described by m features. Hence each known pattern in m-dimensional space is a fuzzy class given by, $A_I = \{A_{i1}, A_{i2}, A_{i3}, \ldots, A_{im}\}$ where $i = 1, 2, \ldots, c$ describe the c-classes. Since some of the features may be more important than others in the pattern recognition process, we introduce normalized weighting factors w_j, where

$$\sum_{j=1}^{m} W_j = 1$$

$$(B, A_i) = \sum_{j=1}^{m} W_j(B_j, A_{ij})$$

Then the equations in the approaching degree concept is modified for each of the known c-patterns $(I = 1, 2, \ldots, c)$ by As before in the approaching degree, sample B is closest to pattern A_j when

$$(B, A_j) = \max_{1 \leq i \leq c} \{(B, A_i)\},$$

$$\mu_{A_i}(x) = \sum_{j=1}^{m} W_j \cdot \mu_{A_{ij}}(x).$$

Note that when the collection of fuzzy sets $B = \{B_1, B_2, B_3, \ldots, B_m\}$ reduces to a collection of crisp singletons, i.e., $B = \{x_1, x_2, x_3, \ldots, x_m\}$ then the above equations get modified as,

And the sample singleton is closest to pattern A_j

$$\mu_A(x) = \max_{1 \leq i \leq c} \{\mu_{A_i}(x)\}.$$

Thus the membership function of pattern A_j is obtained.

9

Fuzzy Logic Projects with Matlab

9.1 Fuzzy Logic Control of a Switched Reluctance Motor

This section presents the use of fuzzy logic control (FLC) for switched reluctance motor (SRM) speed. The FLC performs a PI (Proportional Integral)-like control strategy, giving the current reference variation based on speed error and its change. The performance of the drive system was evaluated through digital simulations through the toolbox Simulink of Matlab program.

The SRM has becoming an attractive alternative in variable speed drives, due to its advantages such as structural simplicity, high reliability, and low cost. An important characteristic of the SRM is that the inductance of the magnetic circuit is a nonlinear function of the phase current and rotor position. So, for the control and optimization of this drive, a precise magnetic model is necessary. To obtain this model is not an easy task, because the magnetic circuit operates at varying levels of saturation under operating conditions. Further, the nonlinear characteristic of this plant represents a challenge to classical control. To overcome this drawback, some alternatives have been suggested in using fuzzy and neuronal systems. This paper proposes to control SR drives using FLC, which is mainly applied to complex plants, where it is difficult to obtain accurate mathematical model or when the model is severely nonlinear. FLC has the ability to handle numeric and linguistic knowledge simultaneously. In this section we present a study by simulation of the use of an FLC for SR drive. The SRM simulated has a structure of six poles on the stator and four on the rotor and power of 1 HP. The nonlinear model of this motor was simulated with the Matlab Simulink package and two tables were used to represent the nonlinearities: $I(\theta, \lambda)$, current in function of rotor position and flux, and $\tau(\theta, I)$, torque in function of rotor position and current. The objective of the FLC is to present a good performance, even if the two tables for a given motor were not accurately determined. The proposed control can be divided into two parts. The first employs FLC and will generate current

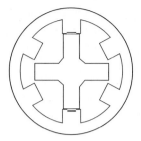

Fig. 9.1. SRM with six poles on the stator and four poles on the rotor

reference variations, based on speed error and its change. The second one has the function of selecting the phase that should be fed to optimize the torque, based on rotor position.

9.1.1 Motor

In a SRM, both stator and rotor have different magnetic reluctance along various radial axis. Figure 9.1 shows the controlled SRM, which has six poles on the stator and four on the rotor.

SRM electromechanical model can be represented by the following equations:

$$V = RI + \frac{\mathrm{d}\lambda}{\mathrm{d}t}, \tag{9.1}$$

$$\tau_e = \frac{\mathrm{d}}{\mathrm{d}\theta} \int_0^i \lambda \, \mathrm{d}i, \tag{9.2}$$

$$\tau = \tau_e - \tau_L = J \frac{\mathrm{d}\omega}{\mathrm{d}t}, \tag{9.3}$$

where V is the stator voltage, R resistance in the winding, λ leakage magnetic flux, τ_e electromechanical torque, τ_L load torque, θ rotor position, ω speed, and J momentum of inertia.

9.1.2 Motor Simulation

Matlab Simulink package was used to simulate the SRM. This choice was taken because this software has a good performance and satisfies all features required. Simulation was based on (9.1)–(9.3) and the tables of torque in function of angle and current, $\tau(\theta, I)$ and current in function of angle and flux linkage, $I(\theta, \lambda)$. These tables, extracted from the numeric data of the motor design by a finite elements program, were used to avoid the time consuming due to partial derivatives equations solution. See, in Fig. 9.2, the block diagram used.

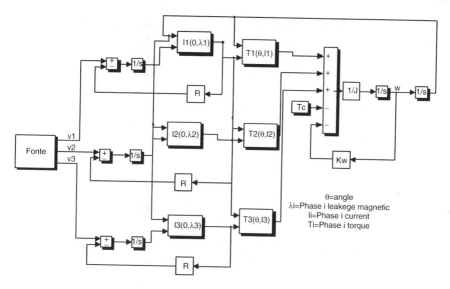

Fig. 9.2. Block diagram of the simulation

9.1.3 Current Reference Setting

In this part, we determine current reference for the three phases currents hysteresis control. The FLC generates current reference changes (ΔI_{ref}), based on speed error ($e\omega = \omega_{ref} - \omega_{actual}$) and its change ($ce\omega = e\omega(k+1) - e\omega(k)$). ΔI_{ref} is integrated to achieve current reference. We will show how the limits for the universes of the antecedents and consequents were initially settled. The $e\omega$ has its minimal value when the motor speed has nominal value, $+180\,\mathrm{rad\,s^{-1}}$, and is inverted to $-180\,\mathrm{rad\,s^{-1}}$. So, we have $e\omega = \omega_{ref} - \omega = (-180) - (+180) = -360\,\mathrm{rad\,s^{-1}}$. The maximum value, $+360$, is obtained in the opposite situation. The maximum torque obtained with the motor nominal current ($5\,\mathrm{A}$) is $1.2\,\mathrm{N\,m}$, thus which we can calculate the maximum absolute value for $ce\omega$:

$$
\begin{aligned}
ce\omega &= e\omega(k) - e\omega(k-1) \\
&= (\omega_{ref} - \omega(k)) - (\omega_{ref} - \omega(k-1)) \\
&= -(\omega(k) - \omega(k-1)) = -\Delta\omega,
\end{aligned}
$$

$$
J\frac{\Delta\omega}{\Delta t} = \tau \Longrightarrow \Delta\omega = \frac{\Delta t}{J}\tau \quad \therefore |ce\omega| = \frac{\Delta t}{J}\tau = \frac{2\times 10^{-3}}{1.3\times 10^{-3}}\times 1.2 \cong 19,
$$

where Δt is the interruption time.

The maximum absolute value for the ΔI_{ref} universe was obtained by trial and error. So, the initial limits for the universe of the antecedents and consequents were the following:

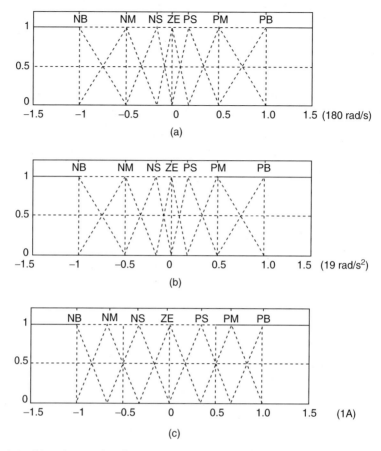

Fig. 9.3. Linguistic rules for current reference determination: **(a)** speed error, **(b)** change of speed error, **(c)** change of current reference

$$e\omega : -360a + 360 \quad (\mathrm{rad\,s^{-1}}),$$
$$ce\omega : -19a + 19 \quad (\mathrm{rad\,s^{-1}\,s^{-1}}),$$
$$\Delta I_{\mathrm{ref}} : -1a + 1 \quad (\mathrm{A}).$$

After some manual changes in these limits to optimize the speed control, we got the following values:

$$e\omega : -180 \quad \mathrm{to} \; +180 \quad (\mathrm{rad\,s^{-1}})$$
$$ce\omega : -19 \quad \mathrm{to} \; +19 \quad (\mathrm{rad\,s^{-1}\,s^{-1}})$$
$$\Delta I_{\mathrm{ref}} : -0.7 \quad \mathrm{to} \; +0.7 \quad (\mathrm{A})$$

Both antecedents and consequent linguistic variables are represented by seven triangular membership functions as shown in Fig. 9.3. The rule database formed with the given inputs to obtain the output is as shown in Table 9.1.

Table 9.1. Rule database

ew/cew	NB	NM	NS	ZE	PS	PM	PB
NB	NB	NB	NB	NB	NM	NS	ZE
NM	NB	NB	NB	NM	NS	ZE	PS
NS	NB	NB	NM	NS	ZE	PS	PM
ZE	NB	NM	NS	ZE	PS	PM	PB
PS	NM	NS	ZE	PS	PM	PB	PB
PM	NS	ZE	PS	PM	PB	PB	PB
PB	ZE	PS	PM	PB	PB	PB	PB

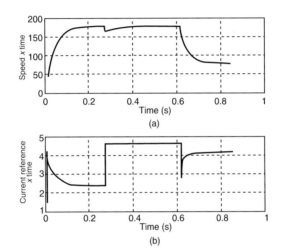

Fig. 9.4. Simulation results: (a) speed x time, (b) current reference x time

Some simulation results are presented on Fig. 9.4, which shows this control performance when there is a change in load and in speed reference. At first, $0.1\,\mathrm{N\,m}$ load is applied to this motor. At $0.27\,\mathrm{s}$, load is increased to $1\,\mathrm{N\,m}$, requiring higher torque. At $0.61\,\mathrm{s}$, speed reference is decreased to $80\,\mathrm{rad\,s}^{-1}$ and in consequence current decreases for deceleration.

9.1.4 Choice of the Phase to be Fed

This part of the control determines which phase should be fed. Its inputs are rotor position and speed.

Consider a phase ideal inductance profile shown in Fig. 9.5. If $\omega > 0$ and $\theta \in$ interval 1, feed the corresponding phase. The presence of current in this increasing inductance region will produce positive torque. If $e\omega > 0$, current should produce electrical torque, τ_e, higher than load one, τ_{load} to accelerate. If $e\omega > 0 < 0$, τ_e, should be lower than τ_{load} for deacceleration. Current reference value (electrical torque) and so acceleration or deacceleration is established by the first part of the control. It is also possible to decelerated the motor feeding

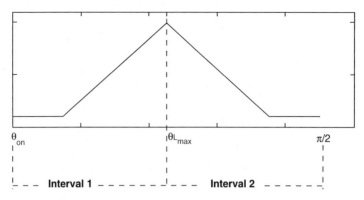

Fig. 9.5. Angle intervals used in the choice of the phase to be fed

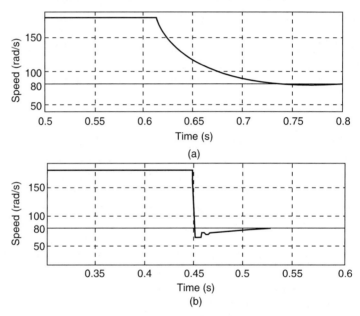

Fig. 9.6. Speed change from 180 to $80 \, \text{rad s}^{-1}$: **(a)** without feeding and **(b)** feeding the phase with decreasing inductance

the phase with decreasing inductance. However, it would cause overshoot. Figure 9.6 shows speed change from 180 to $80 \, \text{rad s}^{-1}$, feeding and not feeding the decreasing inductance phase.

If $\omega > 0, \theta \in$ interval 2 and there will be current in this phase, the source will not supply the phase.

The converter capacitor voltage will disenergize the phase to avoid production of negative torque.

The optimum energization angle θ_{on} is 150 before the ideal inductance profile starts increasing. This choice is taken by the fact that actual inductance

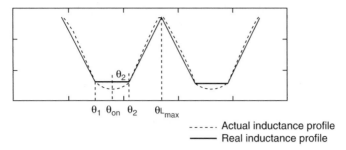

------- Actual inductance profile
———— Real inductance profile

Fig. 9.7. Actual and ideal inductance profiles

profile is not constant between θ_1 and θ_2 (see Fig. 9.7). Inductance decreases between θ_1 and θ_{on} and increases between θ_{on} and θ_2. Therefore, if the phase is fed before θ_{on}, there will be negative torque applied. On the other hand, feeding in θ_{on} will provide positive torque between θ_{on} and θ_2. Its value will be low, but important to compensate the torque fall supplied by another phase that is being disenergized in this same period. Another advantage is that current increases faster due to low inductance in θ_{on} and will have reached the reference when rotor position be in the region of high inductance change ($[\theta_2\theta L_{max}]$).

Finally to conclude, to get dynamics performance predictions of SRMs, including its control, a simulating model has been shown in this paper. The nonlinear modeling has been represented by look-up tables to obtain torque and current. A control has been developed for the SRM speed. This control has two parts. Part 1 determines the reference current, and so electromechanical torque. Part 2 chooses which phase should be fed, based on θ and speed, and is responsible for imposing speed direction. It was shown that inverting speed direction by energizing the phase with decreasing inductance to deaccelerate the motor provided speed overshoot, while the use of load torque on deacceleration made the speed response more smooth. The FLC has demonstrated a good accuracy and has performed well for the speed control of the SRM, surpassing its nonlinearities.

9.2 Modelling and Fuzzy Control of DC Drive

An industrial DC drive (22 kW) with fuzzy controller is simulated. Two models (linear and nonlinear) and two controllers (PID and fuzzy) are investigated. Using fuzzy controller for DC drive operation was successful. Two mathematical models of a DC drive are used. The first model is build as linear transfer function of converter and DC motor. The second model is build using advanced blocks from Power System Blockset (PSB) library. The library is an extension of Matlab/Simulink environment from The MathWorks, Inc. Using fuzzy logic and PSB library model seems to be new and promising approach to control of an electric drive.

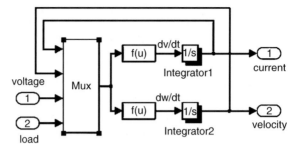

Fig. 9.8. Linear model of DC motor

9.2.1 Linear Model of DC Drive

A linear and nonlinear model of DC drive will be used. The linear model consists of two parts: converter/rectifier and DC motor. A linear model of DC motor (Fig. 9.8) was build using Simulink blocks. There are two inputs (voltage and load) and two outputs (angular motor velocity and current). Its parameters are computed automatically from nominal catalogue data: motor power, voltage, current, speed, etc.). It is very convenient to use nominal motor data as rotor inductance and resistance. DC motor constant and other internal motor parameters are difficult to find converter/rectifier is described as first-order inertia

$$G_{\mathrm{conv}} = \frac{kp(s)}{T_{\mathrm{mip}}s + 1},$$

where kp is gain of converter/rectifier and T_{mip} is mean dead time of converter/rectifier.

The dead time T_{mip} may vary from zero to one-half the period of an AC source (0.01 s for 50 Hz). It is assumed that six-phase thyristor bridge with mean dead time $T_{\mathrm{mip}} = 1.67$ ms is used in the converter. A classic DC drive with two PID controllers is presented on Fig. 9.9. It was assumed to neglect a derivative signal and to use PI operation of a current controller only. Parameters of the current controller were derived from the model parameters using rules of module and symmetry. Then Nonlinear Control Design Blockset (NCD) was used for automatic tuning of the controller parameters to minimize the transient overshot.

Simple transfer function model of motor current versus voltage was used

$$G_{\mathrm{mot}} = \frac{kia}{T_{\mathrm{a}}s + 1},$$

where kia is gain of DC motor and T_{a} the armature circuit time constant.

Similar procedure was used to find parameters of velocity controller. The simulation results (DC motor current and angular velocity vs. time) are presented on Fig. 9.10. This is raw simulation as linear model has very low granularity: AC component of current and switching of currents in thyristor bridge are neglected. Only envelope of transients can be seen on simulation output.

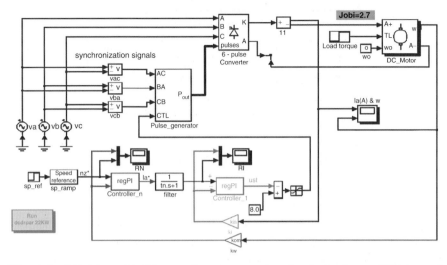

Fig. 9.9. DC drive with PI controllers in current and velocity loops. PSB is used to build advanced drive model (*upper part of block diagram*)

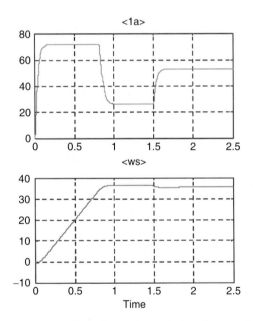

Fig. 9.10. Current (*upper red*) and angular velocity (*lower blue*) of DC motor. Simulink and linear model were used for simulation

9.2.2 Using PSB to Model the DC Drive

An advanced set of linear and nonlinear blocks can be found in PSB. Three
AC sources, three-phase six-pulse converter, pulse generator, and DC motor
are taken from the library. They are used to prepare high-quality model of
three-phase DC drive (see Fig. 9.9).

The three-phase bridge converter is the most frequently used motor control system. Two of six thyristors conduct at any time instant. Gating of each
thyristor initiates a pulse of load current; therefore this is a six-pulse controlled rectifier. The three-phase six-pulse rectifier is also capable of inverter
operation in the fourth quadrant. Electrical phenomena of thyristor bridge
and DC motor are modeled very exactly. Simulation results (Figs. 9.11 and
9.12) are *almost exact* with real *measurement data on industrial object*, but
computation is slow comparing to linear model.

9.2.3 Fuzzy Controller of DC Drive

The fuzzy controller is presented on Fig. 9.13. Advanced model using PSB is
used, but transfer function model can also be useful for preliminary tuning of
controller parameters.

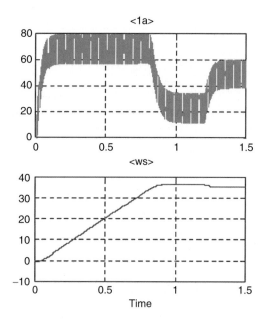

Fig. 9.11. Simulated current signal (*red*) and angular velocity (*blue*) using Simulink
and Power System Blockset (PSB)

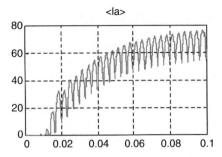

Fig. 9.12. Detail of simulated current signal using Simulink and PSB

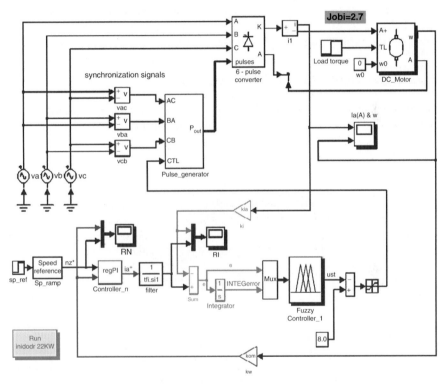

Fig. 9.13. DC drive with fuzzy controller in current loop. PSB is used to build advanced drive model (*upper part of block diagram*)

Linguistic Variables and Rules

There are two fuzzy variables (error and INTEG error) and seven linguistic variables (from *big negative* to *big positive*). The fuzzy controller attributes are:

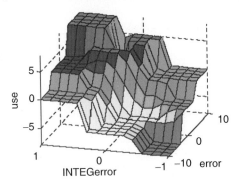

Fig. 9.14. Fuzzy control surface

type: 'mamdani'
andMethod: 'prod'
orMethod: 'max'
defuzzMethod: 'centroid'
impMethod: 'prod'
aggMethod: 'max'
input: [1×2 struct]
output: [1×1 struct]
rule: [1×25 struct]

The membership functions (pimf and gausmf are used) and rules are design tools that give opportunity to model a control surface and controller properties. It is obvious that using this attributes one can more precisely fulfill a quality criterion in full operational range. The control surface (Fig. 9.14) is defined with 25 rules.

9.2.4 Results

Simulation output for fuzzy controller is similar to PI controller output presented – unless one considers how controller reacts for external disturbation. The investigation showed that even simple fuzzy controller used to control DC drive operation (Fig. 9.13) is more precise and faster than of PI controller (Fig. 9.15).

9.3 Fuzzy Rules for Automated Sensor Self-Validation and Confidence Measure

In this section we present a methodology for the development of a generic, automated self-validation technique that can be used to improve the operation

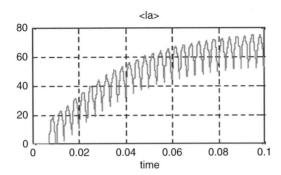

Fig. 9.15. Detail of simulation results (motor current). Fuzzy controller (here) reacts faster than PI

of a controller-based system. The reliability of a controller-based system depends on the validity of the data provided for the development and operation of the controller. The self-validation algorithm described in this paper is based on fuzzy logic rules described by membership functions. The membership functions are created from data set parameters (e.g., the standard deviation and the range of the data set). Raw data are median filtered and then passed through these membership functions to obtain a measure of confidence. The methodology is illustrated using temperature data from an iron-melting cupola furnace.

The most important step before developing an intelligent algorithm for sensor fusion or inferential measurement is to determine what data are to be used for development of the system. In an industrial implementation of a controllerbased system, the key input signals or data may suffer from severe noise and external disturbances. For reliable operation of these systems, we require a valid data set from the sensors. Invalid or erroneous data will result in an incorrect operation of the controller and hence a deterioration in the overall performance quality of the system. In this section we develop a methodology for automated data self-validation; i.e., the only information used to establish the confidence of the particular sensor's data is that sensor's historical behavior.

Previously neural networks were used to validate signal sets. To implement neural networks, large data sets are required to train a network. This is not always a feasible option because of the nonavailability of large data sets. Moreover, a neural network trained by a particular data set is inclined to that particular application. Hence this technique is not generic.

The objective of this section is to develop and implement a generic and automated self-validation system. The presented methodology utilizes fuzzy logic rules for establishing a confidence measure for the sensor data under consideration.

9.3.1 Preparation of Membership Functions

Data Preprocessing

The signal self-validation algorithm is based only actual data taken from the system. We for now assume that the developer has validated the data set used for preparation of the membership functions. Note, however, that the following methodology can be implemented in a recursive manner to establish the confidence of an unknown data set. To remove the effects of impulse noise, the raw data are passed through a median filter.

Curve Fitting

To quantify the variations in sensor readings, a polynomial curve is used to approximate the data set. By using this approximation, the regular "noise" level of the signal should be quantified. The user should chose the order of the polynomial such that it is high enough to give a good correlation of the original data set, but low enough to not be seriously affected by the spikes. The order chosen will depend on the amount of noise in each data set and therefore may be unique to each sensor.

Parameter Calculation

Descriptions

Although the overall shapes of the three fuzzy membership functions are the same for every sensor, the range of values is unique to each sensor. Therefore, certain parameters must be calculated in order to adapt the membership functions to the specific sensor. In our design, the nine parameters listed in Table 9.2 have been selected as necessary to complete the membership functions.

Table 9.2. List of parameters to complete membership functions

Parameter	Description
T_1	Low temperature
T_2	Minimum of ideal range of temperature
T_3	Maximum of ideal range of temperature
T_4	High temperature
R_7	Max/min of ideal rate of change
R_8	High/low rate of change
S_1, S_2, S_3	Standard deviation parameters

Parameters T_1, T_2, T_3, T_4

As discussed, the pre-processed temperature sensor data are fitted to a polynomial curve, T_{fit}, to reduce the amount of noise. The noise can then found by subtracting points on the fitted curve from the original data set. The standard deviation of this error, σ_T, is used to determine temperature range parameters T_1, T_2, T_3, and T_4 as follows:

$$T_1 = \text{minimum } (T_{\text{fit}}) - 3\sigma_T,$$
$$T_2 = \text{minimum } (T_{\text{fit}}) - \sigma_T,$$
$$T_3 = \text{maximum } (T_{\text{fit}}) + \sigma_T, \text{ and}$$
$$T_4 = \text{maximum } (T_{\text{fit}}) + 3\sigma_T$$

Parameters R_7, R_8

A similar method is used to calculate the rate of change parameters R_7 and R_8. The rate of change of the median-filtered original data is found by incrementally dividing the change in temperature by the change in time. The fitted curve rate of change, R_{fit}, is found by taking the mathematical derivative of the coefficients of the fitted curve, T_{fit}. Again, by subtracting points on the fitted curve from the original data, the standard deviation of this error, σ_R, is found. Then R_7 and R_8 are determined as follows:

$$R_7 = \text{maximum } (\text{abs } (R_{\text{fit}})) + \sigma_R,$$
$$R_8 = \text{maximum } (\text{abs } (R_{\text{fit}})) + 3\sigma_R.$$

Parameters S_1, S_2, S_3

The final parameters to be determined are those relevant to the standard deviation; i.e., S_1, S_2, and S_3. The fitted curve of standard deviation, S_{fit}, is found by taking the standard deviation of a window of five coefficients from fitted curve, T_{fit}. Points of S_{fit} are subtracted from the standard deviations of the original data. The standard deviation of this difference, σ_S, is then found. Note: σ_S is *actually* the standard deviation *of* the standard deviations of the difference between the real and fitted curves (σ_T).

$$S_1 = \sigma_T,$$
$$S_2 = \text{maximum } (\text{abs } (S_{\text{fit}})) + \sigma_S, \text{ and}$$
$$S_3 = \text{maximum } (\text{abs } (S_{\text{fit}})) + 3\sigma_S.$$

9.3.2 Fuzzy Rules

Membership Functions

The first membership function analyzes the temperature values. The range is from zero to well above the highest expected temperature. Each of the membership functions is trapezoidal in shape. The parameters for each of the

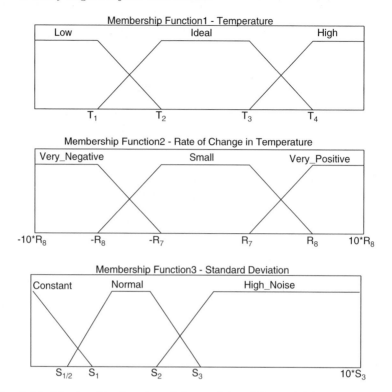

Fig. 9.16. Membership functions (**a**) top, (**b**) middle, and (**c**) bottom figures

functions are shown in Fig. 9.16. Data in the ideal range will have the highest confidence while data outside this range have diminished confidence.

The second membership function analyzes the rate of change in the temperature values and is shown in Fig. 9.16b. It is symmetric about the origin, and the highest/lowest values of the range are well above/below the highest/lowest expected rate of change. Again, the shape is trapezoidal. High rates of change are typically uncharacteristic of the process and thus this data with this behavior are given lower confidence.

The third membership function deals with standard deviation parameters and is shown in Fig. 9.16c. These parameters are computed for constant, normal, and high-noise ranges of the data. The purpose of the constant membership is to detect when sensor values have "flat-lined." This scenario can be indicative of a sensor or signal conditioner failure and thus we reduce the self-confidence as a means of raising a flag to the system monitor.

9.3.3 Implementation

In this section, we demonstrate the fuzzy logic self-validation methodology on real data obtained from the experimental cupola at the Albany Research

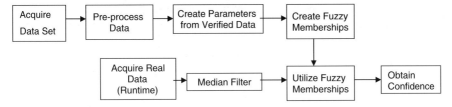

Fig. 9.17. Block diagram of presented methodology

Center (ALRC), OR. The sensor data are contained in a data file that consist of the sample time and the sampled value. Although only temperature sensors have been considered for this illustration, the scheme effectively produces a confidence plot and measure for all the sensors. The methodology was implemented with a combination of Excel, Matlab, and LabView routines. The routines automate the process completely such that the user selects a data file and the program produces the membership functions in a format that can be implemented by the Matlab Fuzzy Logic Toolbox.

Description

The complete procedure of the self-validation scheme can be described with the block diagram of the LabView graphical interface as shown in Fig. 9.17. The data are first acquired and then verified. For our purposes, verifying the data was selecting data from good data records. In general, this can be automated using a recursive self-validation routine and then selecting only data with long sections of highly confident data. The verified data were, however, median filtered to remove any impulse noise.

Parameters for the fuzzy block are computed from the verified data. These parameters are then used to define membership functions. The real data are then median filtered and then passed through the membership functions of the fuzzy block to obtain their self-confidence measure. The results are illustrated with the confidence plots in Section "Confidence Plots".

Confidence Plots

The results of the self-validation scheme are illustrated with the help of the figures. Figure 9.18 shows the confidence plots of good quality and corrupted data. The fuzzy logic membership functions were developed using the good data file and thus its confidence is consistently very high as shown in Fig. 9.19.

Figure 9.18 also presents a deliberately corrupted data set showing various types of inconsistent data behavior including data flat-lining, high rates of change, and being out of ideal range. In each case one notes that the self-confidence of the data is diminished as shown in Fig. 9.19. These figures illustrate the effectiveness of the algorithm in detecting data inconsistent with the historical behavior of the sensor.

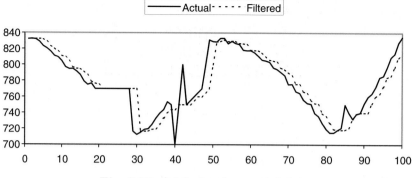

Fig. 9.18. Original and corrupted data

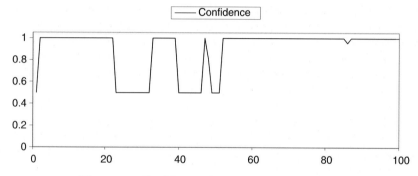

Fig. 9.19. Confidence of corrupted sensor signal

To conclude, in this section, we have described research that has been performed to design a signal self-validation methodology. Self-validation is accomplished by implementing the fuzzy rules described. The membership functions are applied to the median-filtered data set. The validation is based on the confidence measure array that is obtained as the output. In addition to validating the data, this scheme implemented in LabView also provides information about the occurrences of bad data in terms of spikes and warns the engineer of the uncertainty levels in the sensor output. The methodology was demonstrated on temperature data for an iron-melting cupola furnace.

This validation scheme is generic in nature. Though our specific application is data from the cupola furnace, it can be applied any application where time-series data is provided. The fuzzy rules of the developed self-validation scheme are based on the individual data sets coming in from independent sensors. So if any of the sensors fails, the scheme operates despite the fault and hence is a fault tolerant system. Moreover through this scheme we know of the occurrences of all the spikes or disturbances. Hence, if we eliminate these disturbances or random movements the scheme will prove effective in preventing the poor use of the control elements.

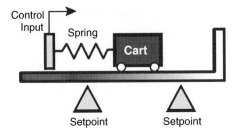

Fig. 9.20. The cart system to be controlled

9.4 FLC of Cart

Fuzzy logic control design is somewhat different from conventional control design methods in that it departs from standard analysis tools such as the Bode frequency response plot and the root locus diagram. In some cases, it may be appropriate to use an entirely fuzzy-based approach. But fuzzy logic can also be used in a hybrid approach with conventional control methods, making the most of both worlds. In this section, we examine how fuzzy logic can simplify gain scheduling between two different PID controllers. The system we will be looking at here is a simple one: a spring-mass-damper system from Dynamics 101 as illustrated in Fig. 9.20. While a basic PID controller will do a fine job of making it behave, fuzzy logic can provide a convenient way to meet stringent control objectives.

In response to a square wave, we want to move the cart back and forth between points A and B. Notice that near point B there is a wall, a hard stop that we want to keep the cart away from. On the other hand, at point A we have considerably more leeway. Let us also assume we want to conserve control power and mechanical wear and tear by using looser, more relaxed control at point A. The design goal is relaxed control at point A, tight control (specifically, fast response with no overshoot) at point B. This situation is similar to the operation of a robot arm in an application where you want precise movement in one position and energy conservation elsewhere. Because the plant is a simple one, both the precise control and the relaxed control can be implemented with a basic PID controller. But to meet both criteria we need some kind of gain scheduling to alternate between the two controllers, each of which has gain parameters tuned for its specific control objective. We can design a fuzzy controller to handle the gain scheduling for us. Let us start with the Simulink model of the system shown in Fig. 9.21.

The cart system is lightly damped. Its dynamics are described in the transfer function block as a function of the frequency domain variable s

$$G(s) = \omega^2/(s^2 + 2\zeta\omega s + \omega^2),$$

where the natural frequency $\omega = 1\,\mathrm{rad\,s^{-1}}$ and the damping is $\delta = 0.1$.

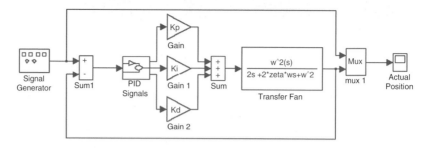

Fig. 9.21. Simulink block diagram of cart system with PID controller in place

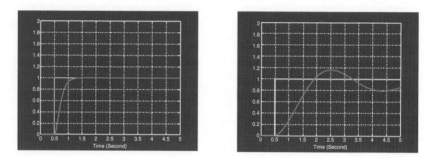

Fig. 9.22. *Left*: closed-loop step response with tight control gains. *Right*: closed-loop step response with loose control gains

Let us assume that we have already specified our gains for both the tightly controlled system and the loosely controlled one. There are any number of ways to choose these gains, and the ones we list later are not necessarily optimal in any sense.

Tight control: $K_p = 60$, $K_i = 4$, $K_d = 14$,

Loose control: $K_p = 5$, $K_i = 1$, $K_d = 2$.

The main design constraint we want to guarantee with the tight control is zero overshoot near setpoint B. On the other hand, the main consideration for the loose control gains is minimizing the control effort (while providing a small degree of damping). Figure 9.22 shows the closed-loop step response for each set of gains.

The point of this particular example is to show how we can use fuzzy logic to work hand-in-hand with conventional control. We do this by making use of what is known as Sugeno fuzzy inference system (FIS) (named for fuzzy logic pioneer Michio Sugeno) to implement a blend of the two different PID controllers. First, we need to make sure we have a good understanding of how a Sugeno system calculates its outputs. In a Sugeno system, the output membership function is a linear function of the inputs. The fuzzy rules for a single input/single output system look like this

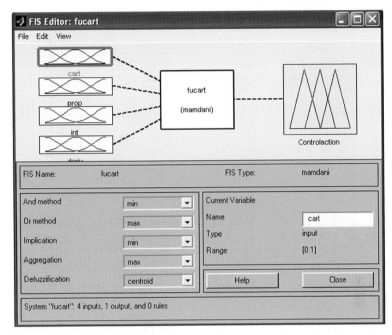

Fig. 9.23. The fuzzy inference system editor in the Fuzzy Logic Toolbox

If input is high, then output = q*input + r,

where q is a gain operating on the input and r is a constant. We need to build a Sugeno system with four inputs: cart position (so we can decide if we are close to point A or point B) and the P, I, and D signals to which we apply the appropriate gains K_p, K_i, and K_d. (The astute reader may notice that the P signal is the same thing as the cart position, but we will continue to refer to both signals for clarity.) Now we can build a rule set with exactly two rules:

1. If cart is near_A then control is loose [so use the gains $K_p = 5$, $K_i = 1$, $K_d = 2$],
2. If cart is near_B then control is tight [so use the gains $K_p = 60$, $K_i = 4$, $K_d = 14$].

The antecedents of these rules (e.g., "if cart is near_A") depend on the membership function for the terms "near_A" and "near_B." The consequents of these rules (e.g., "then control is loose") contain the three gains K_p, K_i, and K_d that we have calculated ahead of time. One output membership function implements all three gains at once. The system switches between two different controllers, so there are two output membership functions and two rules.

The window shown in Fig. 9.23 is the FIS editor, which we use to create our inputs and outputs for the fuzzy controller. We are building a four input/one output system, so we add inputs (cart, prop, int, deriv), and a single output (control action). We will specify these with the membership function editor.

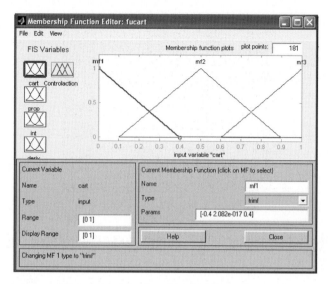

Fig. 9.24. Membership function editor

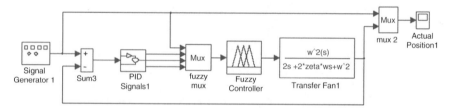

Fig. 9.25. Simulink block diagram of cart system with fuzzy controller in place

The membership function editor shown in Fig. 9.24 is where we define what we mean by the phrase "cart is near_A". Point A corresponds to the numerical value 0 and point B corresponds to the value 1. We have chosen to make a smooth ramp from one to the other. Notice that this means the statement 'cart is near_A' is 100% true when position = 0; it is 50% true when position = 0.5; and it is 0% true when position = 1. The converse is true of the statement 'cart is near_B.'

Once we have built our fuzzy controller using the graphical editors available in the Fuzzy Logic Toolbox, we save its specification in the Matlab workspace as a memory-resident variable. Alternatively, you may use Mat-files to save it to disk. The fuzzy controller is now available to be used in the fuzzy controller block in a Simulink diagram.

Figure 9.25 contains an updated version of our Simulink block diagram. Notice that we have replaced the three PID gains with a fuzzy controller

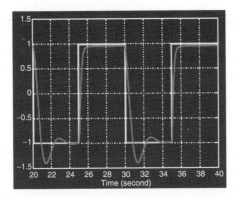

Fig. 9.26. Closed-loop response with the gain-scheduling fuzzy controller

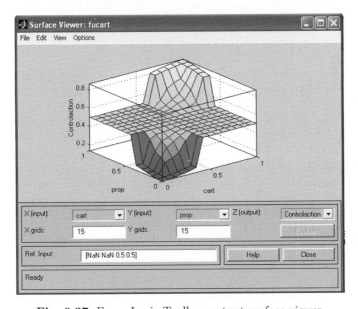

Fig. 9.27. Fuzzy Logic Toolbox output surface viewer

block. Also, we are using the cart position as an extra input so we can blend between the two sets of gains depending on where the cart is.

The plot in Fig. 9.26 shows the simulation of our system as it responds to a square wave. Notice that, just as we expected, the control of the cart near point B is tighter and has no overshoot, while the control is much looser near point A. So the Sugeno fuzzy system has successfully implemented a convenient gain scheduler for us.

In Fig. 9.27 we see the surface plot of the fuzzy controller. This is a plot of show the controller's output signal changes as a function of the cart's position and the proportional signal. Where the three-dimensional shape is mountainous, the control power required is higher and, as expected, corresponds to the region where the cart position is near point B. So we are looking at a map of the required control effort. This kind of visualization can be extremely valuable to a control designer, and it is just one example of the built-in tools available with the Fuzzy Logic Toolbox.

9.5 A Simple Fuzzy Excitation Control System (AVR) in Power System Stability Analysis

This section develops a controller based on fuzzy logic to simulate an automatic voltage regulator (AVR) in transient stability power system analysis. It was simulated a one machine control to check if the fuzzy controller implementation was possible. After that the controller developed was applied to an 18 bus bar system in order to show its behavior, which results were compared to the results obtained with the AVR itself.

From the power system point of view, the excitation system must contribute for the effective voltage control and enhancement of the system stability. It must be able to respond quickly to a disturbance enhancing the transient stability and the small signal stability.

Three principal control systems directly affect a synchronous generator: the boiler, governor, and exciter controls. Assuming that the generating unit has no losses. It is a reasonable assumption when total losses of turbine and generator are compared to total output. Under this assumption all power received, as steam must leave the generator terminals as electric power. The governor controls the steam power amount admitted to the turbine. The excitation system controls the generated EMF of the generator and therefore controls not only the output voltage, but also the power factor and current magnitude as well.

In many present-day systems the exciter is a DC generator driven by either the steam turbine (on the same shaft as the generator) or an induction motor. An increasing number are solid-state systems consisting of some form of rectifier or thyristor system supplied from the AC bus from an alternator exciter.

The voltage regulator is the intelligence of the system and controls the output of the exciter so that the generated voltage and reactive power change in the desired way. In most modern systems the AVR is a controller that senses the generator output voltage (and sometimes the current) then initiates corrective action by changing the exciter control in the desired direction. The speed of the AVR is of great interest in studying stability. Because of the high inductance in the generator field winding, it is difficult to make rapid changes in field current. This introduces a considerable lag in the control function and is one of the major obstacles to be overcome in designing a regulating system. The purpose of this work is the development of a fuzzy controller (software) to simulate the AVR behavior.

9.5.1 Transient Stability Analysis

The first demand of electrical system reliability is to keep the synchronous generators working in parallel and with adequate capacity to satisfy the load demand. If at any time, a generator looses synchronism with the rest of the system, significant voltage, and current fluctuation can occur and transmission lines can be automatically removed from the system by their relays deeply affecting the system configuration. The second demand is maintaining power system integrity. The high voltage transmission system connects the generation sources to the load centers. Interruption of these nets can obstruct the power flow to the load. This usually requires the power system topology study, once almost all electrical systems are connected to each other. When a power system under normal load condition suffers a disturbance there is synchronous machine voltage angles rearrangement. If at each disturbance occurrence an unbalance is created between the system generation and load, a new operation point will be established and consequently there will be voltage angles adjustments. The system adjustment to its new operation condition is called "transient period" and the system behavior during this period is called "dynamic performance."

As a primitive definition, it can be said that the system oscillatory response during the transient period, short after a disturbance, is damped and the system goes in a definite time to a new operating condition, so the system is stable. This means that the oscillations are damped that the system has inherent forces which tend to reduce the oscillations. The instability in a power system can be shown in different ways, according to its configuration and its mode of operation, but it can also be observed without synchronism loss.

9.5.2 Automatic Voltage Regulator

Automatic devices control generators voltages output and frequency, in order to keep them constant according to pre-established values.

These automatic devices are:

– Automatic voltage regulator
– Governor

However any governor due to its action loop, is slower than the AVR. This is associated mainly to its final action in the turbine. The main objective of the AVR is to control the terminal voltage by adjusting the generators exciter voltage. The AVR must keep track of the generator terminal voltage all the time and under any load condition, working in order to keep the voltage within pre-established limits. Based on this, it can be said that the AVR also controls the reactive power generated and the power factor of the machine once these variables are related to the generator excitation level.

The AVR quality influences the voltage level during steady state operation, and also reduces the voltage oscillations during transient periods, affecting

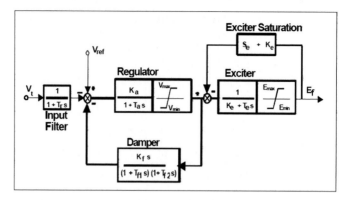

Fig. 9.28. AVR model type II of IEEE [4]: RT-IEEE2

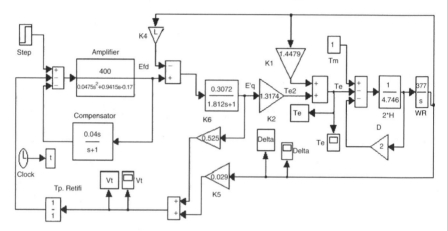

Fig. 9.29. Block diagram of one synchronous machine with AVR simulated in Matlab

the overall system stability. Figure 9.28 illustrates the AVR model used in the program developed.

9.5.3 Fuzzy Logic Controller Results Applied to a One Synchronous Machine System

The block diagram shown in Fig. 9.29 shows a synchronous machine for which output the voltage is controlled by an AVR applied to its excitation system, in the Matlab simulation. All data were taken from reference.

Next step is to replace the AVR device by a fuzzy logic controller in order to check its efficiency in the synchronous machine excitation voltage control. The rule base used by fuzzy controller to simulate an AVR in the Matlab program is shown in Fig. 9.30. The fuzzy controller run with the input and output normalized universe $[-1, 1]$ (Fig. 9.31). The seven linguistic variables

ERROR VARIATION

E R R O R	MN	MN	SN	SN	Z
	MN	MN	SN	Z	SP
	SN	SN	Z	SP	SP
	SN	Z	SP	SP	MP
	Z	SP	SP	MP	MP
	SP	SP	MP	MP	LP
	SP	MP	MP	LP	LP

Fig. 9.30. The output rule base used by fuzzy logic controller in Matlab program simulation

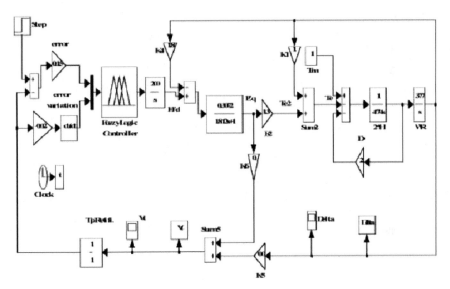

Fig. 9.31. Fuzzy logic controller model

used were generator voltage error and generator voltage error variation, which are:

– LN: large negative
– MN: medium negative
– SN: small negative
– Z: zero
– SP: small positive
– MP: medium positive
– LP: large positive

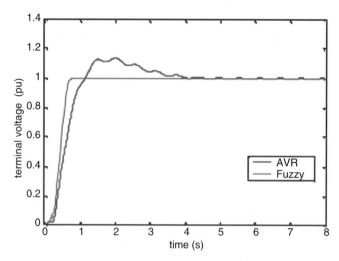

Fig. 9.32. One machine analysis (terminal voltage V_t)

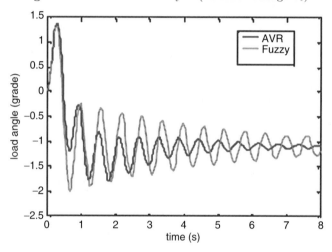

Fig. 9.33. One machine analysis (load angle)

Figure 9.32 shows the terminal voltage of a synchronous machine connected to an infinite bus bar by a transmission line. Figure 9.33 shows the load angle of a synchronous machine.

Next Fig. 9.34 shows the synchronous machine electric torque controlled by the AVR and by the fuzzy controller.

9.5.4 Fuzzy Logic Controller in an 18 Bus Bar System

The 18 bus IEEE system was simulated in an stability analysis program developed by Federal University of Uberlândia called Transufu, for two different disturbance types: short circuit and generation loss.

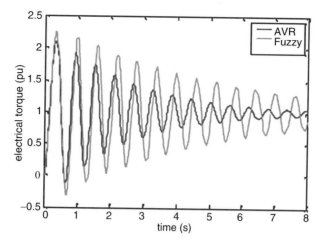

Fig. 9.34. One machine analysis (electrical torque)

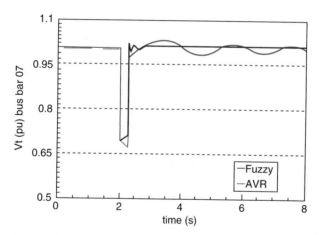

Fig. 9.35. The 18 bus bar system analysis (terminal voltage V_t in bus bar 7)

Four load modeling kinds were studied: constant impedance, constant current, 50% constant current and 50% constant power (mist 1), and finally 21% constant current and 79% constant power (mist 2). The fuzzy controller applied here has seven predicates.

Figure 9.35 shows the terminal voltage at bus bar 7 of the IEEE system for a short circuit simulation at bus bar 12 lasting 300 ms. The load was modeled as 50% I constant and 50% P constant.

Figure 9.36 shows the terminal voltage at the IEEE system bus bar 7 for a generator loss simulation at bus bar 4 at 2 s. The load was modeled as 50% constant current and 50% constant power.

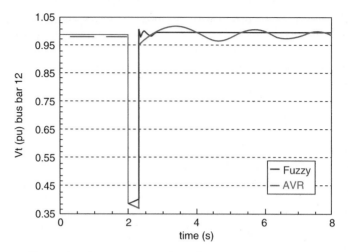

Fig. 9.36. The 18 bus bar system analysis (terminal voltage V_t in bus bar 12)

It could be observed for both studies (Matlab simulation and stability program simulation) an excellent response of the fuzzy controller and with no oscillation, while the AVR response presented a ripple in both studies and some oscillations before reaching the steady state operation point. It is shown that an excellent performance of the fuzzy control over the conventional one for the excitation control of synchronous machines could be achieved.

9.6 A Low Cost Speed Control System of Brushless DC Motor Using Fuzzy Logic

In this section, it focuses on a low cost speed control system using a fuzzy logic controller for a brushless DC (BLDC) motor. In a digital controller of BLDC motor, the control accuracy is of a high level, and it has a fast response time. We used a hall IC signal for the permanent magnet rotor position and for the speed feedback signals, and also for a microcontroller of 8-bit type (80CL580); furthermore, we designed the fuzzy logic controller and implemented the speed control system of BLDC motor. To acquire an accurate FLC algorithm, a simulation with the Matlab program has been made, while the performance of the system, done with an experiment for a unit step response, was also verified.

Recently, a BLDC motor has been rapidly demanded due to preciseness of industrial technology and increase of various kind of control device. Because a BLDC motor is suitable as a servomotor because of its high efficiency and excellent control character. So, we designed a low cost controller for BLDC motor, with high rely on quality. In this section, fuzzy reasoning algorithm was adapted to response well to various kinds of load. Also, the 80CL580, i.e.,

8-bit CPU was used to produce the system for a low cost. Additionally we designed a F/V converter to monitor a digital speed change in analog signal for a efficiency analyzing. So we evaluate a step response characteristic and experimented response characteristic in a instant torque change using torque meter device.

9.6.1 Proposed System

Figure 9.37 shows a block diagram of proposed system.

The system above is composed of BLDC motor, six step inverter, gate drive of inverter, Programmable Logic Device (PLD), and fuzzy reasoning controller. Switching logic, speed multiplexer, and PWM resolution converter are located within the PLD. Initially, the hall IC signal feedback occurred starting the motor by speed command. Logic signals of FET are generated using these three hall IC signals in the switching logic part. This circuit is shown in Fig. 9.38.

Only one phase of driving part is shown in this figure, it is realized using FET gate switching signals that are generated mixing hall IC signals and using AND operation between PWM which is speed control signal and RUN signal which is control of RUN/STOP. Also, only one speed signal is created using three hall IC signals in the speed pulse multiplexer. This circuit is shown in Fig. 9.39.

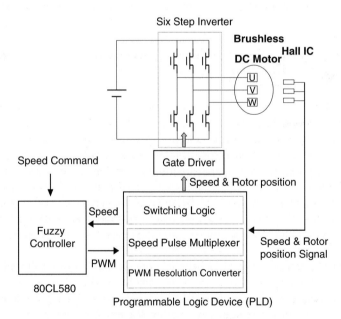

Fig. 9.37. Overall system block diagram

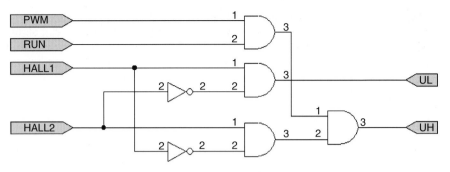

Fig. 9.38. The signal of one phase

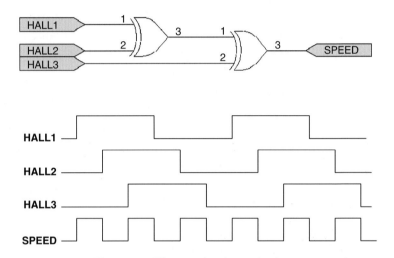

Fig. 9.39. The speed pulse multiplexer

In Fig. 9.39, we extend frequency as three times using XOR logical operation because the resolution of frequency is low related to speed at one hall IC signal. The PWM duty ratio for motor control is generated using the fuzzy reasoning algorithm, and the speed command that was input by speed signal feedback with fuzzy controller. The circuit for a PWM resolution is shown in Fig. 9.40. The resolution range of a PWM duty ratio supplied in an 80CL580 microcontroller is 0–255. As the circuit of Fig. 9.40, we enhanced the PWM duty resolution to two times and improved an accuracy of the speed control.

A specification of the BLDC motor we used in this paper are four pole, six slot, three phase, and a level of 50 W. A speed of the motor is 2,000 rpm to the DC 30 V, a inductance between the phases is 25 mH and resistance between phases is 3.2 Ω. Therefore, to observe the switching time for a drive, the speed frequency of the BLDC motor revolved at N rpm is following:

$$F_{\mathrm{m}} = N \, \mathrm{rpm}/60.$$

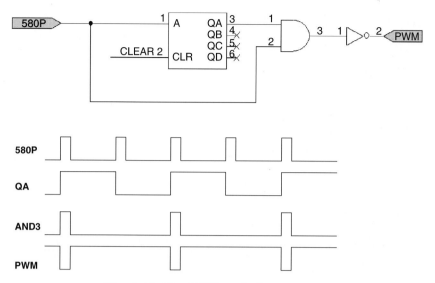

Fig. 9.40. The PWM resolution converter

So, the speed frequency is $6 \times F_m$, and a switching frequency of the one FET is $2 \times F_m$ because a FET turns on/off at one time for one a period of the hall IC signals.

9.6.2 Fuzzy Inference System

Recently the PI and the PID controller have been widely used as a control method of servodriving in industrial control fields. A driving specific property can be estimated well, once a control constant is properly set. But the control constant should be changed in order to maintain an optimum driving state if the driving point or the motor parameters are changed. Recently the fuzzy controller has appeared which is based on knowledge or an experience of expert rather than on a complicate mathematical modeling. The fuzzy controller works well using experimental information even if not having mathematical modeling. Moreover, the fuzzy controller is capable of real time control using fuzzy rule base. In this study, we tried to control the speed of BLDC motor using Mamdani-style FIS, which is composed of each one of input, and of output variable. Generally it is most important to assign the range of input and output membership function in the fuzzy inference engine. We defined input and output variables of FIS in this paper as following:

$$\Delta\varepsilon = C_m - V, \tag{9.4}$$

$$D_{new} = D_{old} + \Delta D. \tag{9.5}$$

The fuzzy input value $\Delta\varepsilon$ is deviation between motor speed command C_m and hall IC signal value V. When V_{rot} is counter step number per one rotation, V

becomes average motor speed of one rotation. The fuzzy output ΔD changes the pulse duty D_{new} which determines motor speed so that stable speed and torque may be maintained in case of starting or load changing of motor. The fuzzy rule base to control the speed of BLDC motor is composed of following five rules:

Rule 1 IF (ε is GN) THEN (ΔD is GN)
Rule 2 IF (ε is SN) THEN (ΔD is SN)
Rule 3 IF (ε is ZE) THEN (ΔD is ZE)
Rule 4 IF (ε is SP) THEN (ΔD is SP)
Rule 5 IF (ε is GP) THEN (ΔD is GP)

$$V = \frac{\sum V_{\text{rot}}}{6}.$$

Each linguistic variable presents degree of fuzzy input and output variables. GN means great negative, SN small negative, ZE zero, SP small positive, and GP great positive. On starting a BLDC motor simulation using FIS, the specific response character of 1,400 rpm speed was controlled with initial condition of C_{m} as 7,600 and D_{old} as 1% and was experimented in a motor load condition. Also, we implemented software simulation using the Matlab Ver. 5.1 (Math Work Co.) for exact performance result of fuzzy logic reasoning and realized fuzzy algorithm in the BLDC motor target board.

Figure 9.41 shows the flow chart of fuzzy reasoning system.

At first, to input $\Delta\varepsilon$ value acquired from hall IC signals to the FIS and to fuzzify the inputs value $\Delta\varepsilon$ is fuzzified. Output values of fuzzy membership function to be reasoned and mapping results to be aggregated according to each rule. Then to defuzzify output result, to be effected on PWM duty and finally speed to be returned of the motor. Figure 9.42 shows the membership function of fuzzy input and Fig. 9.43 shows the membership function of fuzzy output. In Fig. 9.42, control range of input variable $\Delta\varepsilon$ being between -500 and 500, two trapezoidal membership function and, three triangle membership function have been used. In the Fig. 9.43, control range of output variable ΔD being between -100 and 100, two trapezoidal membership function and three triangle membership function have been used.

Figure 9.44 presents a three-dimensional curve mapping from the error value ε to the ΔD amount. x-axis is present error value $\Delta\varepsilon$, y-axis is ΔD, and z-axis is degree of $\Delta\varepsilon$, ΔD.

9.6.3 Experimental Result

In this section, following the experimental set has been composed.

As Fig. 9.45, a controller of the motor, a driver, a RS-232C communication port, a start switch, and LED panel are composed on one board, the response characteristic of the system was confirmed using a frequency to voltage (F/V) converter. Also the torque meter was used to inflict a constant load. The

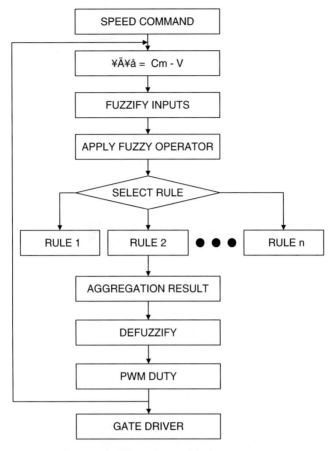

Fig. 9.41. Flow chart of fuzzy reasoning

circuit of the F/V converter module for analysis and its frequency response
are as following Figs. 9.46 and 9.47.

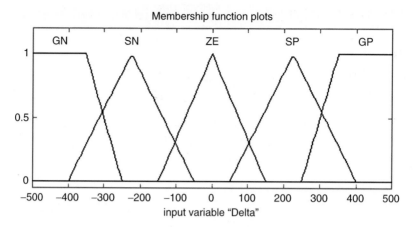

Fig. 9.42. Input membership function

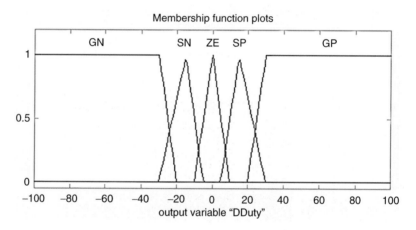

Fig. 9.43. Output membership function

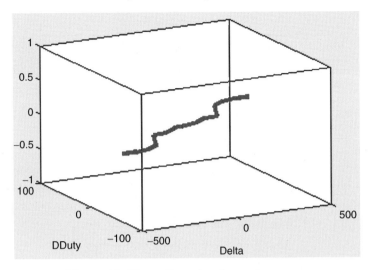

Fig. 9.44. Three-dimensional mapping curve

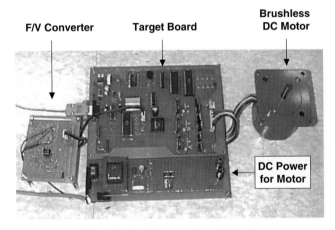

Fig. 9.45. The experimental set

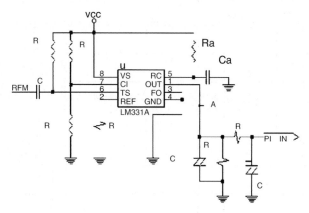

Fig. 9.46. The circuit of F/V converter

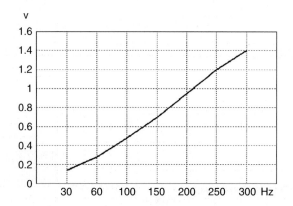

Fig. 9.47. The frequency response of F/V converter

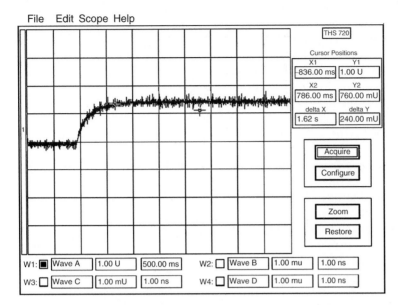

Fig. 9.48. The unit step response to no load

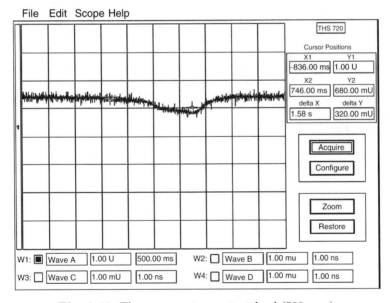

Fig. 9.49. The response to constant load (500 g cm)

After setting a speed command for 1,400 rpm in the program of the target board pushing the start switch on the board, we measured a unit step response. Also we inflict a constant load of 500 g cm in a moment with torque meter during the operation. The characteristic of a unit step response to no load is shown in the Fig. 9.48.

In this figure, the rising time to be increased up to 60% of a maximum speed is about 250 ms. Also, the characteristic of a response to constant load (500 g cm) while the motor is operating of to 1,400 rpm is shown in the Fig. 9.49.

The characteristic curve to a load variation is similar to that to no load.

Hence in this section, we realized the controller for a BLDC motor, which is demanded increasingly using the fuzzy logic, and evaluated a performance of the system with the experimental set. We detected a speed of the motor using only the hall IC signals instead of an expensive encoder. In this section, we presented fuzzy reasoning algorithm to control BLDC motor in order to improve the PI controller, which is hard to get optimum control under the unstable driving situation or different condition of load. As a low cost CPU like 80CL580 was used, execution speed was slightly slow, but we could get the same simulation result compared to the existing speed controller. Hereafter if we take advantage of PI controller and fuzzy reasoning system, better characteristic controller for speed control in BLDC motor may be made.

Appendix A

Fuzzy Logic in Matlab

Fuzzy logic in Matlab can be dealt very easily due to the existing new Fuzzy Logic Toolbox. This provides a complete set of functions to design an implement various fuzzy logic processes. The major fuzzy logic operation includes fuzzification, defuzzification, and the fuzzy inference. These all are performed by means of various functions and even can be implemented using the Graphical User Interface (GUI). Many of the applications can be simulated using the "fuzzy logic controller" Simulink block present in Matlab–Simulink toolbox. The features are:

- It provides tools to create and edit fuzzy inference system (FIS).
- Allows integrating fuzzy systems into simulation with Simulink.
- It is possible to create stand-alone C programs that call on fuzzy systems built with Matlab.

The Toolbox provides three categories of tools:

1. Command line functions
2. Graphical or interactive tools
3. Simulink blocks.

Command Line FIS Functions

addmf   – Add membership function to FIS.
addrule – Add rule to FIS.
addvar – Add variable to FIS.
defuzz – Defuzzify membership function.
evalfis – Perform fuzzy inference calculation.
evalmf – Generic membership function evaluation.
gensurf – Generate FIS output surface.
getfis – Get fuzzy system properties.

mf2mf   – Translate parameters between functions.
mfstrtch – Stretch membership function.
newfis   – Create new FIS. parsrule – parse fuzzy rules.
plotfis   – Display FIS input–output diagram.
plotmf   – Display all membership functions for one variable.
readfis   – Load FIS from disk.
rmmf   – Remove membership function from FIS.
rmvar   – Remove variable from FIS.
setfis   – Set fuzzy system properties.
showfis   – Display annotated FIS.
showrule – Display FIS rules
writefis   – Save FIS to disk.

Graphical User Interface Editors (GUI tools)

fuzzy   – Basic FIS editor.
mfedit   – Membership function editor.
ruleedit – Rule editor and parser.
ruleview – Rule viewer and fuzzy inference diagram.
surfview – Output surface viewer.

Simulink Blocks

Once fuzzy system is created using GUI tools or some other method can
be directly embedded into Simulink using the fuzzy logic controller block as
shown in Fig. A.1.

Important. make sure that the FIS matrix corresponding to the fuzzy system
is both in the Matlab workspace and referred to by name in the dialog box
associated with this fuzzy logic controller.
Although it is possible to use the Fuzzy Logic Toolbox by working strictly
from the command line, in general it is much easier to build a system up
graphically, so that GUI tools are commonly used for building, editing, and
observing FIS.
The process of mapping from a given input to an output using fuzzy logic
involves membership functions, fuzzy logic operators, and IF–THEN rules.

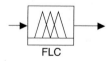

Fig. A.1. Fuzzy logic controller Simulink block

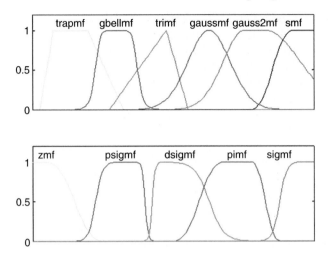

Fig. A.2. Membership functions

Membership functions. this Toolbox includes 11 built-in membership function types, built from several basic functions: piecewise linear functions (triangular and trapezoidal), the Gaussian distribution function (Gaussian curves and generalized bell), the sigmoid curve, and quadratic and cubic polynomial curves (Z, S, and Pi curves) (refer Fig. A.2).

Fuzzy logic operators. according to the fuzzy logical operations, any number of well-defined methods can fill in for the *AND operation* or the *OR operation.* In the Fuzzy Logic Toolbox, two built-in *AND methods* are supported: *min* (minimum) and *prod* (algebraic product). Two built-in *OR methods* are also supported: *max* (maximum), and the *probor* (probabilistic OR, also known as algebraic sum).

Related to *implication method,* two built-in methods are supported, and they are the same functions that are used by the AND method, so that, *min* method truncates the output fuzzy set, and *prod* scales the output fuzzy set.

Related to *aggregation method,* three built-in methods are supported: *max* (maximum), *probor* (probabilistic OR), and *sum* (simply the sum of each rule's output set).

Although centroid calculation is the most popular *defuzzification method,* there are five built-in methods supported: *centroid, bisector, middle of maximum, largest of maximum,* and *smallest of maximum.*

IF–THEN rules: since rules can be edited in three different formats (*verbose, symbolic,* and *indexed*), verbose format makes the system easier to interpret.

Every rule has a *weight* (a number between 0 and 1) which is applied to the number given by the antecedent. Generally this weight is 1 and so it has no effect at all on the implication process. For example, let us enter a sample rule (rule number 1):

Verbose format: 1. IF temperature is warm THEN sky is gray (1)
Symbolic format: 1. (temperature = = warm) => sky = gray (1)
Indexed format: 1,1 (1): 1

i.e., the first "1" corresponds to the input variable, the second corresponds to the output variable, the third displays the *weight* applied to each rule, and the fourth is shorthand that indicates whether this is an OR (2) rule or an AND (1) rule. So a literal interpretation of rule number 1 is: "if input 1 is MF1 (the first membership function associated with input 1) then output 1 should be MF1 (the first membership function associated with output 1) with the weight 1." Notice that as long as the aggregation method is commutative, then the order in which the rules are executed is not important.

Once a FLC is created, it can be saved to disk (FIS-file is created, i.e., juggler.fis) as an ASCII text format so that it can be edited and modified. An FLC can also be saved into Matlab workspace as a matrix variable (FIS matrix) so that it can be modified but its representation is extremely different from FIS-file representation.

Next, main windows corresponding to *GUI tools* are depicted as shown in Figs. A.3–A.8.

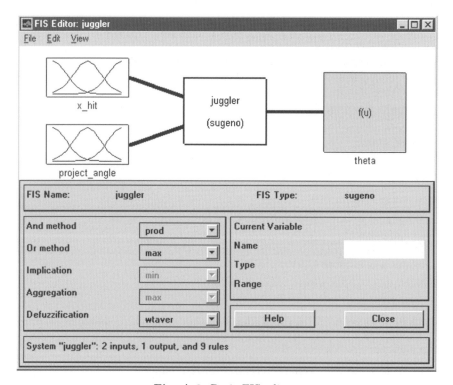

Fig. A.3. Basic FIS editor

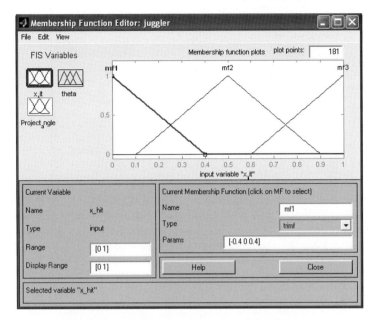

Fig. A.4. Membership function editor (input 1)

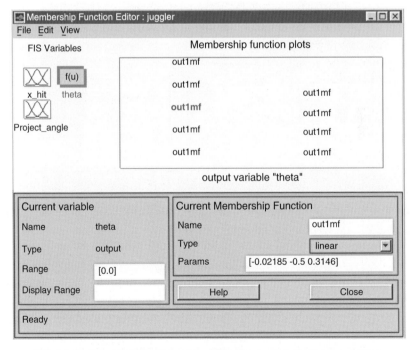

Fig. A.5. Membership function editor (output 1, in Sugeno style)

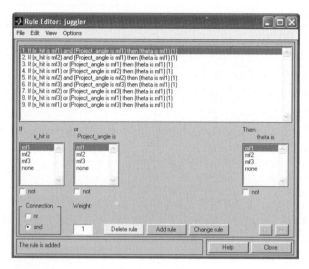

Fig. A.6. Rule editor (sample with nine rule in verbose format)

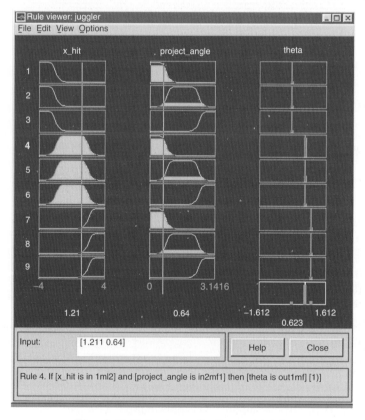

Fig. A.7. Rule viewer (in Sugeno style)

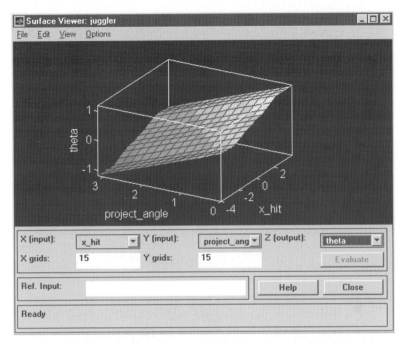

Fig. A.8. Surface viewer (input 1 and input 2 versus output 1)

Fuzzy Controllers: Examples

In order to see some results with FLC techniques, we will look at the example of *water level control* (can be executed by typing sltank in Matlab command window). We can change the valve controlling the water that flows in, but the outflow rate depends on the diameter of the outflow pipe (constant in this example) and the pressure in the tank (which varies the water level: see Fig. A.9). This system has some very nonlinear characteristics. Controller's input is the current error (error = desired water level − actual water level), and its output is the rate at which the valve is opening or not:

Rules should be as follows:

1. IF (level is okay) THEN (valve is no_change) (1)
2. IF (level is low) THEN (valve is open_fast) (1)
3. IF (level is high)THEN (valve is close_fast) (1)

Before editing that rules, membership functions must be defined with membership function editor (see Fig. A.10):

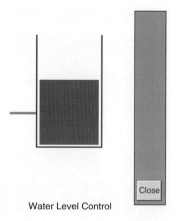

Water Level Control

Fig. A.9. Water level control

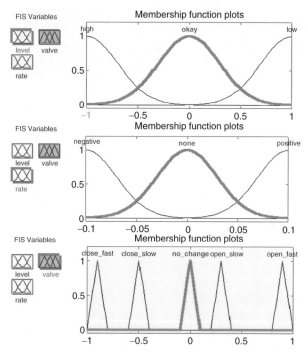

Fig. A.10. Membership function for input and output

Some other rules (with AND connectives) could be added after some delay:

4. IF (level is okay) and (rate is positive) THEN (valve is close_slow) (1)
5. IF (level is okay) and (rate is negative) THEN (valve is open_slow) (1)

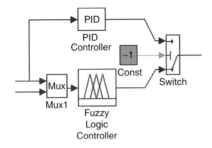

Fig. A.11. Simulink model

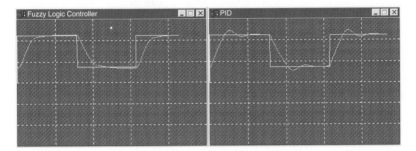

Fig. A.12. Fuzzy logic and PID

This FIS system can be simulated in Simulink and comparison between fuzzy logic control and PID control is done changing cont parameter (−1 means fuzzy control, 0 means PID control) (refer Figs. A.11 and A.12):

Obtained results are as follows:

Thus the "fuzzy logic controller" technique is used efficiently to control the level of liquid in the tank.

Summary

Thus Fuzzy Logic Toolbox in Matlab can be used for various applications. The membership values are assigned easily. The application developed is capable of executing within a fraction of second. The fuzzification and the defuzzification process are performed using command line functions as well by the GUI. The user can obtain the output in any required form.

References

1. Abbod MF, von Keyserlingk DG, Linkens DA, Mahfouf M (2001) Survey of utilization of fuzzy technology in medicine and healthcare. Fuzzy Sets Syst 120:331–349
2. Adlassnig KP (1986) Fuzzy set theory in medical diagnosis. IEEE Trans Syst Man Cybern 16(2):260–265
3. Adlassnig KP, Kolarz G (1982) CADIAC-2: computer-assisted medical diagnosis using fuzzy subsets. In: Gupta MM, Sanchez E (eds) Approximate reasoning in decision analysis. North-Holland, New York, pp. 219–247
4. Akbarzadeh-T M-R (1994) A fuzzy hierarchical controller for a single flexible arm. In: Proceedings of the 1994 international symposium on robotics and manufacturing, ISRAM'94, Maui, Hawaii
5. Akbarzadeh-T M-R, Jamshidi M (1997) Incorporating a priori expert knowledge in genetic algorithms. In: Proceedings of the 1997 IEEE conference on computational intelligence in robotics and automation, Monterey, CA, pp. 300–305
6. Albertos P, Strietzel R, Mart N (1997) Control engineering solutions: a practical approach. IEEE Computer Society, Silver Spring, MD
7. Allahverdi N (2002) Expert systems. An artificial intelligence application. Atlas, Istanbul, 248 p
8. Allahverdi N, Yaldiz S (1998) Expert system applications in medicine and example of design of a pre-diagnosis expert system. In: Proceedings of 2nd Turkish–German joint computer application days, 15–16 October 1998, Konya, pp. 175–192
9. von Altrock C, Krause B, Zimmermann H-J (1992a) Advanced fuzzy logic control technologies in automotive applications. In: Proceedings of 1992 IEEE international conference on fuzzy systems, San Diego, March 1992, pp. 835–843
10. von Altrock C, Krause B, Zimmermann H-J (1992b) On-line development tools for fuzzy knowledge-based systems of higher order. In: Proceedings of the 2nd international conference on fuzzy logic and neural network, Iisuka, July 1992, pp. 269–272
11. Arabshahi P, Choi JJ, Marks RJ, Caudell TP (1992) Fuzzy control of back-propagation. In: First IEEE international conference on fuzzy systems (FUZZ-IEEE'92), San Diego, CA, pp. 967–972

12. Bauer P, Bodenhofer U, Klement EP (1996) A fuzzy algorithm for pixel classification based on the discrepancy norm. In: Proceedings of 5th IEEE international conference on fuzzy systems, New Orleans, LA, September 1996, vol 3, pp. 2007–2012
13. Beierke S, Konigbauer R, Krause B, von Altrock C (1995) Fuzzy logic enhanced control of AC motor using DSP. In: Embedded systems conference California, 1995
14. Bellman RE, Zadeh LA (1970) Decision-making in a fuzzy environment. Manage Sci 17:B-141–B-164
15. Berenji HR (1991) Fuzzy logic controllers. In: An introduction to fuzzy logic applications in intelligent systems. Kluwer, Boston, pp. 69–96
16. Bersini H, Bontempi G, Decaestecker C (1995) Comparing RBF and fuzzy inference systems on theoretical and practical basis. In: Proceedings of International conference on artificial neural networks, ICANN '95, Paris, France, vol 1, pp. 169–174
17. Boegla K, Adlassniga K-P, Hayashic Y, Rothenfluhd TE, Leiticha H (2002) Knowledge acquisition in the fuzzy knowledge representation framework of a medical consultation system. Artif Intell Med 676:1–26
18. Bojadziev G, Bojadziev M (1997) Fuzzy logic for business, finance and management. World Scientific, Singapore
Bonissone P, Badami V, Chiang KH, Khedkar PS, Marcelle K, Schutten MJ (1995) Industrial applications of fuzzy logic at general electric. In: Proceedings of the IEEE, vol 83(3), pp. 450–465
19. Bonissone P, Khedkar P, Chen Y-T (1996) Genetic algorithms for automated tuning of fuzzy controllers: a transportation application. In: 5th IEEE International conference on fuzzy systems (FUZZIEEE'96), New Orleans, LA, pp. 674–680
20. Bouchon-Meunier B, Yager R, Zadeh L (1995) Fuzzy logic and soft computing. World Scientific, Singapore
21. Buckley JJ, Siler W, Tucker D (1986) FLOPS, a fuzzy expert system: applications and perspectives. In: Nogpita CV, Prade H (eds) Fuzzy logics in knowledge engineering. Verlag TUV Rheinland, Germany
22. Burkhardt D, Bonissone P (1992) Automated fuzzy knowledge base generation and tuning. In: First IEEE international conference on fuzzy systems (FUZZ-IEEE'92), San Diego, CA, pp. 179–188
23. Castillo O, Melin P (1994) Developing a new method for the identification of microorganisms for the food industry using the fractal dimension. J Fract 2(3):457–460
24. Castillo O, Melin P (1997) Mathematical modelling and simulation of robotic dynamic systems using fuzzy logic techniques and fractal theory. In: Proceedings of IMACS'97, Berlin, Germany, vol 5, pp. 343–348
25. Castillo O, Melin P (1998) A new fuzzy-fractal-genetic method for automated mathematical modelling and simulation of robotic dynamic systems. In: Proceedings of FUZZ'98, IEEE Press, Anchorage, AK, vol 2, pp. 1182–1187
26. Castillo O, Melin P (1999a) A new fuzzy inference system for reasoning with multiple differential equations for modelling complex dynamical systems. In: Proceedings of CIMCA'99, IOS Press, Vienna, Austria, pp. 224–229
27. Castillo O, Melin P (1999b) Automated mathematical modelling, simulation and behavior identification of robotic dynamic systems using a new fuzzy-fractal-genetic approach. J Robot Auton Syst 28(1):19–30

28. Cavallo A, Setola R, Vasca F (1996) Using Matlab, Simulink and control system toolbox. Prentice-Hall, Englewood Cliffs, NJ

29. Chang HC, Wang MH (1995) Neural network based self organizing fuzzy controller for transient stability of multimachine power systems. IEEE Trans Energy Conver 10(2):339–347

30. Chang CS, Chen JM, Liew AC (1995) An expert system approach for fault diagnosis considering uncertainties. In: Proceedings of international conference on intelligent manufacturing (ICIM'95), Wuhan, China, June, 1995

31. Chang CS, Chen JM et al (1996) Power system fault diagnosis sing fuzzy sets for uncertainties processing. Accepted for publication in ISAP'96, Orlando, USA

32. Chen S, Hwang C (1992) Fuzzy multiple attribute decision making: methods and applications. Springer, Berlin Heidelberg New York

33. Chen G, Pham TT (2001) Introduction to fuzzy sets, fuzzy logic, and fuzzy control systems. CRC, Boca Raton, FL

34. Cho HJ, Park JK, Lee HJ (1994) A fuzzy expert system for fault diagnosis of power systems. In: Proceedings of ISAP'94, Montpellier, France, September, pp. 217–222

35. Choi YS, Krishnapuram R (1997) A robust approach to image enhancement based on fuzzy logic. IEEE Trans Image Process 6(6):808–825

36. Cohen ME, Hudson DL (1988) The use of fuzzy variables in medical decision making. In: Gupta MM, Yamakawa T (eds) Fuzzy computing. Elsevier Science, North Holland, pp. 263–271

37. Cordon O, Herrera F, Hoffmann F, Magdalena L (2001) Genetic fuzzy systems – evolutionary tuning and learning of fuzzy knowledge bases, vol 19 of Advances in fuzzy systems – applications and theory. World Scientific, Singapore

38. Cox E (1984) The fuzzy systems hand book – a practitioners guide to building using and maintaining fuzzy system. AP Professional Publishers, Cambridge

39. Craiger J, Goodman D, Weiss RJ, Butler AB (1996) Modeling organizational behavior with fuzzy cognitive maps. Int J Comput Intell Org 1:120–123

40. Davis L (1991) Handbook of genetic algorithms. Van Nostrand Reinhold, New York

41. Dawson JG, Gao Z (1994) Fuzzy logic control of linear systems with variable time delay. In: Proceedings of 1994 IEEE international symposium on intelligent control, August 1994, pp. 5–10

42. Devillers J (1996) Neural networks in QSAR and drug design. Academic, New York

43. Dhayagude N, Gao Z, Mrad F (1996) Fuzzy logic control of automated screw fastening. J Robot Comput Aid Manuf 12(3):235–242

44. Drake P, Mazuelos D (1994) An introduction to fuzzy logic control. In: Fuzzy logic'94 conference proceedings tutorials, San Diego, CA, September 13–15, pp. 5–44

45. Driankov D, Hellendoorn H, Reinfrank M (1993) An introduction to fuzzy control. Springer, Berlin Heidelberg New York

46. Driankov D, Hellendoorn H, Reinfrank M (1996) An introduction to fuzzy control. Springer, Berlin Heidelberg New York

47. Dubois D, Prade H (1980) Fuzzy sets and systems: theory and applications. Academic, New York

48. Dubois D, Prade H (1985) A review of fuzzy set aggregation connectives. Inform Sci 36:85–121

49. Dubois D, Prade H (1990) An introduction to possibilistic and fuzzy logics. In: Shafer G, Pearl J (eds) Readings in uncertain reasoning. Morgan Kaufmann, San Francisco
50. Dubois D, Prade H (1998) Soft computing, fuzzy logic, and artificial intelligence, soft computing a fusion of foundations, methodologies and applications, vol 2(1). Springer, Berlin Heidelberg New York, pp. 7–11
51. El-Metwally KA, Malik OP (1996) Applications of fuzzy stabilizers in a multimachine power system environment. In: IEEE proceedings of generation, transmission and distribution, May, vol 143(3), pp. 263–268
52. El-Metwally K, Hancock G, Malik O (1996) Implementation of a fuzzy logic PSS using a microcontroller and experimental 3–13 test results. IEEE Trans Energy Conver 1(1):91–96
53. Fahmy HI, Douligeris C (1995) END: an expert system designer. IEEE Netw Mag 9(6):18–27
54. Franke K, Koeppen M, Nickolay B (2000) Fuzzy image processing by using Dubois and Prade fuzzy norm. In: Proceedings of 15th international conference on pattern recognition (ICPR), Barcelona, Spain, pp. 518–521
55. Franke K, Zhang YN, Koeppen M (2002) Static signature verification employing a Kosko-neuro-fuzzy approach. In: Pal NR, Sugeno M (eds) Advances in soft computing-AFSS 2002, LNAI 2275. Springer, Berlin Heidelberg New York, pp. 185–190
56. Fujiyoshi M, Shiraki T (1992) A fuzzy automatic-combustion-control-system. In: Proceedings of the 2nd international conference on fuzzy logic and neural network, Iisuka, July 1992, pp. 469–472
57. Fukui C, Kawakami J (1986) An expert system for fault section estimation using information from protective relays and circuit breakers. In: IEEE Transactions on power delivery, October 1986, vol PWRD-1(4)
58. FuzzyTECH User's Manual (1995) INFORM Software Corp., 2001 Midwest Rd, Oak Brook, IL 60521, USA
59. Gao Z, Trautzsch TA, Dawson J (2000) A stable self-tuning fuzzy logic control system for industrial temperature control problems. In: IEEE industrial application society 2000 annual meeting and world conference on industrial applications of electrical energy, October 8–12
60. Garibaldi JM, Ifeachor EC (1996) The comparison of a crisp and fuzzy expert system with practising and expert clinicians. In: Proceedings of the 2nd international conference on neural networks and expert systems in medicine and healthcare, Plymouth, UK, pp. 229–237
61. Garibaldi JM, Ifeachor EC (1999) Application of simulated annealing fuzzy model tuning to umbilical cord acid–base interpretation. IEEE Trans Fuzzy Syst 7(1):72–84
62. Garibaldi JM, Westgate JA, Ifeachor EC, Greene KR (1994) The development of an expert system for the analysis of umbilical cord blood at delivery. In: Proceedings of the international conference on neural networks and expert systems in medicine and healthcare, Plymouth, UK, pp. 394–402
63. Garibaldi JM, Westgate JA, Ifeachor EC, Greene KR (1997) The development and implementation of an expert system for the analysis of umbilical cord blood. Artif Intell Med 10(2):129–144
64. Gebhardt J (1995) New industrial applications of the fuzzy-PLC proceedings of the 3. In: European congress on fuzzy and intelligent technologies (EUFIT 95), Aachen, 08/95

65. Gebhardt J, Müller R (1993) Application of fuzzy logic to the control of a wind energy converter. In: First European congress on fuzzy and intelligent technologies (EUFIT 93), Aachen, 09/93

66. Geyer-Schulz A (1995) Fuzzy rule-based expert systems and genetic machine learning, vol 3 of Studies in fuzziness. Physica-Verlag, Heidelberg

67. Goto K, Yamaguchi T (1991) Fuzzy associative memory application to a plant modeling. In: Proceedings of the international conference on artificial neural networks (ICANN-91), Espoo, Finland 1991, pp. 1245–1248

68. Grabisch M, Nicolas JM (1994) Classification by fuzzy integral: performance and tests. Fuzzy Sets Syst 65:255–271

69. Gupta R, Ragade R, Yager R (eds) (1979) Advances in fuzzy set theory and applications. North-Holland, Amsterdam

70. Hairi A, Malik OP (1997) A self-learning fuzzy stabilizer for a synchronous machine. Int J Eng Intell Syst 5(3):157–162

71. Hairi A, Malik OP (1999) Fuzzy logic power system stabilizer based on genetically optimized adaptive network. Fuzzy Sets Syst 102(1):31–40

72. Halim M, Ho KM, Liu A (1990) Fuzzy logic for medical expert systems. Ann Acad Med 19(5):672–683

73. Handschin E, Hoffmann W, Reyer F, Stephanblome T, Schlucking U et al (1996) A new method excitation control based on fuzzy set theory. IEEE Trans Power Syst 9(1):533–539

74. Hariri A, Malik OP (1996a) Self-learning adaptive-network-based fuzzy logic power system stabilizer. In: Proceedings of international conference on intelligent systems applications to power systems, Orlando, FL, 28 January–2 February, 1996, pp. 299–303

75. Hariri A, Malik OP (1996b) A fuzzy logic based power system stabilizer with learning ability. IEEE Trans Energy Conver 11(6):721–727

76. Hassan M, Malik O (1993) Implementation and laboratory test results for a fuzzy logic based self tuned power system stabilizer. IEEE Trans Energy Conver 8(2):221–228

77. Hassan M, Malik O, Hope GS (1991) Fuzzy logic based stabilizer for a synchronous machine. IEEE Trans Energy Conver 6(3):407–413

78. Hathaway RJ, Bezdek JC, Pedrycz W (1996) A parametric model for fusing heterogeneous fuzzy data. IEEE Trans Fuzzy Syst 4(3):270–281

79. Herrmann CS (1995) A hybrid fuzzy-neural expert system for diagnosis. In: IJCAI, Morgan Kaufman, Montreal, Canada

80. Hilloowala R, Sharaf A (1996) A rule based fuzzy logic controller for a PWM inverter in a stand alone wind energy conversion scheme. IEEE Trans Ind Appl 32(1):57–65

81. Hiyama T (1994) Robustness of fuzzy logic power system stabilizers applied to multimachine power system. IEEE Trans Energy Conver 9(3):451–459

82. Hiyama T (1996) Real time control of micro machine system using microcomputer based fuzzy logic power system stabilizers. IEEE Trans Energy Conver 9(4):724–731

83. Hiyama T, Sameshima T (1991) Fuzzy logic control scheme for on-line stabilization of multi-machine power system. Fuzzy Sets Syst 39:181–194

84. Hiyama T, Oniki S, Nagashima H (1993) Experimental studies on microcomputer based fuzzy logic stabilizer. In: Proceedings of the 2nd international forum on application of neural network to power systems, pp. 212–217

424 References

85. Hiyama T, Kugimiya M, Satoh H (1994) Advanced PID type fuzzy logic power system stabilizer. IEEE Trans Energy Conver 9(3):514–520
86. Hiyama T, Oniki S, Nagashima H (1996a) Evaluation of advanced fuzzy logic PSS on analog network simulator and actual installation on hydro generators. IEEE Trans Energy Conver 11(1):125–131
87. Hiyama T, Mishiro M, Kihara H, Ortmeyer T (1996b) Coordinated fuzzy logic control for series capacitor modules and PSS to enhance stability of power system. IEEE Trans Power Deliv 10(2):1098–1104
88. Hiyama T, Miyazaki K, Satoh H (1996c) A fuzzy logic excitation system for stability enhancement of power systems with multimode oscillations. IEEE Trans Energy Conver 11(2):449–454
89. Hiyama T, Mishiro M, Kihara H, Ortmeyer T (1996d) Fuzzy logic switching of thyristor controlled braking resistor considering coordination with SVC. IEEE Trans Power Deliv 10(4):2020–2026
90. Hoang TC, El-Sharkawi M (1996) High performance speed and position tracking of induction motors using multi-layer fuzzy control. IEEE Trans Energy Conver 11(2):353–358
91. Hoang P, Tomsovic K (1996) Design and analysis of an adaptive fuzzy power system stabilizer. IEEE Trans Energy Conver 11(2):455–461
92. Hsu Y-Y, Lu FC, Chien Y et al (1991) An expert system for locating distribution system faults. IEEE Trans Power Deliv 5(2):366–371
93. Ifeachor EC, Outram NJ (1995) A fuzzy expert system to assist in the management of labour. In: Proceedings of the international ICSC symposium on fuzzy logic. Zurich, Switzerland, pp. 97–102
94. Irani KB, Khabbaz NG (1982) A methodology for the design of communication networks and the distribution of data in distributed supercomputer systems. IEEE Trans Comput C-31(5)
95. Jacobs RA (1988) Increased rates of convergence through learning rate adaptation. Neural Netw 1:295–297
96. Jang J-SR (1992a) ANFIS: adaptive-network-based fuzzy inference systems. IEEE Trans Syst Man Cybern 23(3)
97. Jang J-SR (1992b) Self-learning fuzzy controller based on temporal backpropagation. IEEE Trans Neural Netw 3(5):714–723
98. Jang J-SR (1993) ANFIS: adaptive-network-based fuzzy inference system. IEEE Trans Syst Man Cybern 23:665–684
99. Jang J-SR, Sun CT, Mizutani E (1997) Neuro-fuzzy and soft computing. Prentice-Hall, Englewood Cliffs, NJ
100. Jarventausta P, Verho P, Partanen J (1994) Using fuzzy sets to model the uncertainty in the fault location process of distribution networks. IEEE Trans Power Deliv 9(2):954–960
101. Johansen TA (1994) Fuzzy model based control: stability, robustness, and performance issues. IEEE Trans Fuzzy Syst 2:221–234
102. Kasabov NK (1996) Foundation of neural networks, fuzzy systems, and knowledge engineering. MIT, Cambridge, MA
103. Kaufmann A, Gupta MM (1985) Introduction to fuzzy arithmetic. Van Nostrand, New York
104. Kaufmann A, Gupta MM (1988) Fuzzy mathematical models in engineering and management science. North-Holland, Amsterdam
105. Kezunovic M, Fromen CW et al (1993) An expert system for transmission substation event analysis. IEEE Trans Power Deliv 8(4):1942–1949

106. Kimura T, Nishimatsu S et al (1992) Development of an expert system for estimating fault section in control center based on protective system simulation. IEEE Trans Power Deliv 7(1):167–172

107. Klir GJ, Yuan B (1995) Fuzzy sets and FuzzyLogic: theory and applications. Prentice-Hall, New York

108. Koeppen M, Franke K (1999) Fuzzy morphology revisited. In: Proceedings of 3rd international workshop on softcomputing in industry (IWSCI), Muroran, Japan, pp. 258–263

109. Kohonen T (1995) Self-organizing maps. Springer, Berlin Heidelberg New York

110. Kosko B (1986) Fuzzy cognitive maps. Int J Man Mach Stud 24:65–75

111. Kosko B (1992) Neural networks and fuzzy systems: a dynamical systems approach to machine intelligence. Prentice-Hall, Englewood Cliffs, NJ

112. Kosko B (1997) Fuzzy engineering. Prentice-Hall, Englewood Cliffs, NJ

113. Krause B, von Altrock C, Limper K, Schäfers W (1994) A neuro-fuzzy adaptive control strategy for refuse incineration plants. Fuzzy Sets Syst 63(3)

114. Krause B, von Altrock C, Pozybill M (1996) Intelligent highway by fuzzy logic: congestion detection and traffic control on multi-lane roads with variable road signs. In: Proceedings of EUFIT'96, Aachen, Germany

115. Kruse R, Gebhardt J, Klawonn F (1994) Foundations of fuzzy systems. Wiley, New York

116. Lang JSR, Gulley N (1995) Fuzzy Logic Toolbox user's guide. The MathWorks, Natick, MA

117. Langari R, Berenji HR (1992) Fuzzy logic in control engineering. In: Handbook of intelligent control. Van Nostrand, New York

118. Lee CC (1990) Fuzzy logic in control systems: fuzzy logic controller: Parts I and II. IEEE Trans Syst Man Cybern 20(2):404–435

119. Lee MA, Takagi H (1993a) Integrating design stages of fuzzy systems using genetic algorithms. In: Proceedings of the 2nd international conference on fuzzy systems (FUZZ-IEEE'93), 28 March–1 April, 1993. IEEE, New York, pp. 612–617

120. Lee MA, Takagi H (1993b) Integrating design stages of fuzzy systems using genetic algorithms. In: Proceedings of the 1993 IEEE international conference on fuzzy systems, San Francisco, CA, pp. 612–617

121. Lee C-S, Kuo Y-H, Yau P-T (1997) Weighted fuzzy mean filter for image processing. Fuzzy Sets Syst 89(2):157–180

122. Levi R, Rivers M, Hickey K (1994) An intelligent instrument for high voltage equipment insulation evaluation. In: Proceedings of the 1994 international conference on intelligent system application to power systems, France, September 1994, pp. 471–476

123. Lin Y, Cunninghan GJ (1994) Building a fuzzy system from input–output data. J Intell Fuzzy Syst 2(3):243–250

124. Lin C-T, Lee CSG (1991) Neural-network-based fuzzy logic control and decision system. IEEE Trans Comput 40(12):1320–1336

125. Lin CT, Lee G (1996) Neural fuzzy systems: a neuro-fuzzy synergism to intelligent systems. Prentice-Hall, Upper Saddle River, NJ

126. Lorenz A, Blum M, Ermert H, Senge Th (1997) Comparison of different neuro-fuzzy classification systems for the detection of prostate cancer in ultrasonic images (http://www.lp-it.de/neuro-fuzzy-classification.pdf)

127. Lpez de Mantaras R, Agusti J, Plaza E, Sierra C (1991) MILORD: a fuzzy expert system shell. In: Kandel A (ed) Fuzzy expert systems. CRC, Boca Raton, FL
128. Luo RC, Kay MG (1989) Multisensor integration and fusion in intelligent systems. IEEE Trans Syst Man Cybern 19:901–931
129. Makkonen A, Koivo HN (1995) Fuzzy control of a nonlinear servo motor model. J Intell Fuzzy Syst 3:145–154
130. Mamdani EH (1974) Application of fuzzy algorithms for control of simple dynamic plant. Proc Inst Elect Eng 121:1585–1588
131. Mamdani EH (1981) Advances in the linguistic synthesis of fuzzy controllers, fuzzy reasoning and applications. Academic, London, pp. 325–334
132. Mamdani EH (1993) Twenty years of fuzzy control: experiences gained and lessons learnt. IEEE Int Conf Fuzzy Syst 339–344
133. Mamdani EH, Assilian S (1975) An experiment in linguistic synthesis with a fuzzy logic controller. Int J Man Mach Stud 7(1):1–13
134. Mamdani EH, Gaines BR (eds) (1981) Fuzzy reasoning and its applications. Academic, London
135. Marquardt DW (1963) An algorithm for least squares estimation of non-linear parameters. J Soc Ind Appl Math 11:431–441
136. Mayer F, Morel G, Iung B, Leger JB (1996) Integrated manufacturing system metamodelling at the shop-floor level. In: Proceedings of advanced summer institute ASI'96, Toulouse, France, pp. 232–239
137. McDonald JR, Burt GM, Young DJ (1992) Alarm processing and fault diagnosis using knowledge based systems for transmission and distribution network control. IEEE Trans Power Syst 7(3):1292–1298
138. Mendel JM (1995) Fuzzy logic systems for engineering: a tutorial. Proc IEEE 83(3):345–377
139. Mitchell T (1997) Machine learning. McGraw-Hill, New York
140. Mitsumoto M, Zimmermann H-J (1992) Comparison of fuzzy reasoning methods. Fuzzy Sets Syst 8:253–285
141. Monostori L; Egresits Cs; Kádár B (1996) Hybrid AI solutions and their application in manufacturing. In: Proceedings of IEA/AIE-96, the 9th international conference on industrial and engineering applications of artificial intelligence and expert systems, 4–7 June 1996. Gordon and Breach Publishers, Fukuoka, Japan, pp. 469–478
142. Naunin D, Beierke S, Heidrich P (1991) Transputer control of asynchronous servo drives. EPE Florence
143. Negoita C (1985) Expert systems and fuzzy systems. Benjamin Cummings, Menlo Park, CA
144. Nguyen HP, Kreinovich V (2001) Fuzzy logic and its applications in medicine. Int J Med Inform 62:165–173
145. N.N., fuzzyTECH 4.2 (1996) User manual and reference manual, INFORM
146. N.N., fuzzyTECH fuzzyLAB Manual (1994) Microchip Technologies, Arizona
147. Noroozian M, Andersson G, Tomsovic K (1996) Robust, near time optimal control of power system oscillations with fuzzy logic. IEEE Trans Power Deliv 11(1):393–400
148. Ono H, Ohnishi T, Terada Y (1989) Combustion control of refuse incineration plant by fuzzy logic. Fuzzy Sets Syst 32:193–206
149. Ortmeyer T, Hiyama T (1995) Frequency response characteristics of the fuzzy polar power system stabilizer. IEEE Trans Energy Conver 10(2):333–338

150. Pal SK (1998) Soft computing tools and pattern recognition. IETE J Res 44(1–2):61–87
151. Pal SK, Mitra S (1992) Multilayer perception, fuzzy sets and classification. IEEE Trans Neural Netw 3(5):683–696
152. Palm R (1992) Sliding mode fuzzy control. In: Proceedings of the 1st IEEE international conference on fuzzy systems, San Diego, pp. 519–526
153. Park YM, Moon UC, Lee KY (1995) Self-organization of fuzzy logic controller for dynamic systems using a fuzzy auto-regressive moving average (FARMA) model. IEEE Trans Fuzzy Syst 3:75–82
154. Park YM, Moon UC, Lee KW (1996) A self-organizing power system stabilizer using fuzzy autoregressive moving average (Farma) model. IEEE Trans Energy Conver 11(2):442–448
155. Parodi A, Bonelli P (1993) A new approach to fuzzy classifier systems. In: Forrest S (ed) Proceedings of the ICGA'97, Los Altos, CA. Morgan Kaufmann, San Francisco, pp. 223–230
156. Pedrycz W (1989) Fuzzy control and fuzzy systems. Wiley, New York
157. Pelaez CE, Bowles JB (1995) Applying fuzzy cognitive maps knowledge representation to failure modes effects analysis. In: Proceedings of annual reliability and maintainability symposium, pp. 450–455
158. Pelaez CE, Bowles JB (1996) Using fuzzy cognitive maps as a system model for failure models and effects analysis. Inform Sci 88:177–199
159. Protopapas CA, Machias AV et al (1991) An expert system for substation fault diagnosis and alarm processing. IEEE Trans Power Deliv 6(2):648–655
160. Ramaswamy P, Edwards RM, Lee KY (1993a) An automated tuning method of a fuzzy logic controller for nuclear reactors. IEEE Trans Nucl Sci 40:1253–1262
161. Ramaswamy P, Riese M, Edwards RM, Lee KY (1993b) Two approaches for automating the tuning process of fuzzy logic controllers. In: Proceedings of the IEEE 32nd conference on decision and control, San Antonio, TX, 15–17 December 1993, pp. 1753–1758
162. Rao DH (1998) Fuzzy neural networks. IETE J Res 44(4–5):227–236
163. Ronco AL, Fernandez R (1999) Improving ultrasonographic diagnosis of prostate cancer with neural networks. Ultrasound Med Biol 25(5):729–733
164. Ross TJ (1993) In: Jamshidi M, Vadiee N, Ross TJ (eds) Fuzzy logic and control. PTR Prentice-Hall, Englewood Cliffs, NJ
165. Ross TJ (1995) Fuzzy logic with engineering applications. McGraw-Hill, New York
166. Russo M (1998) FuGeNeSys – a fuzzy genetic neural system for fuzzy modeling. IEEE Trans Fuzzy Syst 6(3):373–388
167. Saridis G (1989) Analytic formulation of the principle of increasing precision with decreasing intelligence for intelligent machines. Automatica 25:461–467
168. Schneider M, Kandel A, Langholz G, Chew G (1996) Fuzzy expert system tools. Wiley, New York
169. Schneider M, Shneider E, Kandel A, Chew G (1998) Automatic construction of FCMs. Fuzzy Sets Syst 93:161–172
170. Seker H, Odetayo M, Petrovic D, Naguib RNG (2003) A fuzzy logic based method for prognostic decision making in breast and prostate cancers. IEEE Trans Inform Technol Biomed (in press)
171. Simpson PK (1992) Fuzzy min–max neural networks – part I: classification. IEEE Trans Neural Netw 3(5):776–786

172. Stylios CD, Groumpos PP (1998a) The challenge of modeling supervisory systems using fuzzy cognitive maps. J Intell Manuf 9:339–345
173. Stylios CD, Groumpos PP (1998b) Using fuzzy cognitive maps to achieve intelligence in manufacturing systems. In: Proceedings of the 1st international workshop on intelligent manufacturing systems, Lausanne, Switzerland, pp. 85–95
174. Sugeno M (1985) Industrial applications of fuzzy control. Elsevier, Amsterdam
175. Sugeno M, Kang GT (1988) Structure identification of fuzzy model. Fuzzy Sets Syst 28(1):15–33
176. Sugeno M, Takagi T (1985) Fuzzy identification of systems and its application to modelling and control. IEEE Trans Syst Man Cybern 15:116–132
177. Sugeno M, Yasukawa T (1993) A fuzzy-logic-based approach to qualitative modeling. IEEE Trans Fuzzy Syst 1(1):7–31
178. Taber R (1991) Knowledge processing with fuzzy cognitive maps. Expert Syst Appl 2:83–87
179. Takagi H, Hayashi I (1991) NN-driven fuzzy reasoning. Int J Approx Reason 191–212
180. Takagi T, Sugeno M (1985) Fuzzy identification of systems and its applications to modeling and control. IEEE Trans Syst Man Cybern 15(1):116–132
181. Tanaka K, Sugeno M (1992) Stability analysis and design of fuzzy control systems. Fuzzy Sets Syst 45(2):135–156
182. Tang Y, Xu L (1994) Fuzzy logic application for intelligent control of a variable speed drive. IEEE Trans Energy Conver 9(4):679–685
183. Tang Y, Xu L (1996) Vector control and fuzzy logic control of doubly fed variable speed drives with DSP implementation. IEEE Trans Energy Conver 10(4):661–668
184. Thole U, Zimmermann H-J, Zysno P (1975) On the suitability of minimum and product operators for the Intersection of fuzzy sets. Fuzzy Sets Syst 2:173–186
185. Thomas DE, Armstrong-Helouvry B (1995) Fuzzy logic control – a taxonomy of benefits. Proc IEEE 83(3):407–421
186. Tobi T, Hanafusa T (1991) A practical application of fuzzy control for an air-conditioning system. Int J Approx Reason 5:331–348
187. Togai M, Watanabe H (1986) An inference engine for real-time approximate reasoning: toward an expert system on a chip. IEEE Exp 1:55–62
188. Toliyat H, Sadeh J, Ghazi R (1996) Design of augmented fuzzy logic power system stabilizers to enhance power systems stability. IEEE Trans Energy Conver 1(1):97–103
189. Tsoukalas LH, Uhrig RE (1997) Fuzzy and neural approaches in engineering. Wiley, Singapore
190. Tsukamoto Y (1979) An approach to fuzzy reasoning methods. In: Gupta M, Ragade R, Yager R (eds) Advances in fuzzy set theory and applications. North-Holland, Amsterdam, pp. 137–149
191. Tunstel E (1996) Mobile robot autonomy via hierarchical fuzzy behavior control. In: 6th international symposium on robotics and manufacturing, 2nd world automation congress, May 1996, pp. 837–842
192. Tunstel E, Lippincott T, Jamshidi M (1997) Behavior hierarchy for autonomous mobile robots: fuzzy-behavior modulation and evolution. Int J Intell Automat Soft Comput 3(1):37–49
193. Turksen IB (1988) Approximate reasoning for production planning. Fuzzy Sets Syst 26:23–37

194. Ungar LH (1995) A bioreactor benchmark for adaptive network-based process control. In: Neural networks for control. MIT, Cambridge, MA, pp. 387–402

195. Vachtsevanos G, Kim S (1997) The role of the human in intelligent control practices. In: Proceedings of the 12th IEEE international symposium on intelligent control, Istanbul, Turkey, pp. 15–20

196. Vadiee N, Jamshidi M (1994) The promising future of fuzzy logic. IEEE Expert 9:37

197. Valiquette B, Torres GL (1991) An expert system based diagnosis and advisor tool for teaching power system operation emergency control strategies. IEEE Trans Power Syst 6(3):1315–1322

198. Varley RF (1996) Fuzzy logic controller extends direct drive torque motor performance. In: Proceedings of intelligent motion systems conference

199. Von Altrock C (1997) Fuzzy logic and neurofuzzy applications in business and finance. Prentice-Hall, Englewood Cliffs, NJ

200. Von Altrock C, Krause B, Zimmermann H-J (1992) Advanced fuzzy logic control technologies in automotive applications. In: IEEE conference on fuzzy systems, ISBN 0-7803-0237-0, pp. 831–842

201. Von Altrock C, Arend H-O, Krause B, Steffens C, Behrens-Rommler E (1994) Customer-adaptive fuzzy control of home heating system. In: IEEE conference on fuzzy systems in Orlando

202. Wang L-X (1993) Stable adaptive fuzzy control of nonlinear systems. IEEE Trans Fuzzy Syst 1(2):146–155

203. Wang L-X (1994a) A supervisory controller for fuzzy control systems that guarantees stability. IEEE Trans Automat Control 39(9):1845–1848

204. Wang L-X (1994b) Adaptive fuzzy systems and control: design and stability analysis. Prentice-Hall, Englewood Cliffs, NJ

205. Wang L-X (1997) A course in fuzzy systems and control. PTR Prentice-Hall, Upper Saddle River, NJ

206. Watanabe H, Yakowenko WJ, Kim YM, Anbe J, Tobi T (1994) Application of a fuzzy discrimination analysis for diagnosis of valvular heart disease. IEEE Trans Fuzzy Syst 2(4):267–276

207. Yager RR, Zadeh LA (eds) (1991) An introduction to fuzzy logic applications in intelligent systems. Kluwer, Boston

208. Yager RR, Zadeh LA (1992) An introduction to fuzzy logic applications in intelligent systems. Kluwer, Boston

209. Yasunobu S, Myamoto S (1985) Automatic train operation by predictive fuzzy control. In: Industrial applications of fuzzy control. North-Holland, Amsterdam

210. Yongli Z, Yang YH, Hogg BW, Zhang WQ, Gao S (1994) An expert system for power systems fault analysis. IEEE Trans Power Syst 9(1):503–509

211. Zadeh LA (1950) Thinking machines – a new field in electrical engineering. Columbia Eng 3:12–13, 30, 31

212. Zadeh LA (1965) Fuzzy sets. Inform Control 8:338–353

213. Zadeh LA (1973) Outline of a new approach to the analysis of complex systems and decision processes. IEEE Trans Syst Man Cybern SMC-3:28–44

214. Zadeh LA (1976) A fuzzy-algorithmic approach to the definition of complex or imprecise concepts. Electronics Research Laboratory Report

215. Zadeh LA (1992) The calculus of fuzzy if–then rules. AI Exp 7(3):22–27

216. Zadeh LA (1971) Toward a theory of fuzzy systems. In: Aspects of network and system theory. Rinehart and Winston, New York, 469–490

217. Zadeh LA (1974) On the analysis of large scale systems. In: Systems approaches and environment problems, pp. 23–27

218. Zadeh LA (1975) The concept of a linguistic variable and its application to approximate reasoning. Inform Sci 8:199–249. Department of Computer Science, Tijuana Institute of Technology, P.O. Box 4207, Chula Vista, CA 91909, USA

219. Zadeh LA (1977) Fuzzy sets and their applications to classification and clustering. In: Van Ryzin J (ed) Classification and clustering. Academic, New York, pp. 251–299

220. Zadeh LA (1993) The role of fuzzy logic and soft computing in the conception and design of intelligent systems. In: Klement EP, Slany W (eds) FLAI. Springer, Berlin Heidelberg New York, p 1

221. Zadeh LA, Yager RR (eds) (1991) Uncertainty in knowledge bases. Springer, Berlin Heidelberg New York

222. Zadeh LA, Fu KS, Tanaka K, Shimura M (1975) Fuzzy sets and their applications to cognitive and decision processes. Reports, Special Publications, Theses, Academic, New York

223. Zhao ZY, Tomizuka M, Isaka S (1993) Fuzzy gain scheduling of PID controllers. IEEE Trans Syst Man Cybern 23(5):1392–1398

224. Zheng L (1992) A practical guide to tune proportional and integral (PI) like fuzzy controllers. In: 1st IEEE international conference on fuzzy systems (FUZZ-IEEE'92), San Diego, CA, pp. 633–640

225. Zimmermann H-J (1991) Fuzzy set theory and its applications, 2nd edn. Kluwer, Boston MA, ISBN 0-7923

226. Zimmermann HJ, Zysno P (1980) Latent connectives in human decision making. Fuzzy Sets Syst 4:37–51

227. http://www.fuzzytech.de/e/e_a_htw.html

228. http://journals.eecs.qub.ac.uk/codata/Journal/contents/1_1/1_1pdf/DS111.Pdf

229. http://pami.uwaterloo.ca/tizhoosh

230. http://morden.csee.usf.edu/~hall/adrules/segment.html

231. http://www.ncbi.nlm.nih.gov/entrez/query.fcgi?cmd=Retrieve&db=pubmed&dopt=Abstract&list_uids=9919828

232. http://Circuit Cellar - Digital Library - 103 Constantin von Altrock.htm

233. http://www.fuzzytech.com/e/e_a_tfc.html

234. http://www.fuzzytech.com/e/e_a_pfd.html

235. http://www.fuzzytech.com

236. http://chemindustry.intota.com

237. http://www.erudit.de/erudit/events/esit99/12529_p.pdf

238. http://www.smallsat.org/proceedings/12/ssc98/12/sscxii7.pdf

239. http://www.tech.plym.ac.uk/spmc/people/jtilbury/pdf/Tilbury-NNESMED1998.pdf

240. http://www.mathworks.com